Chemical Sensors and Microinstrumentation

ACS SYMPOSIUM SERIES **403**

Chemical Sensors and Microinstrumentation

Royce W. Murray, EDITOR
University of North Carolina

Raymond E. Dessy, EDITOR
Virginia Polytechnic Institute and State University

William R. Heineman, EDITOR
University of Cincinnati

Jiri Janata, EDITOR
University of Utah

W. Rudolf Seitz, EDITOR
University of New Hampshire

Developed from a symposium sponsored
by the Division of Analytical Chemistry
at the 196th National Meeting
of the American Chemical Society,
Los Angeles, California,
September 25–30, 1988

American Chemical Society, Washington, DC 1989

Library of Congress Cataloging-in-Publication Data

Chemical sensors and microinstrumentation.

(ACS symposium series, ISSN 0097–6156; 403)
"Developed from a symposium sponsored by the Division of Analytical Chemistry at the 196th National Meeting of the American Chemical Society, Los Angeles, California, September 25–30, 1988."

Includes bibliographies and indexes.

1. Chemical detectors—Congresses. 2. Engineering instruments—Congresses.

I. Murray, Royce W. II. American Chemical Society. Division of Analytical Chemistry. III. American Chemical Society. Meeting (196th: 1988: Los Angeles, Calif.). IV. Series.

TP156.C3C48 1989 681′.2 89–15149
ISBN 0–8412–1661–4

PRINTED IN THE UNITED STATES OF AMERICA

ACS Symposium Series

M. Joan Comstock, *Series Editor*

Foreword

The ACS SYMPOSIUM SERIES was founded in 1974 to provide a medium for publishing symposia quickly in book form. The format of the Series parallels that of the continuing ADVANCES IN CHEMISTRY SERIES except that, in order to save time, the papers are not typeset but are reproduced as they are submitted by the authors in camera-ready form. Papers are reviewed under the supervision of the Editors with the assistance of the Series Advisory Board and are selected to maintain the integrity of the symposia; however, verbatim reproductions of previously published papers are not accepted. Both reviews and reports of research are acceptable, because symposia may embrace both types of presentation.

Contents

Preface xi

1. **Chemical Sensors and Microinstrumentation: An Overview** 1
   Royce W. Murray

SENSORS BASED ON ELECTRICAL MEASUREMENTS

Introduction: Electrochemical Sensors 22
William R. Heineman

2. **New Anion- and Gas-Selective Potentiometric Sensors** 26
   M. E. Meyerhoff, D. M. Pranitis, H. S. Yim, N. A. Chaniotakis, and S. B. Park

3. **Electrical Properties of Phospholipid Bilayer Membranes Measured with a Light Addressable Potentiometric Sensor** 46
   George B. Sigal, Dean G. Hafeman, J. Wallace Parce, and Harden M. McConnell

4. **Electropolymerized Films in the Construction of Biosensors** 65
   Robert J. Geise and Alexander M. Yacynych

5. **Electrochemical Sensors Based on Polymer Films Immobilized by Gamma Irradiation** 78
   Louis A. Coury, Jr., Christos Galiatsatos, Edward W. Huber, Sandra Sponaugle, and William R. Heineman

6. **Pulse Voltammetric Biosensing Device for Real Time Determination of a Stationary Droplet Glucose Sample** 96
   Yoshihito Ikariyama, Naoki Shimada, and Shigeru Yamauchi

7. **Bioelectrochemistry at Microelectrodes** 105
   H. Allen O. Hill, Napthali P. Klein, A. Surya N. Murthy, and Ioanna S. M. Psalti

8. Microelectrodes To Probe Spatially Heterogeneous Concentrations 114
R. Mark Wightman, Leslie J. May, John Baur, David Leszczyszyn, and Eric Kristensen

9. A Fiber-Optic Electrode for Optoelectrochemical Biosensors 129
Masuo Aizawa, Masaru Tanaka, and Yoshihito Ikariyama

SENSORS BASED ON MASS OR THERMAL MEASUREMENTS

Introduction: Mass and Thermal Microsensors 140
Raymond E. Dessy

10. Fiber-Optic- and Polymer Film-Based Enthalpimeters for Biosensor Applications 143
Raymond E. Dessy, Lloyd Burgess, Larry Arney, and James Petersen

11. Vapor Detection with Surface Acoustic Wave Microsensors 157
Hank Wohltjen, David S. Ballantine, Jr., and N. L. Jarvis

12. Selective Measurement of Vinyl Acetate Vapor Using a Coated Surface Acoustic Wave Oscillator: Applications to Industrial Hygiene 176
Edward T. Zellers

13. Liquid-Phase Sensors Based on Acoustic Plate Mode Devices 191
A. J. Ricco, S. J. Martin, T. M. Niemczyk, and G. C. Frye

14. Monitoring Thin-Film Properties with Surface Acoustic Wave Devices: Diffusion, Surface Area, and Pore Size Distribution 208
G. C. Frye, S. J. Martin, A. J. Ricco, and C. J. Brinker

15. Elastic Properties of Thin Polymer Films Investigated with Surface Acoustic Wave Devices 222
David S. Ballantine, Jr., and Hank Wohltjen

16. Sensors Based on Biomolecules Immobilized on the Piezoelectric Quartz Crystal Microbalance: Detection of Glucose Using Hexokinase 237
Steven J. Lasky and Daniel A. Buttry

SENSORS BASED ON OPTICAL MEASUREMENTS

Introduction: Optical Sensors 248
W. Rudolf Seitz

17. **Design, Preparation, and Applications of Fiber-Optic Chemical Sensors for Continuous Monitoring** 252
David R. Walt, Christiane Munkholm, Ping Yuan, Shufang Luo, and Steven Barnard

18. **Poly(vinyl alcohol)-Based Indicators for Optical pH and Mg(II) Sensing** 273
Ma Wangbai, Zhang Zhujun, and W. Rudolf Seitz

19. **Construction and Characterization of Optical Sensors with Porous Polymer Film** 283
Scott M. Stole, Thomas P. Jones, Lai-Kwan Chau, and Marc D. Porter

20. **Fiber-Optic-Based Biocatalytic Biosensors** 303
Mark A. Arnold

21. **Design Considerations for Antibody-Based Fiber-Optic Chemical Sensors** 318
Michael J. Sepaniak, Bruce J. Tromberg, Jean-Pierre Alarie, James R. Bowyer, Arthur M. Hoyt, and Tuan Vo-Dinh

22. **Acetylcholine Receptor-Mediated Optical Transduction from Lipid Membranes** 331
U. J. Krull, R. S. Brown, K. Dyne, B. D. Hougham, and E. T. Vandenberg

23. **New Developments and Applications of Fiber-Optic Sensors** 345
S. M. Angel, M. N. Ridley, K. Langry, T. J. Kulp, and M. L. Myrick

24. **Optical Waveguide Chemical Sensors** 364
John F. Giuliani

25. **Multidimensional Fluorescence with a Simple Fiber-Optic-Based Probe** 380
Kevin S. Litwiler and Frank V. Bright

INDEXES

Author Index 398

Affiliation Index 399

Subject Index 399

Preface

CHEMICAL SENSORS AND MICROINSTRUMENTATION, the symposium on which this volume is based, began with proposals from the American Chemical Society Division of Analytical Chemistry for two separate symposia, one on chemical sensors and one on microinstrumentation. However, as our discussions progressed, the substantial intermingling and dependence of modern chemical sensors on miniaturization ideas in physical and chemical transducers emerged with such clarity that a combined symposium was recognized as the more rational approach.

The science and technology of miniaturization can be seen in all the forms of chemical sensor measurements included in the symposium: electrical, optical, mass, and thermal. Electrical measurements profit from the development of miniaturized voltammetric electrodes (microelectrodes). Optical measurements using optical fibers and planar waveguides became feasible as a result of technical research and developments in the telecommunications industry. Mass measurements depend on the availability of piezoelectric materials, which additionally offer a new future for analytical chemistry with scanning tunneling microscopy. Mass measurements also profit from application of microlithographic methods for preparing patterned electrodes needed for surface acoustical wave generation and detection. It is true that a good deal of modern chemical sensor research is built upon the analytical chemist "thinking small," and the symposium on Chemical Sensors and Microinstrumentation was replete with such notions.

Acknowledgments

The organizers and editors were delighted at the response from leading scholars in chemical sensor research to the invitation to participate in the symposium and in this volume. We believe that an excellent cross section of the significant current research in chemical sensors was represented by the participants and writers, and we express our gratitude to them. The research reported here also reflects attention to incorporating new chemical and physical concepts into the bases for making chemical measurements. We view this as the fundamental form of analytical chemical research and a form that serves the future of chemical sciences, and we are pleased to report it here.

ROYCE W. MURRAY
Kenan Laboratories of Chemistry
University of North Carolina
Chapel Hill, NC 27599–3290

RAYMOND E. DESSY
Department of Chemistry
Virginia Polytechnic Institute
and State University
Blacksburg, VA 24061

WILLIAM R. HEINEMAN
Department of Chemistry
University of Cincinnati
Cincinnati, OH 45221–0172

JIRI JANATA
Department of Materials Science
University of Utah
Salt Lake City, UT 84112

W. RUDOLF SEITZ
Department of Chemistry
University of New Hampshire
Durham, NH 03824

April 6, 1989

Chapter 1

Chemical Sensors and Microinstrumentation

An Overview

Royce W. Murray

Kenan Laboratories of Chemistry, University of North Carolina, Chapel Hill, NC 27599–3290

A chemical sensor is a measurement system designed to exhibit an experimental response relatable to the quantity of a sample chemical species or class of chemical species. It may be designed to respond to the target species as present in a gaseous, liquid, or solid sample of indeterminate volume, or to a target contained within a defined space such as within a small volume, on the surface of the sample, or on a microscopic part of the surface of the sample. Also known as a chemical transducer, the chemical sensor can be based on any characteristic of the sample species that serves to distinguish it from its matrix, including its chemical reactivity, optical or electrical properties, or mass. The student of analytical chemistry will recognize that these statements encompass the field traditionally known as instrumental analysis. The more recent popularity of the descriptor chemical sensor subtlely reflects the quite important fact that the chemical/instrument interface (i.e., the sensor) is today more commonly the limiting aspect of the chemical analysis capability than is the instrument itself.

The range of sample characteristics and manner of their detection, is much larger than can be realistically addressed in the space of a single chapter. We will confine this chapter mainly to the chemical sensor research areas discussed in other chapters in this volume, dividing them into electrical, optical, and mass and thermal measurements. Our focus will furthermore be on the generic chemical and physical phenomena upon which such measurements can be based, as opposed to the alternative organization that would address chemical sensors in the context of their application (i.e, auto exhaust sensor, clinical diagnostic sensor, environmental sensor) or of the kinds of samples detected (i.e, CO sensors, humidity sensor, biosensor, etc.), as used in a previous ACS Symposium Series volume on Chemical Sensors (D. Schuetzle, R. Hammerle, Eds., ACS Sympos. Ser. 309, 1986).

0097–6156/89/0403–0001$06.00/0

Throughout current chemical sensor design research, one can readily see the impact of new chemical materials, of microfabrication technology that arranges chemical materials in miniaturized forms, and of microelectronics and communications devices and technology. Microelectronics technology has given us the science of microlithography, which in particular allows the spatially defined arranging of electrically sensitive elements and contacts to our chemical systems on scales of a few microns. Patterns of metal films that detect surface acoustical waves (1,2), molecular conductivity (3), and electrochemical charge transfers (4-6) are products of the analytical chemists' use of microlithographic science. Telecommunications technology has given us optical fibers, which yield new dimensions of contacting optically sensitive chemical systems with sample media (7,8). The science of polymer composites has led to the availability of _ca_. 5 micron radius carbon fibers which have found use as microelectrodes (9) in _in vivo_ sensing of neurotransmitters. This is of course an old story (10); measurement science has traditionally benefited from and re-contributed to technological developments and products that at their roots were driven by consumer or military economics and goals.

Dealing with molecules, chemists are also good at miniaturizing the physical dimensions of materials they work with. This has been particularly evident in exploitation of thin polymeric films in modified electrodes (11,12), quartz crystal microbalances (13,14), and surface acoustical wave (1,2), optical fiber (7,8), and potentiometric (15,16) chemical sensors. Small polymer film thicknesses are especially significant in manipulating the response times and selectivities of chemical sensors via transport rate effects. Combining chemical reagents with ultrathin polymer films (and with ultrathin organized molecular films) has become an enormously profitable tactic in sensor design, as well as a stimulus for analytical chemists to draw upon synthetic aspects of polymer chemistry.

CHEMICAL SENSORS BASED ON ELECTRICAL PHENOMENA

The most important electrical phenomena in chemical sensing are electrical conductivity, interfacial potentials (i.e., potentiometry), and electrochemical reaction currents (i.e, faradaic or amperometric methods). All depend on charge exchange and transport processes in some manner, and all have been active research areas in recent years.

Conductivity. Electrical conductivity measurements are appealingly simple and sensitive, and reflect the charge carrying ability (carrier population and mobility) of a test medium interposed between two electrodes biased by DC or AC potentials. The key for chemical sensoring is to design some element of chemical selectivity into the values of carrier population or mobility. Measurements in liquid ionic solutions (the classical method of conductimetry) have little intrinsic chemical selectivity and are mainly useful as detectors of ionic populations after a chromatographic separation. Solid state conductances can have greater significance for chemical sensing if a carrier generation process can be tied to a chemical process specific to a target chemical. For example, binding of a target chemical from its vapor into a thin molecular film, in a chemically selective way that produces an increase in charge carriers, is a useful design goal. Barger et al (3) have for example reported how the conductivity of a thin phthalocyanine film is enhanced by traces of axially-coordinating organic vapor contacting the film. This work was done with a lithographically defined conductance cell consisting of interdigitated metal fingers, the phthalocyanine film being deposited over the metal finger array and in the insulating gaps in between the fingers.

Potentiometry. Chemical sensors based on measurements of interfacial electrochemical potentials (potentiometry) mainly involve potential differences that exist across membranes that separate two solutions. The origin of the membrane potential rests on binding of a charged species at the membrane surface and/or transport of the species through it. If the binding and transport are selective to a given charged species, then membrane potential varies with $(RT/nF)\ln[a_1/a_2]$ where a_1 and a_2 are the concentrations on the sample and reference sides of the membrane, respectively. If the membrane is not entirely selective, this relation is changed (15) to

$$E_{cell} = \text{constant} + (0.059/z)\log[(a_1 + b^{Za/Zb}K_{a,b})/a_2]$$

where b is the activity of the interfering sample species and $K_{a,b}$ is the potentiometric selectivity coefficient. $K_{a,b}$ is the usual way to express the ideality of the membrane selectivity and is zero for total selectivity toward a.

Clearly potentiometric sensor design is substantially based on chemical design of membranes that minimize $K_{a,b}$ coefficients. The potentiometric literature (15,17) is substantial in the areas of H^+, alkali metal ion, Ca^{2+}, and F^- sensors and has involved membranes prepared from glasses, crystalline materials, composites, and functionalized polymers. The situation is less satisfactory for potentiometric sensors for anions and for more complex molecular ions, and some current work has been directed at the problems of designing chemical binding specificity into polymer films. For metal-coordinating ionic bases, the metalloporphyrins (16,18) offer a rich coordination chemistry that is manipulable by choice of metal and porphyrinic framework. The principles of host-guest (19) molecular design are also potentially important in this area. These efforts, as do most involving designed molecular specificity, carry a substantial chemical synthetic involvement, or collaboration with synthetic chemists (a useful moral).

Faradaic Methods. Chemical sensors based on electrochemical reactions tend to be more complex than potentiometric sensors because their operation depends on a greater diversity of kinetic events: heterogeneous electron transfer kinetics, mass transport kinetics, and the kinetics of chemical reactions coupled to the heterogeneous electrode reaction. At the same time, the sensitivity of kinetics to molecular character and the utility of controlling reaction rates with applied electrode potentials creates a useful scope of sample selectivity. A value of oxidation or reduction current relatable to the sample concentration is the usual faradaic method output.

A substantial portion of recent electrochemical sensor work has been directed at measurements of (or in) biological systems. The active research areas include detection of neurotransmitters (catecholamines) and of reactions involving enzymes and antibodies. The latter classes of study seek to take advantage of the sharp chemical selectivities that natural systems have evolved.

Several ways that enzymes have been incorporated into electrode reaction schemes are illustrated in Fig. 1. The classical method (Fig. 1A) uses a solution of sample substrate, enzyme, and enzyme cofactor (mediator oxidant or reductant) and detects with the electrode, a product of the enzyme reaction (either that of the substrate or of the mediator). For the important case of glucose determination, the H_2O_2 product of the aerobic turnover of the enzyme can be detected by its oxidation current at a Pt electrode, or the consumption of dioxygen can be detected through diminution of its reduction current (20).

Most enzymes do not readily undergo direct electron transfers with conducting electrode surfaces (Fig. 1B); the enzyme active site is either buried and inaccessible to a macroscopically huge electrode surface, or the enzyme may be steered by adsorptive or electrostatic forces into contacting the electrode at a region on its outer surface unfavorable for electron transfer. Adding fast electron transfer mediators (21,22) can be an effective way to deal with the active site accessibility problem (Fig. 1C). In the determination of glucose (22) for example, the electrochemical oxidation of ferrocene generates ferricenium, an oxidant which reacts with glucose oxidase to produce the active form that consumes glucose and simultaneously regenerates the ferrocene. The catalytic regeneration enhances the electrode current which is in turn related to the glucose concentration. A recent innovative step(23) is to chemically bind the mediator to the enzyme in a way allowing the mediator to react directly with both the electrode and the enzyme active site (Fig. 1D). While this approach requires developing and characterizing the mediator-enzyme attachment chemistry, it apparently can provide an efficient electron coupling between electrode and enzyme turnover of substrate. Fourthly, a useful variant to all of the above schemes is to attach the enzyme directly to the electrode surface. Recent work in this direction (24,25) aims at improving the attachment stability and the kinetics and selectivity of transport of charge between electrode and enzyme and of substrate between sample solution and immobilized enzyme. Finally, it is important to understand

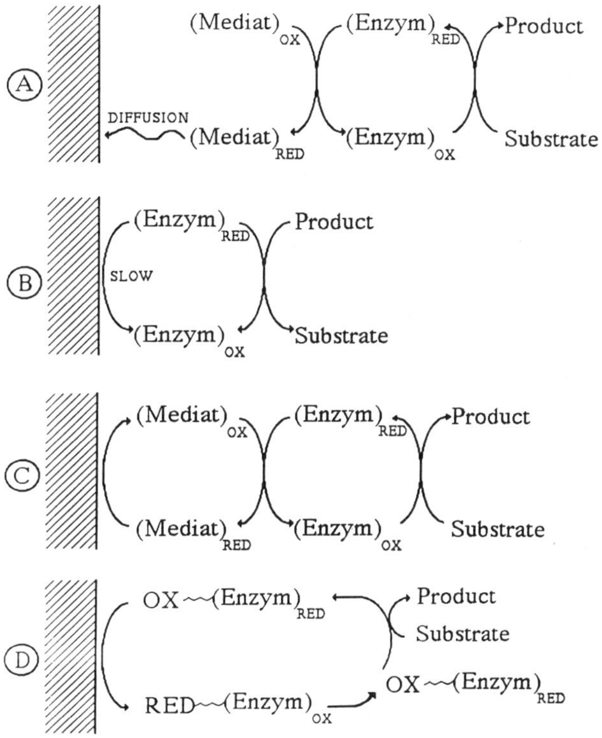

Figure 1. Schemes for coupling enzyme reactions to electrodes. A. Enzyme reacts in bulk solution with oxidant $(\text{mediat.})_{ox}$ and substrate, detect change in bulk concentration of the oxidant or its product $(\text{mediat.})_{red}$; B. Direct reaction of enzyme with electrode (generally slow); C. Mediation of electron transfer between enzyme and electrode; D. Same as C but mediation bound to enzyme.

that effective electrochemical observation of an enzyme-driven reaction requires that the enzyme reaction rate exceed that with which the electron transfer mediator (in either free or enzyme bound form) diffuses into/out of the depletion layer around the electrode. It has not always been recognized that the electrochemical method and electrode geometry should be selected with this timescale requirement in mind.

Polymer films on chemically modified electrodes offer additional ways (11) to enhance the chemical selectivity of electrochemical sensors. The polymer film can be designed chemically to sequester (extract) a target sample species from the contacting medium, concentrating it at the electrode surface prior to initiation of an electrochemical reaction that measures the amount of extracted target (26). Investigations of neurotransmitters in cerebral fluids have exploited polymer films as transport barriers to avoid or minimize reactions of electroactive interferants. For example, the electrode reaction of ascorbic acid overlaps that of dopamine (a target neurotransmitter) which is a serious problem since dopamine is present at much lower concentrations. A film of the polymer Nafion (a perfluorosulfonate polymer) coated over the electrode concentrates dopamine (a cation at physiological pH) within the polymer at the electrode surface, whereas ascorbic acid (an anion) is charge-expelled. The result (27) is a great improvement in selectivity of this sensor. Other polymers (designed or selected for the particular purpose) have also been used to address the dopamine/ascorbic acid problem (28) as well as that of electroactive interferants in enzyme-based schemes (25).

Miniaturization of electrodes and of electrochemical cells has had considerable impact on electrochemical sensor design as well as on fundamental electrochemical studies (9,29). Microelectrodes in particular have opened up new avenues for designing electrochemical sensors. Microelectrodes are physically small electrodes (microdisks have radii of 1-5 microns, microbands have widths down to _ca_ 5 nm). Such dimensions permit analysis of small spaces, including _in situ_ neurotransmitter determination in the brain (9) and even in single nerve cells (30). Microelectrodes also permit a new form of microscopy, scanning electrochemical microscopy (31) as discussed later. An important aspect of electrode smallness is that the currents that flow from electrochemical reactions are also small. This would seem to be a disadvantage, since smaller currents mean more difficulties with overcoming noise and retaining sensitivity to a given sample concentration. However, the small currents also allow access of electrochemical experiments to media normally far too resistive to accommodate electrochemical experiments at all. Thus, possibilities for the design of electrochemical sensors that work in hydrocarbon solvents (32), frozen solvents (33), and in solid polymers (34) have recently opened up.

Miniaturization of the entire electrochemical cell can be accommodated with microlithographic procedures (4). The traditional difficulty of the complexity of maintaining clean electrodes, etc., in routine electrochemical analyses, may be circumvented by the capacity to employ electrochemical cells that are disposable. Small electrochemical cells can also miniaturize the required sample volume; the glucose sensor scheme of Figure 1A can for example be reduced to a 10 microliter volume requirement in an appropriately designed cell (35).

Lastly, electrochemical gas sensors can be based on ceramic ZrO_2 materials that at high temperatures act as oxide ion carriers. These sensors can operate both on potentiometric and amperometric (36) principles and have been of great interest for dioxygen partial pressure sensors in harsh environments like internal combustion engines.

USES OF PIEZOELECTRIC MATERIALS IN CHEMICAL SENSORS

Piezoelectric materials have the property of undergoing a change in physical shape under the influence of an applied electrical field. This property has been exploited in a wide range of technologies ranging from quartz crystal oscillators in timepieces to strain gauges in seismometers. Piezoelectric materials have been applied to chemical sensor design in two ways, as microscopic devices that serve to generate a precisely controlled physical motion of a sensor probe over very small distances, and in mass-sensitive oscillators.

Microscopic Sensors. Analytical situations frequently present themselves in small spaces, volumes, and on surfaces. We have already noted that microelectrodes (9), being small, sense only a small local volume of contacting medium (of dimension approximately equal to the electrode radius) and are thus useful for analysis of small spaces. This aspect of microelectrodes has been combined by Bard (31) with the precise positioning of the microelectrode with respect to a sample surface that is possible with a piezoelectric device, to devise a new scanning electrochemical microscopy technique. The experiment can image, at the few micron level, both sample topology and chemical composition by taking advantage of the interaction between the sample surface and the flux of electroactive materials within the diffusion layer of the microelectrode.

Another important application of piezoelectric positioning is the recently invented (37) scanning tunnelling microscopy (STM) experiment. A fine probe is brought within a few Angstroms of a conducting sample surface, where under a voltage bias, electrons tunnel between the (ideally) single atom on the probe tip and atoms of the sample surface. This experiment is of enormous importance to chemistry, including analytical chemistry, because of its ability to image the atomic structure of surfaces and to observe the dynamic ways in which they change in response to various stimuli. Single molecules can be imaged; the resolution of single molecules on surfaces is presently poor but some information may be obtainable on molecular shape and further developments may improve resolution. Chemical sensor applications of STM have not yet appeared, but they surely will in time as the number and variety of STM researchers increases. Imagine for example designing and attaching a molecular site to a surface which will selectively bind a target molecule, and subsequently interrogating that site for the presence of bound target using an STM probe. The author regards this as an entirely realistic hypothesis of future single molecule analysis.

Mass Sensitive Applications. Uses of piezoelectric materials in mass-sensitive oscillators encompass several configurations (Fig. 2). Mechanical oscillations propagated perpendicularly between the parallel faces of a thin quartz crystal piezoelectric element are used in the quartz crystal microbalance (13,14), mechanical up-and-down undulations at radio frequency propagated along the surface of a crystal (surface acoustical waves, Rayleigh waves) are used in SAW devices (1,2), and sideways undulations similarly propagated (shear horizontal acoustic plate mode, SH-APM, horizontally polarized shear waves) are used in a newly introduced experiment (38). Typically, various chemically sensitive films are placed on the surfaces of the piezoelectric crystal (or on the metal film electrodes on the crystal), and the sensor response is to some change in the film brought about by contact with a medium containing the target molecule. The simplest version of such a response is through the change in mass of the film; increased film mass lowers the frequency at which the resonant RF oscillation of the piezoelectric crystal is excited. Theory for this response shows(2,38,39) that the frequency change is proportional to the mass change and to the square of the resonant frequency, and importantly, that the proportionality constant is calculable in an absolute manner. The absolute character of the relation between frequency change and mass change requires, however, that the film is thin enough or mechanically rigid enough that its own viscoelastic characteristics, and any change in those characteristics produced by binding of the target species, are not important. This condition is sometimes not met (40-42), but frequency changes that are usefully correlatable with binding of a target molecule to the film can often nonetheless be obtained. Some of the current research with piezoelectric sensors is aimed at the understanding and exploitation of those circumstances.

Quartz Crystal Microbalance. This experiment depends on the manner in which the resonant frequency of an oscillating quartz crystal piezoelectric element responds to a change in mass on the interface perpendicular to the oscillatory motion.

Typically the oscillation is excited between facing metal film electrodes on opposite side of the crystal as shown in Figure 2(top). A change in mass at one of the interfaces produces a proportional oscillatory frequency decrease.

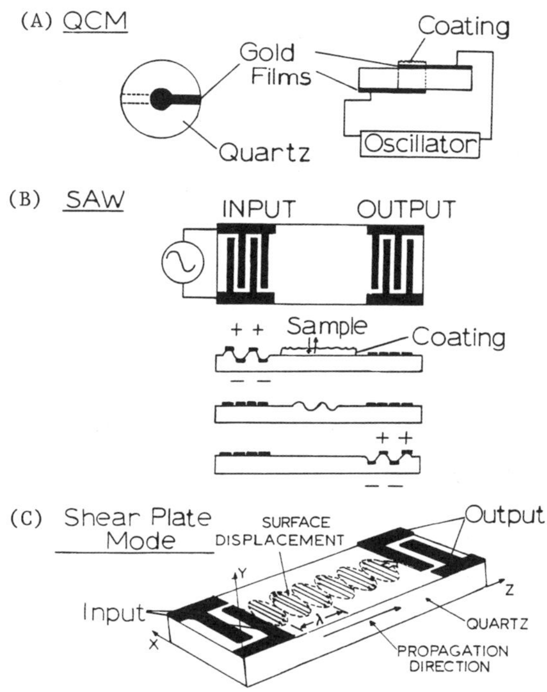

Figure 2. Schemes for using piezoelectric quartz crystals. A. Quartz crystal microbalance configuration, standing shear wave between facing Au electrode contacts; B. Surface acoustical mode configuration, surface undulation caused by bias between metal fingers; C. Horizontal shear plate mode.

Quartz crystal microbalances work best in contact with gases or vacuum and are used as deposition monitors in thin film deposition apparatus. They have been employed for chemical vapor sensing purposes by coating one of the crystal interfaces with a film that selectively binds the target molecule, detecting the consequent change in the mass of the film.

More recent work (13,14,40,43,44) has solved some of the problems of viscous loading when the piezoelectric crystal is contacted by liquid samples, and new measurements with QCM devices are now being made in solutions. Much of this work has exploited one of the metal film electrodes as the working electrode of an electrochemical cell; the experiment detects alterations in films deposited on that electrode. Reductive electrodeposition of a sample ion as a metal film or as an insoluble salt on the working, mass sensitive electrode produces a decrease in the oscillator frequency that is sensitively detectable on a monolayer scale. Electroactive polymer films coated on the working electrode lose or gain weight as the film is oxidized or reduced to incorporate or lose counterions and/or solvent, or potentially as the film binds a target molecule from the contacting solution. Applicability of these processes in useful chemical sensors depends of course, as is the case in all measurements that are themselves not chemically selective (i.e., mass), on the chemical selectivity of the process.

The reactions of antibodies and enzymes are attractively selective chemical reactions, and responses of their reactions in binding to their substrates have begun to be investigated. A potential difficulty, probably an intrinsic one, is in immobilizing the antibody or enzyme species on the piezoelectric surface within a medium that is sufficiently thin and rigid to avoid viscoelastic effects. Buttry for example (40) has found responses to glucose binding to the hexokinase enzyme entrapped in a poly(acrylamide) film that are far larger than expected simply on a mass change basis. The frequency change response is, nonetheless, proportional to glucose concentration and thus potentially useful.

Surface Acoustical Wave Devices. As illustrated in Fig. 2(middle), application of an AC potential bias between a pair of adjacent, closely spaced (by 1/4 of the acoustical wavelength) electrodes will produce (2,39) an undulatory surface oscillation, or surface wave, that is propagated toward a similar pair of detecting electrodes at a rate determined primarily by the properties of the quartz piezoelectric material (_ca_. 10^3 m/s) but also by any mass that contacts the quartz crystal in the region between the two sets of electrodes. Changes in the surface wave propagation rate, and in the amplitude of the surface oscillation, can be detected as a changes in the resonant oscillation frequency.

The typical SAW sensor application is for gaseous samples and is based on selective interaction of the target gas with a film coating placed in the mass sensitive region. If the film coating is thin, isotropic and non-conducting, a change in its mass resulting from a target species binding or partitioning into it will lower the resonant frequency proportionally to the mass change and as noted above, an explicit, absolute relation can be written to connect frequency and mass changes. This relation, in the simplest terms, requires that the film thickness be small compared to the acoustical wavelength, which while dependant on the oscillation frequency, puts the desired film thickness in the ultrathin film (< 100 NM) range. If the film is thicker, there is the possibility that its elastic properties may influence the frequency-mass relation; this is less likely with a rigid film than with a soft one. The absolute frequency-mass relationship is altered if the film has appreciable elasticity or if the target molecule binding alters the film elasticity.

Typical SAW chemical sensors operate in the 30-160MHz frequency range. Since the magnitude of the SAW mass-frequency shift depends on the square of the frequency of the oscillation mode, the trend for enhanced mass sensitivity has been toward higher oscillation frequencies (39). The SAW measurement has excellent mass sensitivity at any frequency; example values cited later in this volume are 840 Hz frequency change per microgram on a 1.5 cm^2 area at 30 MHz and 365 Hz frequency change per nanogram on a 8 mm^2 area at 158 MHz.

A principal requirement in exploiting this substantial sensitivity of SAW sensors is choosing a coating or film on the piezoelectric surface that interacts with the target component in the sample gas matrix. Ideally, the film-sample interaction involves a large equilibrium constant (providing sensitivity), is selective toward the desired sample species, occurs rapidly, and is readily reversible. These requirements can be contradictory; strong, selective binding is often not easily reversed. This can lead coating design in two directions, one compromising reversibility, and the other compromising selectivity. The analysis of vinyl acetate vapor by observing the rate of its substitutional binding to a Pt complex (45) selected from the appropriate coordination chemistry literature exemplifies the former approach. This kinetic analysis is quite selective for vinyl acetate, even among other potential other interferants, but the surfaces of the Pt complex particles embedded in the polymer film coating must periodically be regenerated by substitutional replacement of bound vinyl acetate with ethylene.

A different film selection approach relies on the differences between partition coefficients of the pertinent components of the sample matrix into film materials chosen from gas-liquid chromatography column substrates based on their retention characteristics (41,42). Vapor/polymer partition coefficients typically have great selectivity only between classes of chemical compounds, and only moderate selectivity within a given class, so this approach requires careful choice of the film if the sample matrix contains chemically similar materials. A chemometric approach may offer some relief in this situation;

however, based on the fact that SAW piezoelectric elements are small and several can be incorporated into a single analytical device. Each of the several elements is coated with a different film material, biased for favored partitioning of a selected component of the sample matrix(39). Chemometric manipulations of the data are then required to deconvolve interferences from the desired sample response of each individual element.

Like QCM, SAW measurements have additional chemical uses besides chemical analysis. Investigation of how viscoelastic properties of polymer films vary with temperatures spanning their glass transition temperatures (41), of surface area measurements in thin porous sol-gel glass films by N_2 absorption mass changes (46), and of the rates of diffusion of sorbed vapors with polymer films (46), are example applications described in chapters later in this volume.

OPTICAL FIBER AND WAVEGUIDE BASED CHEMICAL SENSORS

Technological development in the telecommunications and related areas have made available a wide variety of miniaturized optical structures such as tiny glass and silica rods (fibers), hollow cylinders, and more planar geometries like thin films or thin film channels resting on flat substrates (Fig. 3). The optical fibers are fabricated by drawing glass fibers such that the 4-100 micron diameter drawn fiber has a step function or gradient of refractive index change between the central fiber core and its outer cladding. Often used in combination with a plastic cladding, these optical fibers exhibit very efficient and low loss internal reflection of light and provide a basis for transmission of optical signals over substantial distances. Names like optical fibers, lightguides, and optical waveguides are employed to describe these materials, which have provided the chemist with new ways to contact measurement systems with media containing objects of chemical analysis.

There are four (at least) significant elements of using optical waveguides or fibers. Firstly, just as in optical communications, an optical fiber can serve as a means to transport a spectrometric light source to a point of optical coupling to the sample (as in optical absorbance or fluorescence), and by appropriate design can also transport optical responses attributable to absorption, fluorescent, or chemiluminescent events back to a photodetector. This feature allows for remote analysis, analysis of materials within small spaces, and, importantly, for novel ways to combine optical response-producing reagents with the sample medium. Such combinations involve either supplying the optical-response-producing reagent as a film of solution or as an immobilized film of material at the interface between the optical waveguide or fiber and the sample medium (a solution or a vapor). We will call such films the <u>indicator phase</u>.

A second element of using optical waveguides or fibers concerns how the light transmitting medium is coupled to the chemical medium or the indicator phase. Three general

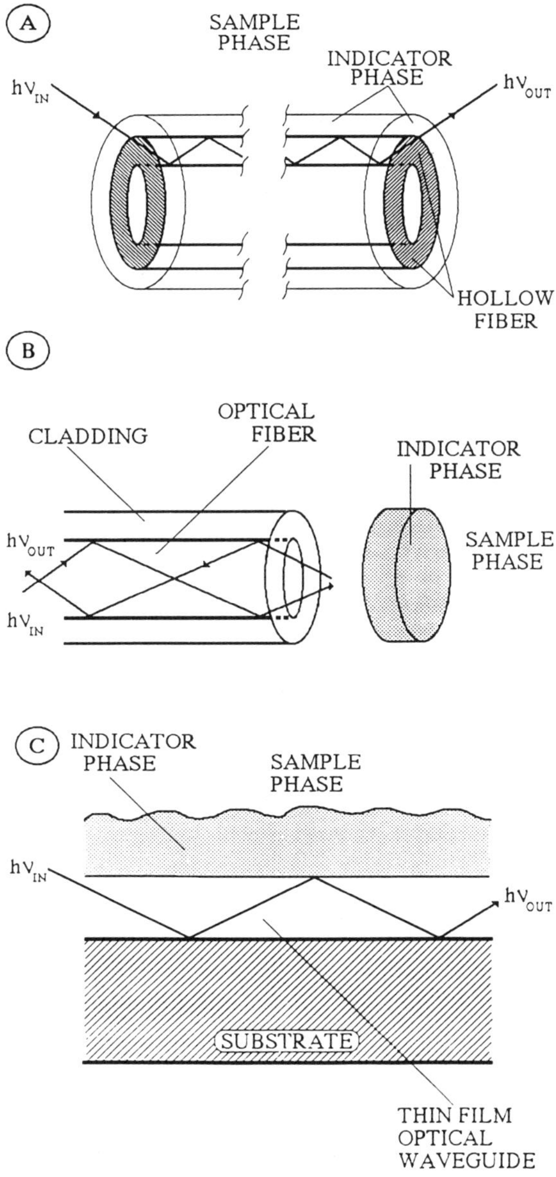

Figure 3. Schemes for coupling indicator phases with optical elements. B. Optical fiber coated on end; A. Hollow optical fiber coated on cylindrical surface; C. Flat optical waveguide coated with indicator phase.

approaches are illustrated in Fig. 3. The optics-chemistry coupling can either be done at the end of the optical structure or longitudinally along its length. The former, the more common approach thus far, places the distal end of an optical fiber (or fiber bundle) into contact with the indicator phase which in turn is contacted by the source of the sample species (Fig. 3B). In this case, the quantity of light absorbing or emitting material observed by the fiber depends on a number of factors including the depth of light penetration into and/or the dimension of the indicator phase and the optical input and/or collection characteristics of the fiber/indicator phase interface. If the optics-chemistry coupling is done longitudinally, this involves either removing the cladding from a commercial optic fiber, or using hollow fiber or planar geometries (Fig. 3A,C), and coating the indicator phase along the length of the optically transmitting phase. Longitudinal coupling is actually a form of internal reflection spectroscopy (47), where the interaction between the indicator phase and the internally reflected light depends on the penetration of the radiation field into the chemical phase (the evanescent wave). In the indicator phase, the radiation field dies away exponentially over a distance of the order of one wavelength and the experiment thus samples a quite thin film of that phase.

The manner of eliciting and detecting an optical response is another significant aspect of using optical fibers or waveguides as chemical sensors. Fluorescence is an often preferred approach from the viewpoint of sensitivity to the sample target. The intrinsically small dimensions of fiber optic and waveguide experiments mean that the observable quantities of sample contained within the indicator phase are rather small, which make sensitive measurement approaches especially appealing. Chemiluminescent response are similarly attractive. For fluorescence measurements, the fiber optics design must provide for admitting the exciting wavelengths and collecting the emitted wavelengths; this can be done either by using some combination of two different fibers (bifurcated) (48), or by using a single fiber (49,50) and by exploiting differences between the collimation of the exciting source (typically a laser) and the emanating emitted light at the instrument end of the fiber. Absorbance measurement is the usual mode for longitudinal (waveguide) coupling (51), and can also be used with optical fibers, with attention to molar extinction considerations and providing some means of reflecting (52) the incident radiation after passage through the indicator phase back into the optical fiber assembly.

The fourth element of using optical waveguides and optical fibers for chemical sensing concerns the nature of the indicator phase itself, in particular how it is formed and what reagents it contains. This chemical consideration is generally acknowledged as constituting the primary challenge in the development of this area of chemical sensing. Considering that the first application of a fiber optic element for sensing (pH) was as recent as 1980 (53), recent reviews (7,8) and the chapters in this volume indicate that substantial progress has been made in addressing the chemical problem, and a reasonable diversity of sample environments and of species detectable is beginning to emerge.

Chemical and Physical Design of Indicator Phases. General considerations of indicator phase designs include what reagent chemicals are supplied within it, how they selectively and sensitively produce an observable optical response when placed into contact with the sample-containing phase, on what timescale and how stably and reversibly is the response obtained, and in what manner are the significant reagent chemicals immobilized within the indicator phase on the surface of the optical fiber, the hollow fiber, or the planar optical waveguide.

Monolayers. The response timescale factor involves how rapidly the sample species penetrates the indicator phase volume as well as the chemical reaction rate of the optical-response-producing reaction. Clearly the indicator phase volume should be small to offer any chance of a reasonably fast response; a monomolecular layer is the limiting case in this regard. Because of sensitivity considerations, planar waveguides would appear to be the most suitable geometry for use with reagent monolayers; a sensitivity enhancement for absorbances by monolayers of 10^3 has been reported (54).

Trapping Membranes. The simplest route to immobilizing chemicals at an interface is to trap them under a membrane. When used with gas (but not solution) permeable membranes this tactic has been long been effective for design of optically and electrochemically based gas sensors. In contact with solutions, however, it is difficult to trap reagents next to the optical sensors unless they have molecular dimensions so substantial that molecular sieving membrane materials (such as a dialysis membrane) can contain them. This category includes however, enzymes and antibodies which because of their great chemical selectivity and sensitivity, and wide range of conceivable substrates, clearly form a fertile area for application of fiber optic methodology. In the case of enzyme applications, the sample substrate is allowed to diffuse through the trapping membrane into a small volume (40 nanoliter to 1 microliter) of enzyme solution trapped next to the fiber optic end. The enzyme reaction may produce a strongly absorbing product, produce or consume a fluorescing component (such as NADH), or invoke a chemiluminescent reaction; all of these have been exploited (52). The binding of antigens or labeled antigens to antibodies can be fluorescence detected as competitive/binding, direct assay, or "sandwich" interactions. The Ab/Ag interaction can be quite strong and while this poses delightful selectivity and sensitivity, it carries the burden of being not readily reversed. This problem is addressed in a chapter in this volume (49) by incorporating microcapillaries into the fiber optic assembly to allow for flushing out the indicator cell volume, replacing the antibody reagents, and conducting control experiments. This promising cell design would seem to have many advantages in circumventing the vagaries of dealing with chemical immobilizations in biosensor design.

Immobilization By Polymer Binding. A popular approach to small molecule reagent immobilization is to covalently bind it to a thin polymer film covering the fiber optic end. Hydrogel polymers are preferred for analysis of constituents of aqueous media, and poly-acrylamide (55), poly-vinylalcohol (48), and partially hydrolyzed cellulose acetate (56) films have been described.

These polymer films are assembled in different ways. Poly-acrylamide can be cross-linked and copolymerized in place with an acrylamide derivative of the optical reagent to form a polymer film containing a large concentration of reagent. This polymer was used to introduce (55) polymer bound fluoresceinamine in a pH sensing optical fiber; this reagent's fluorescence occurs competitively with protonic dissociation from the fluoresceinamine excited state so fluorescence intensity varies in a sigmoid fashion centered on the pKa of the excited state. Using a film of poly-vinylalcohol to which fluoresceinamine and a Mg^{2+}-selective reagent were both covalently bound and which was cross-linked in place as a cast solution film, a Mg^{2+} optical fiber sensor was derived (48) from the energy transfer quenching of the fluoresceinamine emission accompanying formation of the colored Mg^{2+} complex. Cellulose acetate can be cast in place on an optical fiber and partially hydrolyzed to yield (56) a polymeric material that widely known dye-fastening chemistry can be applied to. The diversity of covalent binding chemistry possible with these three materials should yield a very wide scope of reagent immobilization chemistry. Of course it is important that the possibilities for alteration of the electronic state energies and lifetimes accompanying chemical binding reactions be recognized, and basic characterization studies of the polymer film are a necessary part of the research in this area (56).

Partitioning Polymers. Used in contact with sample vapors, films of suitable polymer cast over optical fiber ends or onto longitudinal waveguide surfaces (51) can be made to respond to partitioning of the sample vapor into the polymer in several ways. If an absorbing or fluorescing probe molecule is co-dissolved in the polymer and is selected as a solvent polarity probe, then the alteration in its fluorescence wavelengths or intensity can be sensed as the partitioning sample alters the effective polymer solvent polarity. Alternatively, the nature of a polymer/waveguide interface may be altered by partitioning sample so as to alter how effectively internal reflection propagates the radiation. This type of response is not likely to exhibit a strong chemical selectivity, but as noted above in discussion of SAW devices, multiple responses from multiple waveguide sensors coated with differently partitioning polymers may with chemometric manipulation prove useful nevertheless. That waveguides can be designed in a multiplex mode is demonstrated in one of the chapters in this volume (51).

SOME NEW APPROACHES

Some of the chapters in this volume present novel approaches to chemical sensor design which do not fit conveniently into the preceding outline.

Thermal Sensors. Chemical sensors that detect the heats of chemical reactions possess selectivity to target samples only to the extent that the chemical reactions themselves are selective. Enzyme reactions are useful in this regard, and Dessy et al (57) exploit an enzymic reaction in a thermal detector based on a fiber optic interferometer. The detecting optical fiber is used in a longitudinal mode, coated with enzyme and exposed to sample in a flow injection analysis system. The thermal effects of the sample-enzyme reaction change the rate of light propagation through the fiber which is in turn detected by a interferometric phase shift relative to a reference fiber. This mode of thermal detection would appear to be capable of substantial sensitivity as this experiment is further developed.

The pyroelectric effect can also be employed as a sensor of an enzymic reaction. Dessy et al (57) placed two poly(vinylidene fluoride) films into contact, with one exposed surface coated with enzyme and contacting a flow injection sample stream. The potential resulting from the thermal bias across the films could be related to the enzyme substrate concentration.

Lipid Bilayers. Two chapters in this volume deal with prospective applications of lipid bilayers to chemical analysis, both having a long term goal of immobilizing such membranes on a sensing element surface. McConnell et al (58) describe a measurement system designed to read out the transmembrane potential, and its conductance and capacitance in a novel manner. The membrane's response to a target sample would for example depend on its interaction with the membrane and with ion carrier or channel sites incorporated into it. The membrane is in series with a insulator film covered Si electrode whose depletion layer capacitance is modulated by photogenerated electron-hole pairs. The project aims at contacting the lipid bilayer directly to the insulator film in which case the response of the membrane read out would be confined to that portion addressed by the focussed light spot.

The chapter by Krull et al (59) use membranes in vesicle form ("membrane bags"). Fluorophore-labeled receptors are incorporated into the vesicles which, drawing upon known receptor binding chemistry, exhibit alterations in fluorescence intensity in the presence of the agonist carbamylcholine. This project aims at incorporating this scheme into membranes on the surfaces of optical fibers.

ACKNOWLEDGMENT. The author of this chapter wishes to thank the co-organizers of the Chemical Sensors and Microinstrumentation Symposium, Professors Dessy, Janata, Seitz, and Heineman for providing many useful remarks, and some draft written materials, to aid in the preparation of this chapter.

REFERENCES:

1. Wohltjen, H.; Dessy, R. Anal. Chem., (1979), 51, 1458, 1465.
2. Wohltjen, H. Sensors and Actuators, (1984), 5, 307.
3. Wohltjen, H.; Barger, W.R.; Snow, A.W.; Jarvis, N.L. IEEE Trans. Electron Devices, (1984), ED-32, 1170.
4. Longmire, M.; Morita, M.; Murray, R.W. Anal. Chem., (1988), 60, 2770.
5. Shu, C.-F.; Wrighton, M.S. J. Phys. Chem., (1988), 92, 5221.
6. Sanderson, D.G.; Anderson, L.B. Anal. Chem. (1985), 57, 2388.
7. Seitz, W.R. Anal. Chem. (1984), 56, 16A; CRC Crit. Rev. Anal.Chem. (1988), 19, 135.
8. Angel, S.M. Spectroscopy (1987), 2, 38.
9. Wightman, R.M. Science (1988), 240, 415.
10. "Opportunities in Chemistry", National Academy Press, Washington, D.C. 1985
11. Ewing, A.E.; Darst, R.; Murray, R.W. Anal. Chem. (1987), 59, 379A.
12. Murray, R.W. Chemically Modified Electrodes in "Electroanalytical Chemistry", Bard, A.J., Ed.; Vol. 13, M. Dekker, NY 1984.
13. Melroy, O.R.; Kanazawa, K.K.; Gordon, J.G.; Buttry, D.A.; Langmuir, (1986), 2, 697.
14. Kanazawa, K.K., Gordon, J.G. Anal. Chem., (1985), 51, 1770.
15. Ammann, D.; Marf, W.E.; Anker, P.; Meier, P.C.; Pretsch, E.; Simon, W. Ion Selective Electrode Rev., (1983), 5, 3 -92.
16. Chaniotakis, N.A.; Chasser, A.M.; Meyerhoff, M.E.; Groves, J.T.; Anal. Chem., (1988), 60, 185.
17. Marf, W.E. "The Principles of Ion Selective Electrodes and of Membrane Transport", Elsevier, Amsterdam, 1981.
18. Meyerhoff, M.E.; Pranitis, D.M.; Yim, H.S.; Chaniotakis, N.A.; Park, S.B., this Volume.
19. Graf, E.; Kintzinger, J.-P.; Lehn, J.-M.; LeMoigne, J.; J. Am. Chem. Soc. (1978), 97, 229.
20. Guilbault, G.G.; Lubrano, G.L. Anal. Chem., (1972), Acta 60, 254; 64, (1973) 439; 97, (1978), 229.
21. Hill, H.A.O.; Klein, N.P.; Murthy, A.S.N.; Psalti, I.S.M.; this Volume.
22. Armstrong, F.A.; Hill, H.A.O.; Walton, N.J.; Acc. Chem. Res. (1988), 21, 407.
23. Degani, Y.; Heller, A. J. Phys. Chem. (1987), 91, 1285; J. Amer. Chem. Soc. (1988), 110, 2618.
24. Umana, M.; Waller, J. Anal. Chem. (1986), 58, 2979.
25. Geise, R.J.; Yacynch, A.M. this Volume.
26. Guadalupe, A.R.; Abruna, H.D. Anal. Chem. (1985), 57, 142.
27. Nagy, G.; Gerhardt, G.A.; Oke, A.F.; Rice, M.E.; Adams, R.N.; Moore, R.B.; Szantirnay, M.N.; Martin, C.R. J. Electroanal Chem. (1985), 188, 89.
28. Coury, L.A.; Galiatsatos, C.; Huber, E.W.; Sponaugle, S.; Heineman, W.R., this Volume.
29. Chidsey, C.E.D.; Murray, R.W. Science (1986), 231, 25.

30. Chien, J.B.; Saraceno, R.A.; Ewing, A.G. in "Redox Chemistry and Interfacial Behavior of Biological Molecules", Plenum Publishing Corp., (1988), pp417.
31. Bard, A.J.; Fan, F.-R.; Kwak, J.; Lev, O. Anal. Chem. (1989), 61, 132.
32. Geng, L.; Ewing, A.G.; Jernigan, J.C.; Murray, R.W. Anal. Chem. (1986), 54, 852.
33. Bond, A.; Fleischmann, M.; Robinson, J. J. Electroanal. Chem. (1984), 180, 257.
34. Geng, L.; Reed, R.A.; Kim, M.-H.; Wooster, T.; Oliver, B.N.; Egekeze, J.; Kennedy, R.; Jorgenson, J.W.; Parcher, J.F.; Murray, R.W. J. Am. Chem. Soc. (1989), 111, 1614.
35. Ikariyama, Y.; Shimada, N.; Yamauchi, S.; this Volume.
36. Usui, T.; Asada, A.; Nakazawa, M.; Osanai, H. J. Electrochem. Soc. (1989(, 136, 534.
37. Golovchenko, J.A. Science (1986), 232, 48.
38. Ricco, A.J.; Martin, S.J.; Niemczyk, T.M.; Frye, G.C.; this Volume.
39. Wohltjen, H.; Ballantine, D.S.; Jarvis, N.L.; this Volume.
40. Lasky, S.J.; Buttry, D.A.; this Volume.
41. Ballantine, D.S.; Wohltjen, H.; this Volume.
42. Grate, J.W.; Snow, A.; Ballantine, D.S.; Wohltjen, H.; Abraham, M.H.; McGill, A.; Sasson, P. Anal. Chem. (1988), 60, 869.
43. Deakin, M.R.; Melroy, D.R. J. Electrochem. Soc. (1989), 136, 349.
44. Orata, D.; Buttry, D.A. J. Am. Chem. Soc. (1987), 109, 3574.
45. Zellers, E.T.; this Volume.
46. Grye, G.C., Martin, S.J.; Ricco, A.J.; Brinker, C.J.; this Volume.
47. Harrick, N.J. "Internal Reflection Spectroscopy", 2nd Ed., Hanick Scientific Corp., N.Y., 1979.
48. Wangbai, M.; Zhujun, Z.; Seitz, W.R.; this Volume.
49. Sepaniak, M.J.; Tromberg, B.J.; Alarie, J.-P.; Bowyer, J.R.; Hoyt, A.M.; Vo-dinh, T.; this Volume.
50. Skogerboe, K.J.; Yeung, E.S. Anal. Chem. (1987), 59, 1812.
51. Guiliani, J.F.; this Volume.
52. Arnold, M.A.; this Volume.
53. Peterson, J.I.; Goldstein, S.R.; Fitzgerald, R.V.; Buckhold, D.K. Anal. Chem. (1980), 52, 864.
54. Stephens, D.A.; Bohn, P.W. Anal. Chem. (1989), 61, 386.
55. Walt, D.R.; Munkholm, C.; Yuan, P.; Luo, S.; Barnard S., this Volume.
56. Stole, S.M.; Jones, T.P.; Chau, L.-K.; Porter, M.D.; this Volume.
57. Dessy, R.E.; Burgess, L.W.; Arney, L.; Petersen, J., this Volume.
58. Sigal, G.B.; Hafeman, D.G.; Parce, J.W.; McConnell, H.M., this Volume.
59. Krull, V.J.; Brown, R.S.; Dyne, K.; Hougham, B.D.; Vandenberg, E.T., this Volume.
60. Angel, S.M.; Nugent, M.; Langry, K.; Kulp, T.J., this Volume.

RECEIVED April 6, 1989

SENSORS BASED ON ELECTRICAL MEASUREMENTS

Introduction: Electrochemical Sensors

Electrochemistry involves the measurement of electrical signals associated with chemical systems that are incorporated in an electrochemical cell. The cell consists of two or more electrodes that function as transducers between the chemical system and an electrical system in which voltage or current can be measured and/or controlled. For sensor applications, one of the electrodes is an indicator electrode that is designed to interact selectively with the species being monitored and another electrode is a reference electrode whose half-cell potential should remain constant during the measurement process. The magnitude of the voltage or current signal from the cell is related to the activity or concentration of that species in the sample.

Electrochemical cells offer certain advantageous features for use as sensors. (i) Electrochemical measurements can be made on extremely small volumes with relative ease. This property arises from the fact that electrochemistry is an interfacial process. The important electrode processes occur at the electrode surface, not in the surrounding solution. Consequently, electrochemical measurements can be made on exceedingly small volumes of sample with miniaturized electrodes. For example, measurements have been made on nanoliter and smaller volumes with relatively simple instrumentation. This ready adaptability to miniaturization is especially desirable for *in vivo* monitoring. (ii) The signal from an electrochemical cell is electrical. So, no conversion to an electrical signal for the measurement process is required as in spectroscopy. (iii) Detection limits, typically ranging from 10^{-9} to 10^{-6} M, are adequate for the detection of many of the analytes of interest. (iv) The relatively simple and inexpensive instrumentation lends itself to portability and disposable devices.

Electrochemical cells also exhibit certain disadvantages, which have restricted their implementation as sensors. Perhaps the main disadvantage of electrochemistry is its inherent lack of selectivity in comparison with analytical techniques such as chromatography and atomic emission spectroscopy, which exhibit excellent selectivity. Nonetheless, sensors with a high degree of selectivity have been devised by a variety of approaches. A second disadvantage is the necessity of the reference electrode. It is important that this electrode maintain a constant half-cell potential during the course of the measurement: a drift in the half-cell potential over time causes erroneous results. This requirement can be problematical when monitoring over a long period of time.

Electrochemical sensors comprise two basic types: potentiometric and amperometric. In potentiometric sensors the potential between an indicator electrode and a reference electrode is measured under the condition of essentially zero current. Since no current passes through the cell while the potential is measured, potentiometry is an equilibrium method that does not perturb the sample. The application of a potential to an electrochemical cell and the measurement of current due to an oxidation or a reduction process at the electrode surface is the basis of the amperometric sensor. A small amount of analyte is consumed by the sensor measurement process.

Perhaps the most important aspect in electrochemical sensor development has been achievement of the degree of selectivity, for the analyte of interest, that is commensurate with the complexity of the sample for which the sensor is to be used. This point has been addressed successfully for both potentiometric and amperometric sensors. In both cases, effective sensors for specific analytes have been developed, produced commercially, and applied to "real world" samples. A second area of investigation has been the miniaturization of electrochemical sensors. Measurements have been made on nanoliter and smaller volumes for both sensor types.

Some of the early potentiometric sensors were based on the use of an indicator electrode which constitutes the other half of the redox couple of the analyte (e.g., the determination of silver ion with a silver wire electrode). Electrodes of this type lack the selectivity required to make measurements in complex samples. A significant improvement in selectivity was realized by the ion selective electrode in which the potential across a semi-permeable membrane is measured. In this case, the selectivity of the sensor is determined by the selectivity of the interaction of the analyte and the membrane. Excellent ion selective electrodes have been developed for H^+, Na^+, K^+, Ca^{2+}, F^- and other ions. These sensors are now routinely used for the determination of electrolytes in biological fluids such as serum samples and for environmental monitoring. Electrodes for the determination of gases such as CO_2, NH_3, H_2S have been devised by combining gas permeable membranes with pH electrodes. The scope of ion selective electrodes has been expanded dramatically by combining the electrode with enzymes. Such electrodes have not realized their commercial potential in part because of inadequate shelf-life due to denaturation of the enzyme.

One of the first amperometric sensors was the Clark oxygen electrode, which has proved to be an excellent device. Its selectivity is imparted by the use of a gas permeable membrane which prevents most interfering species from reaching the electrode surface. This electrode has been used for in vivo monitoring of dissolved oxygen, and the environmental monitoring of PO_2 in streams and lakes. Enzyme catalyzed reactions have also been used to impart selectivity to amperometric devices. For example, glucose oxidase immobilized on a platinum electrode enables the determination of glucose by means of the following reaction

$$\text{glucose} + O_2 \longrightarrow \text{gluconic acid} + H_2O_2$$

in which either the decrease in oxygen or the appearance of hydrogen peroxide can be monitored electrochemically. This sensor has been available commercially in both forms for the determination of glucose in serum samples, beverages, etc. The enzyme is immobilized between appropriate membranes to provide additional selectivity.

Thus, electrochemical sensors have an established track record of commercial development and application. However, the challenge for the future is to dramatically widen the scope of application. This will necessitate improvements in selectivity and detection limit to enable a wider range of analytes to be measured in complex sample matrices. Since these advances will likely come in part via selective biological reactions involving enzymes, antibodies, and receptors, techniques for immobilizing these biological materials to provide commercially viable (easily manufactured, adequate shelf-life, reproducible, etc.) devices will be important. The successful development of implantable devices for continuous _in vivo_ monitoring will require miniaturization, exceptional stability, _in vivo_ calibration techniques, and avoidance of clotting and tissue growth on the device.

The following chapters on electrochemical sensors deal primarily with two important aspects of sensor development, selectivity and miniaturization. Both classes of sensors, potentiometric and amperometric, are represented. Meyerhoff et al. describe a strategy that enables selectivity to be achieved for anion and gas selective potentiometric sensors based on the incorporation of metal ligand binding complexes within polymeric membranes. Highly selective sensors for sulfite, salicylate and thiocyanate are based on specific reactions with metalloporphyrins. This approach has enabled deviation from the Hofmeister anion pattern, which has previously dictated anion selectivity. The use of bilayer membranes as a means of achieving selectivity for sensors is a topic that is currently receiving a good deal of attention since it provides a means for incorporating the selectivity of, for example, biological receptors in a device. McConnell et al. describe a novel light addressable potentiometric sensor for measuring the electrical properties of phospholipid bilayer membranes. Yacynych et al. and Heineman et al. describe techniques for immobilizing polymer films on electrode surfaces as a means of imparting selectivity. Yacynych et al. explore the use of electro-polymerized films as a means of eliminating electroactive interferences and electrode fouling as exemplified with a glucose sensor based on reticulated vitreous carbon. Heineman et al. use gamma-irradiation as a means of immobilizing water soluble polymers and enzymes on graphite electrodes. The application of miniaturized sensors is the topic of papers by Ikariyama and Yamauchi, Hill et al. and Wightman et al. Ikariyama and Yamauchi describe a microfabricated biosensor for glucose determination in a droplet (10 μL) of sample. Hill et al. consider the use of microelectrodes for mediator-enzyme based biosensors. Wightman et al. demonstrate electrodes with micrometer dimensions to probe spatially heterogeneous chemical concentrations. This approach is potentially important for the measurement of secretion of chemical substances from living cells. Improved selectivity is sought by combining optical electrochemical techniques in the novel fiber optic electrode for an opto-electrochemical sensor described by Aizawa et al.

William R. Heineman
Department of Chemistry
University of Cincinnati
Cincinnati, OH 45221–0172

RECEIVED April 28, 1989

Chapter 2

New Anion- and Gas-Selective Potentiometric Sensors

M. E. Meyerhoff, D. M. Pranitis[1], H. S. Yim, N. A. Chaniotakis, and S. B. Park

Department of Chemistry, University of Michigan, Ann Arbor, MI 48109

Although highly selective membrane electrodes are now used routinely for direct measurements of various cations (e.g. H^+, K^+, Na^+, Ca^{2+}), there are relatively few analogous sensing devices that can be utilized to determine specific anions and gases. In this paper, recent progress in the design of simple anion and gas selective potentiometric chemical sensors based on solvent/polymeric membranes will be summarized. In the case of anion sensing, the incorporation of various metal-ligand complexes, including metalloporphyrins, within polymeric membranes has resulted in the development of highly selective sensors for sulfite, salicylate and thiocyanate. The sulfite selective membrane may be further used in conjunction with an outer gas permeable membrane to design SO_2 selective sensing systems. Existing cation and the newer anion polymeric membrane electrodes can also be employed to fabricate novel differential gas sensing cells that exhibit enhanced sensitivity toward the analyte gas. The principles of this differential approach are demonstrated via the design of a new ammonia selective sensor.

Advances in the development of ion-selective electrodes over the past 20 years have focused primarily on the design of membrane-based sensors for monitoring

[1]Current address: Lever Brothers Company, Edgewater, NJ 07020

0097-6156/89/0403-0026$06.00/0

physiologically important cations. Perhaps the most useful potentiometric ion-selective devices are now fabricated with solvent/polymeric membranes. Selectivity of these electrodes for one ion over others is dictated by the relative interactions of ions in the sample solution with active components of the organic, polymeric membrane phase.

In practice, the selectivity of polymer membrane electrodes can be evaluated using the well known Nicolsky equation:

$$E_{cell} = K + (0.059/z_i)\ \log\left(a_i + \sum_j k_{i,j}^{pot}\ a^{z_i/z_j}\right) \quad (1)$$

where a_i is the activity of the analyte ion with charge z_i; a_j is the activity of an interferent ion with charge z_j, $k_{i,j}^{pot}$ is the potentiometric selectivity coefficient, and E_{cell} is expressed in volts.

Either neutral or charged carrier molecules (e.g. crown ethers, natural antibiotics, synthetic ionophores, etc.) can be doped into polymeric membranes to achieve desired selectivities. However, despite successes in the fabrication of selective and stable cation electrodes (1-5), the development of analogous devices for anions has thus far been limited by the inability to identify appropriate anion-selective ionophores.

In this paper, results are presented from recent studies regarding the use of metal-ligand complexes as anion selective membrane components. This anion sensor work is one component of a longer term effort to develop new and improved potentiometric gas sensors. Indeed, Figure 1 illustrates how appropriate polymeric ion-selective membranes can be used in either static (A) or flow-through gas sensing configurations (B). By sensing ionic forms of the gases in a recipient buffer (rather than pH changes in a non-buffered layer as in conventional Severinghaus style gas sensors), enhanced selectivity over other acidic and basic gases can be achieved. This concept has been successfully applied for the selective detection of ammonia in solution as well as in air (6-8) using an internal nonactin-based polymeric membrane sensitive to ammonium ions. Extending this concept to other gases, particularly SO_2, NO_2, and CO_2, requires the development of suitable anion selective polymeric membranes.

While it is most advantageous to use membrane electrodes sensitive to ionic forms of the analyte gases, certain fabrication advantages are also realized merely by

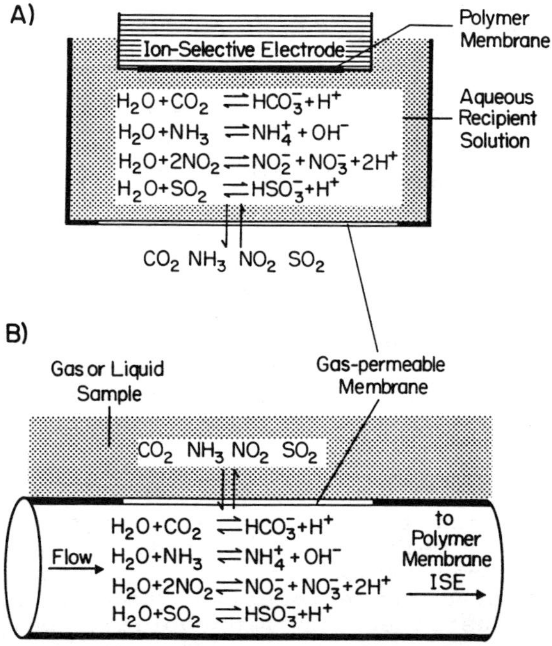

Figure 1. Schematic of static (A) and continuous flow (B) gas sensing system using solvent/polymeric ion-selective membrane electrodes as detectors.

using polymer rather than glass pH selective membranes as internal transducers in more classical Severinghaus designs (9, 10). Moreover, as described below, the combination of using these two distinctly different gas detection schemes (gas-ion detection and pH detection) simultaneously in a single differential gas sensor configuration can offer a unique enhancement in gas sensitivity.

Anion Selective Membrane Electrodes

The literature is full of reports that describe the response properties of anion selective solvent/polymeric membrane electrodes (4, 5, 11, 12). In almost every instance, the active membrane components of such electrodes are lipophilic quaternary ammonium or phosphonium salts. Since the organic cations of these salts are very hydrophobic, the positively charged sites remain in the organic membrane phase. Assuming no direct interaction between the lipophilic cationic sites and counter anions in the membrane phase, phase boundary potentials are generated at the sample/membrane interface via the partitioning of the more hydrophilic counter anions into the aqueous sample. Thus, for a fixed anion activity in the sample, the solubility of the given anion in the organic membrane phase will dictate the magnitude of the phase boundary potential generated. With regard to selectivity, this mechanism always results in the so-called Hofmeister pattern ($ClO_4^- > SCN^- > I^- > Br^- > Cl^- > HCO_3^-$) regardless of the structure of quaternary ammonium or phosphonium salt incorporated in the membrane. In physical terms, the Hofmeister pattern correlates directly with the free energies of hydration for the various anions.

To deviate from Hofmeister behavior, the polymeric membrane must be doped with species that directly interact with, or preferentially solvate, selected anions. To date, such deviations have been observed in only a few instances. For example, carbonate anion selectivity can be induced by incorporating trifluoroacetophenone derivatives as the membrane active component (13). In this case, the carbonate ion is a strong enough nucleophile to react with the electrophilic carbonyl group of the ketone, forming a stable dianion adduct in the membrane phase. Organotin compounds have been shown to exhibit selectivity toward thiocyanate, chloride, and phosphate, depending on the structure of the species doped into the membrane (14-16). Membranes containing a lipophilic derivative of vitamin B_{12} have exhibited unusually strong response to nitrite and thiocyanate, apparently arising from preferential coordination of these anions with the cobalt (III) center of the corine structure (17).

Thiocyanate and Salicylate Selective Membranes Using Metalloporphyrins as Ionophores. Studies with metalloporphyrins have suggested that such species can also yield anion sensitive membranes with anti-Hofmeister selectivity sequences (18, 19). Potentiometric anion responses of the membranes doped with metalloporphyrins may be attributed to the direct interaction of sample anions with axial coordination sites of the central metal. Thus, controlling the equilibrium constant for anion coordination should influence the potentiometric anion selectivity sequences observed. Variations in the coordination affinities of anions may be achieved by altering the central metal ion and/or changing structural appendages on the porphine ring.

To demonstrate these principles, the potentiometric anion selectivities of membrane electrodes prepared by doping the various metalloporphyrins shown in Figure 2 into plasticized PVC membranes (at approx. 1 wt% of porphyrin) have been examined. Table I summarizes the results of such studies.

Table I. Potentiometric Selectivity Coefficients, log $k^{pot}_{Cl,x}$ of Different Metalloporphyrin - PVC Membranes Relative to Chloride[a]

Anion	TDMA-Cl (1)	Mn[TPP]Cl (2)	Mn[TPPP]Cl (3)	Mn[TBrNP]Cl (4)	$Sn[TPP]Cl_2$ (5)
Cl^-	0.0	0.0	0.0	0.0	0.0
Br^-	0.6	0.3	1.0	0.2	0.1
I^-	2.8	1.5	2.3	1.6	0.1
ClO_4^-	4.6	1.5	2.8	0.4	0.4
IO_4^-	4.0	1.8	2.5	0.6	0.2
SCN^-	3.0	1.3	5.2	3.5	1.3
Sal^{-b}	2.7	2.1	3.5	2.0	3.8

[a]Evaluated in 0.05 M MES(2-Morpholinoethanesulfonic acid)buffer, pH 5.5
[b]Sal^- = salicylate

For comparison purposes, selectivity data is also presented for membranes prepared with a typical quaternary ammonium exchanger, tridodecylmethylammonium chloride (TDMA-Cl) (1). All the porphyrin species tested yield selectivities which clearly deviate from that observed with TDMA-Cl. More importantly, two compounds, Mn[TPPP]Cl (3) and $Sn[TPP]Cl_2$ (5), show extraordinary selectivity toward thiocyanate and salicylate, respectively. In the case of Mn[TPPP]Cl, adding conjugated and bulky substituents in the form of twelve phenyl groups to the basic tetraphenylporphyrin [TPP] structure dramatically enhances the response and selectivity toward thiocyanate

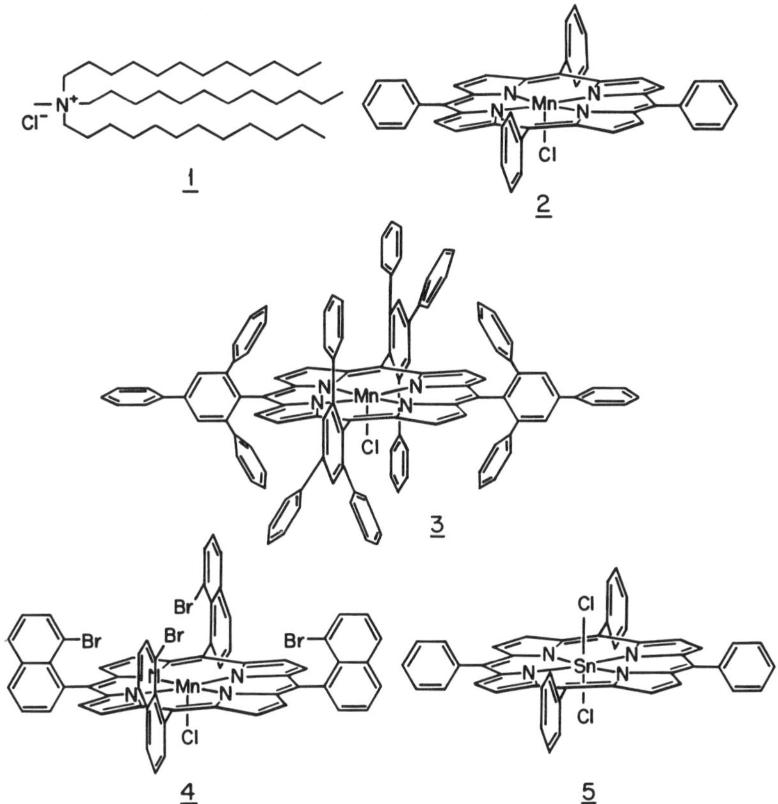

Figure 2. Structures of some compounds incorporated into poly(vinyl chloride) membranes for the development of anion sensors; (1) tridodecylmethyl ammonium chloride (TDMA-Cl); (2) chloro (5,10,15,20 - tetraphenyl-porphyrinato) manganese (III) (Mn[TPP]Cl); (3) chloro (5,10,15,20) -tetra(triphenyl) - phenylporphyrinato) manganese (III) (Mn[TPPP]Cl); (4) α, α, α, α - chloro (5,10,15,20) - tetrakis (8-bromo-1-naphthyl) porphyrinato) maganese (III) (Mn[TBrNP]Cl); (5) dichloro (5,10,15,20 - tetraphenyl-porphyrinato) tin (IV) ($Sn[TPP]Cl_2$).

(relative to Mn[TPP]Cl (2)). In fact, the resulting membrane electrode responds in a Nernstian fashion to thiocyanate in the range from 0.01 mM to 10 mM (see Figure 3). The high selectivity over chloride and salicylate make this sensor potentially useful for detecting SCN^- in biological samples (e.g., urine and saliva) where elevated levels of thiocyanate correlate with excessive cigarette smoking (20).

The data obtained for the Sn[TPP]Cl_2-based membranes clearly illustrate how a change in the central metal ion of the porphyrin can influence potentiometric anion selectivities. As shown in Table I, incorporation of this compound into membranes yields a selectivity pattern which is markedly different than that for Mn[TPP]Cl. Salicylate (o-hydroxybenzoate) remains the preferred anion, but the relative response toward this ion over other anions is enhanced significantly. The increased response to salicylate is due to its stronger interaction as an axial ligand, probably due to the oxophilic nature of the Sn(IV) center. Since the charge on the tin center is +4, interaction of anion ligands at either axial coordination site mandates that this metalloporphyrin function as a charged carrier type ionophore in the membrane phase.

As shown in Figure 4, the selectivity of the Sn[TPP]Cl_2-based membrane over other aromatic organic compounds is quite high. This fact coupled with the membrane's high selectivity over chloride (see Table I) makes the resulting membrane electrode potentially useful for monitoring levels of free salicylate in physiological samples. In this regard, preliminary results have been obtained for the determination of salicylate in spiked and unspiked human urine samples (Table II). As can be seen, there is reasonably good correlation between the values obtained with the new salicylate selective membrane electrode and the conventional colorimetric method now used in hospitals.

Unfortunately, at this point, interpreting results for the electrode measurement of salicylate in blood samples is complicated by the fact that a large fraction of the total salicylate is bound to proteins (22, 23). The Sn(TPP)Cl_2-based membrane electrode detects "free" salicylate while the conventional colorimetric procedure (i.e. the Trinder method (21) or variations thereof) measures total salicylate concentration (free plus bound). The former is the physiologically active form of the compound (24). Thus, the salicylate selective electrode could provide a new analytical tool for scientists who are examining the pharmacological effects of aspirin and other

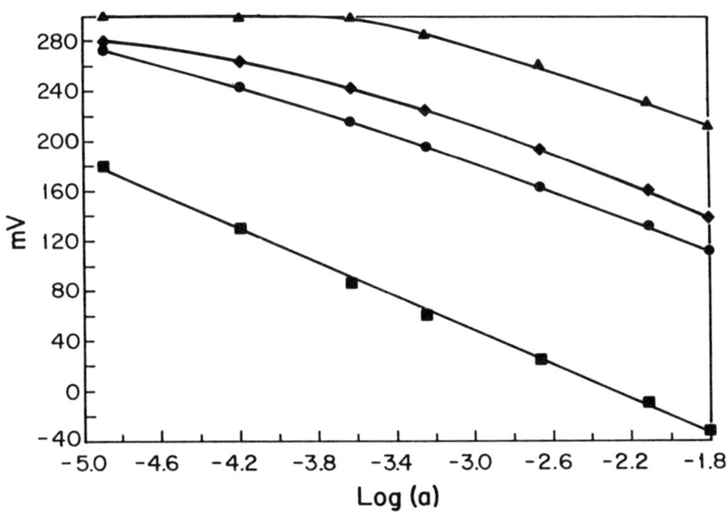

Figure 3. Potentiometric response of Mn[TPPP]Cl-PVC membrane toward various anions in MES buffer, pH 5.5 (■) SCN^-; (●) $ClO4^-$; (◆) I^-; (▲) Br^-.

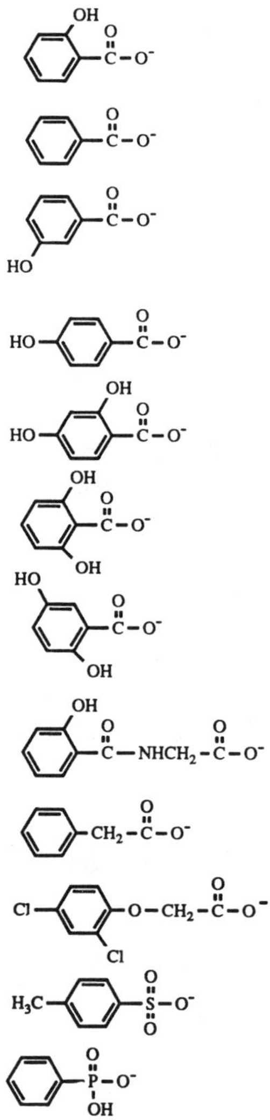

Anion	Log $k^{pot}_{sal/x}$
Salicylate	0.0
Benzoate	-1.4
3-Hydroxybenzoate	-1.1
4-Hydroxybenzoate	-1.0
2,4-Dihydroxybenzoate	-1.4
2,6-Dihydroxybenzoate	-0.6
2,5-Dihydroxybenzoate (Gentisate)	-1.3
Salicylurate	-1.5
Phenylacetate	-1.8
2,4-Dichlorophenoxyacetate (2,4-D)	-1.1
p-Toluenesulfonate	-2.9
Phenylphosphonate	-2.9

Figure 4. Potentiometric anion selectivity coefficients, relative to salicylate, of Sn[TPP] Cl_2-PVC membrane toward various aromatic anion species.

Table II. Comparison of Results Obtained for Measurement of Salicylate Concentrations in Spiked and Non-Spiked Human Urine Samples by $Sn(TPP)Cl_2$-Based Membrane Electrode and Conventional Colorimetric Method

Sample[a]	Electrode[b, c] (mM)	Colorimetric (mM)
1A	0.32	0.35
1B	1.05	1.02
1C	1.79	1.97
2A	0.24	0.07
2B	0.84	0.88
2C	2.02	2.04
3A	0.26	0.22
3B	0.86	0.80
3C	2.02	1.90
4A	0.18	0.29
4B	0.68	1.02
4C	1.53	1.68
5A	0.24	0.36
5B	0.78	1.02
5C	1.78	1.97

[a]A-samples are unspiked; B-samples are A-samples spiked with salicylate to achieve a change in concentration of 0.79 mM; C-samples are A-samples spiked to achieve change in concentration of 1.9 mM salicylate.

[b]average of 3 determinations

[c]samples diluted 1:10 in MES, pH 5.5 buffer for measurements.

salicylates in living subjects. This application cannot be pursued at this time due to the electrode's pH response. As shown in Figure 5, when the pH of the sample is in the physiological range, detection limits toward salicylate are poor. Only after the sample is diluted in low pH buffer (e.g., pH 5.5) can sub-millimolar levels of salicylate be quantitated accurately. Unfortunately, such a change in pH is likely to disturb the salicylate-protein binding equilibria. Recent studies have suggested that the total water content of the membrane plays an important role in the observed potentiometric pH response (18,25). Thus, by changing the plasticizer and/or by switching to polymer matrices other than PVC, a significant reduction in pH response of the membrane may be possible.

Sulfite/Bisulfite Selective Electrode Based on Bis-(Diethyldithiocarbamato Mercury (II). The incorporation of metal-ligand complexes within polymeric membranes for the development of new anion sensors is not limited to the use of metalloporphyrins. Indeed, recent studies with the Hg (II) complex of diethyldithiocarbamate ($Hg(DDC)_2$) have yielded a new membrane electrode with relatively high specificity for sulfite ions (Pranitis, D.M.; Meyerhoff, M.E. Anal. Chim. Acta, in press). The membrane is prepared by doping the mercury complex at 1.8 wt% in a PVC matrix plasticized with dibutylphthalate. Potentiometric response of the membrane to step changes in sulfite ion activity in phosphate buffer, pH 10.0, occur in ≤ 2 min. When the equilibrium potentials are plotted vs. the logarithm of sulfite activity in the sample solution, a calibration curve such as that shown in Figure 6 is obtained. Near Nernstian behavior is observed in the activity range of 0.1 to 1000 μM sulfite.

Selectivity of the $Hg(DDC)_2$-based membrane toward sulfite relative to other anions correlates with what one might expect for anionic species interacting with the Hg(II) center of the complex. As shown in Table III, iodide, bromide, thiosulfate, and thiocyanate are major interferents. However, the membrane exhibits little or no response to a wide range of other anions including sulfate, phosphate, chloride, and nitrate, making it potentially useful as an analytical sensor for measurements in certain samples, e.g., in monitoring added sulfite levels in food and beverages.

Table III. Potentiometric Selectivity Coefficients of $Hg(DDC)_2$ -Based Membrane[a]

Anion	$\log k^{pot}_{SO_3^{2-},j}$
sulfite	0
chloride, sulfonate (MES), acetate, nitrite, citrate, nitrate, perchlorate, cyanate, salicylate, sulfate, bicarbonate/carbonate	< -4
bromide	0
thiocyanate	0
thiosulfate	0.5
iodide	7

[a]Selectivity coefficients measured with matched-potential, separate-solution method in 10 mM phosphate buffer, pH 10.0.

The mechanism by which the $Hg(DDC)_2$ compound acts as an ionophore for sulfite is not yet clear. Initial studies indicate that the mercury(II) species is essential for sulfite response. When $Zn(DDC)_2$ is used in place of

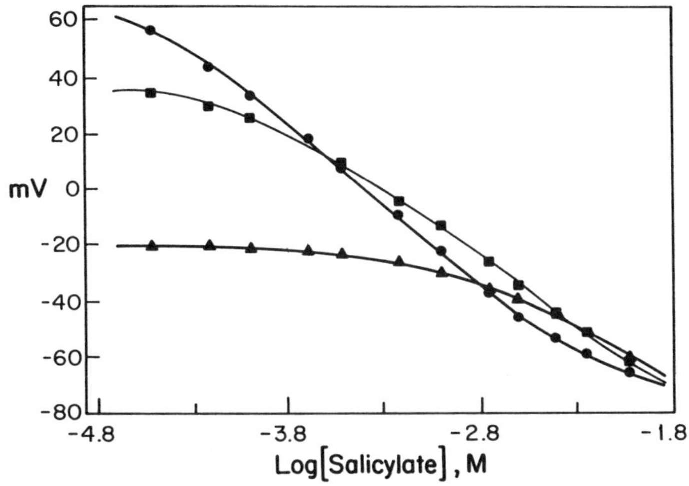

Figure 5. Potentiometric response of Sn[TPP] Cl_2-based membrane electrode toward salicylate in different MES buffers: (▲) pH 7.2; (■) pH 6.2; (●) pH 5.5.

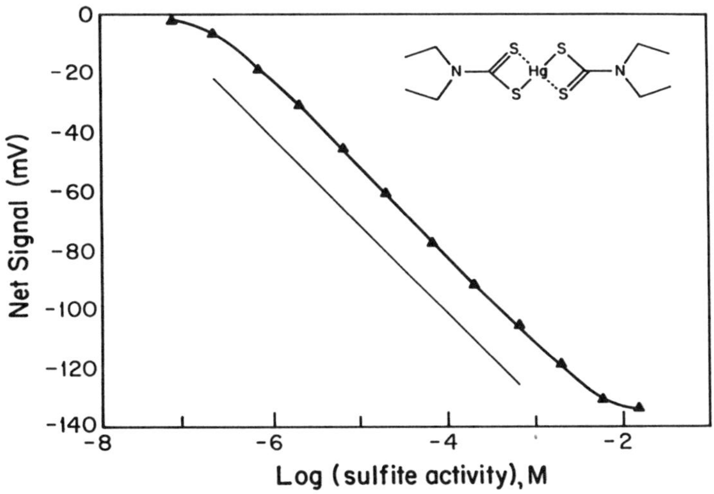

Figure 6. Typical calibration curve toward sulfite for $Hg(DDC)_2$-PVC membrane electrode (solid line is theroetical Nernstian slope line).

the mercury complex, no sulfite sensitivity is observed. However, when diphenylmercury(II) is incorporated into polymeric membranes, significant sulfite response is observed although it is inferior to that found with $Hg(DDC)_2$. It appears that sulfite interacts with the Hg(II) by displacing a thiocarbonyl sulfur of the DDC ligand, yielding a negatively charged divalent complex. If so, in contrast to the $Sn(TPP)Cl_2$ - salicylate system above, the $Hg(DDC)_2$ species should be classified as a neutral carrier type ionophore.

New Potentiometric Gas Sensing Systems

As stated previously, the development of new anion selective membrane electrodes can also lead to the design of improved potentiometric gas sensing systems. At present, almost all commercial gas sensors rely on the use of glass membrane pH electrode to detect pH changes in a thin film of electrolyte sandwiched between the glass membrane and an outer gas permeable membrane (typically microporous Teflon). This is the so-called Severinghaus design (26). For example, an SO_2 sensor may be fabricated by using an internal electrolyte consisting of sodium bisulfite and sodium chloride (27). Equilibration of dissolved SO_2 into the thin layer of electrolyte results in a pH change logarithmically proportional to the concentration of SO_2 in the sample. However, the selectivity of such a sensor is limited. In this case, other volatile acidic species, such as acetic acid, NO_x, and even CO_2 will cause positive interference with the measurement of SO_2 (28).

Sulfur Dioxide Sensing Based on Sulfite Selective Membrane Electrode. The $Hg(DCC)_2$-based sulfite selective electrode described above can be used as a transducer to devise new SO_2 selective gas sensing systems according to the detection schemes depicted in Figure 1. The addition of the outer gas permeable membrane enables measurements to be made in the presence of ions which would normally be major interferents to the membrane electrode (e.g. I^-,Br^-, $S_2O_3^{-2}$ etc.). Initial studies have focused on incorporating the sulfite selective electrode in a flow-through gas sensing arrangement such as the one illustrated in Figure 7. It has been found that the sulfite electrode exhibits a mixed response to both sulfite and bisulfite, which allows for the use of a pH 6.0 buffered recipient stream. Upon acidifying samples containing sulfite/bisulfite, sulfur dioxide gas is formed in proportion to the concentration of the sulfites present. This gas diffuses through the membrane of the dialyzer where it is trapped as a bisulte/sulfite mixture in the recipient buffer stream. Potentiometric response of the downstream sulfite selective electrode is proportional to the logarithm of sulfite/bisulfite concentration in the original acidified sample.

This approach greatly enhances the selectivity for measuring sulfites as SO_2 relative to sensing systems that detect SO_2 via pH changes in a recipient electrolyte solution. Indeed, as illustrated in Figure 8, sulfite can be detected to levels below 10 μM even in the presence of 0.1M acetic acid. While the acetic acid does permeate the gas membrane into the recipient stream, acetate ions formed in the recipient stream go undetected by the $Hg(DCC)_2$-based sulfite selective sensor. This type of enhanced selectivity would be extremely valuable when determining sulfites in wine-vinegar or other samples that contain volatile acidic species.

Differential Gas Sensing Membrane Electrode-Based Detectors with Enhanced Gas Sensitivity. One frequent complaint regarding the use of potentiometric gas and ion-selective membrane electrodes for analytical purposes is lack of precision owing to the logarithmic response of such devices. Thus, uncertainty in measured potentials of ±1mV will result in ±4% precision for sensors with slopes of 59 mV/decade and ±8% for those devices based on response to divalent ions (e.g., the above sulfite and SO_2 gas sensors). One novel approach suggested in the literature for enhancing response slopes of potentiometric sensors is to use several membrane electrode cells in series (28, 29). This arrangement results in response slopes n times the Nernstian value, where n is the number of two electrode cells (working and reference) in series. Unfortunately, this approach results in a more complex system where the number of electrodes required is increased, as is the number of separate sample compartments.

Recently, a novel two electrode differential potentiometric cell for enzyme electrode systems has been described that provides enhanced substrate sensitivities compared to conventional cells composed of a single enzyme electrode and reference (Cha, G.S.; Meyerhoff, M.E. Electroanalysis, in press). The cell employs two working enzyme electrodes, one which responds to the analyte in the positive potential direction via detection of cations, and one which responds to the same analyte but in a negative direction owing to anion detection. A similar approach can be applied in the design of new two electrode gas-selective sensors with enhanced gas sensitivity.

Such a differential gas sensing system is composed of two working gas sensors, each with a different inner ion-selective polymer membrane electrode as the transducer. For example, a differential ammonia sensing arrangement involves the use of two ammonia sensing probes whose

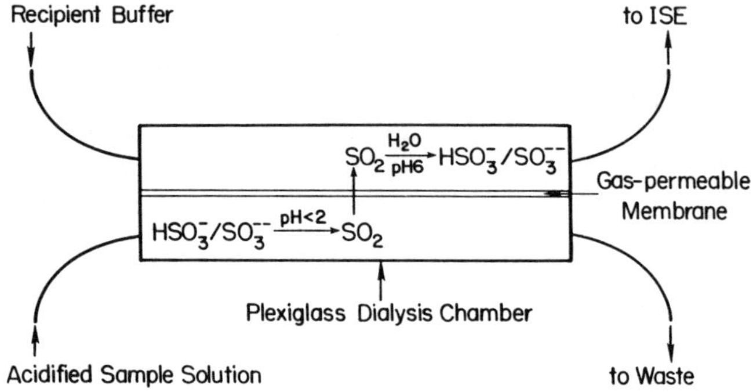

Figure 7. Flow-through gas sensing arrangement used to evaluate $Hg(DDC)_2$-PVC membrane electrodes for selective detection of sulfites as SO_2.

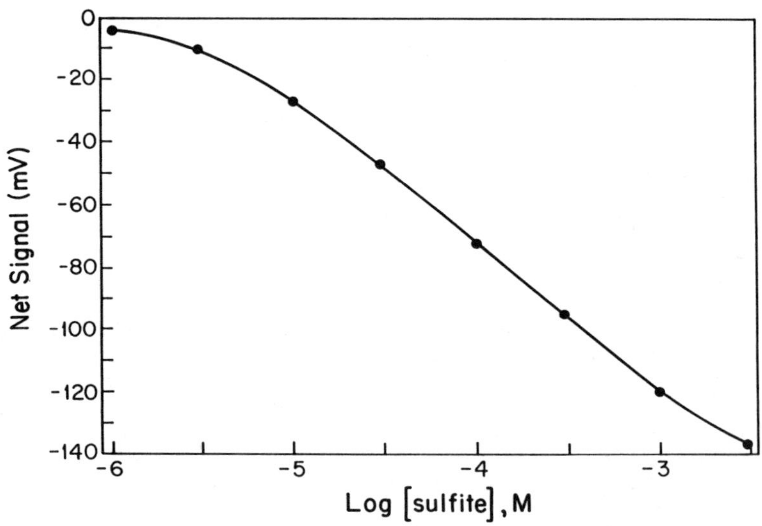

Figure 8. Typical response of SO_2 gas sensing system toward sulfite standards prepared in a background of 0.1 M sodium acetate. Samples acidified with 0.1 M H_3PO_4; recipient stream: 0.1 M MES buffer, pH 6.0.

internal filling solutions are connected via a salt bridge (see Figure 9). One half cell responds to ammonia gas by detecting an increase in pH in a thin film of electrolyte (i.e., NH_4Cl) sandwiched between a polymeric pH sensitive membrane (prepared with tridodecylamine as membrane active species (9)) and an outer gas permeable membrane. The second half cell detects ammonia gas in the sample by responding to changes in ammonium ion activities in a thin layer of buffer sandwiched between a nonactin-based ammonium ion-selective polymeric membrane and another outer gas permeable film (6). Accordingly, the overall measured potential for this two working electrode cell is the difference in potential between the ammonium ion-selective electrode and the polymeric pH electrode:

$$E_{cell} = E_{NH_4^+} - E_{pH} \qquad (2)$$

or

$$E_{cell} = K + 0.059 \log a_{NH_4^+} - 0.059 \log a_{H^+} \qquad (3)$$

where E_{cell} is in volts, $a_{NH_4^+}$ and a_{H^+} are the activities of ammonium ions and protons in the thin films of internal solutions held between the outer gas permeable membranes and the respective ion-selective membranes, and K is the sum of all constant potentials in the cell (e.g., junction potentials at salt bridge, inner Ag/AgCl potentials of each membrane electrode, etc.). Diffusion of gaseous ammonia into the thin films results in the equilibrium hydrolysis of the ammonia:

$$NH_3 + H_2O \leftrightarrow NH_4^+ + OH^- \qquad (4)$$

with an equilibrium constant of

$$K_{NH_3} = a_{NH_4^+} \cdot a_{OH^-} / P_{NH_3} \qquad (5)$$

Therefore, if the pH of the film in contact with the ammonium electrode is buffered, the $a_{NH_4^+}$ in the film is directly proportional to the partial pressure of ammonia gas,

$$a_{NH_4^+} = K_{NH_3} \cdot P_{NH_3} / a_{OH^-} \qquad (6)$$

On the other hand, for the pH electrode half cell, since $a_{OH^-} = K_w/a_{H^+}$, then the activity of protons in the thin film is given by,

$$a_{H^+} = a_{NH_4^+} \cdot K_w/P_{NH_3} \cdot K_{NH_3} \quad (7)$$

However, in the pH half cell, $a_{NH_4^+}$ is kept high and relatively constant by using NH_4Cl as the thin film electrolyte. Thus, substituting equations (6) and (7) into equation (3) and combining all the constant terms together yields the following expression for the overall differential cell potential,

$$E_{cell} = K' + 0.118 \log P_{NH_3} \quad (8)$$

As can be seen, such a cell should respond to the partial pressure of ammonia with a theoretical slope of 118 mV/decade.

Figure 10 illustrates the results obtained for such a differential gas sensor cell arrangement. Varying concentrations of ammonia were generated by adding ammonium chloride standards to a 0.01 M NaOH sample solution. The slope of the differential sensor is 93.6 mV/decade in the range of 10^{-6}-10^{-2}M NH_3. This less than theoretical value is expected based on the difficulties in effectively isolating a thin film of solution at the tip of each sensing half cell from the bulk solution required for electrolytic contact (30), and the fact that the pH and ammonium polymer membranes typically exhibit sub-Nernstian behavior (e.g., slopes of 52-56 mV decade). In addition, earlier theoretical predictions on the ammonia response of the half cell detecting ammonium ions also point to lower than theoretical slopes owing to small pH changes in the buffered film (31).

Naturally the selectivity of the new differential gas sensor design will be dependent on the combined gas selectivities of each gas sensing half-cell. Thus, enhanced sensitivity comes at the expense of poorer selectivity owing to the response of the pH sensing half-cell to volatile amines. However, for certain types of samples (e.g. physiological fluids) this should not pose a problem.

Conclusions

As described above, significant progress in the design of anion and gas selective membrane electrodes has been made. While further work is needed to understand fully the response mechanisms and to improve the performance of the new thiocyanate, salicylate, and sulfite selective membrane electrodes, each of these sensors appears to offer adequate selectivity for use in real sample measurements. In addition, by carefully

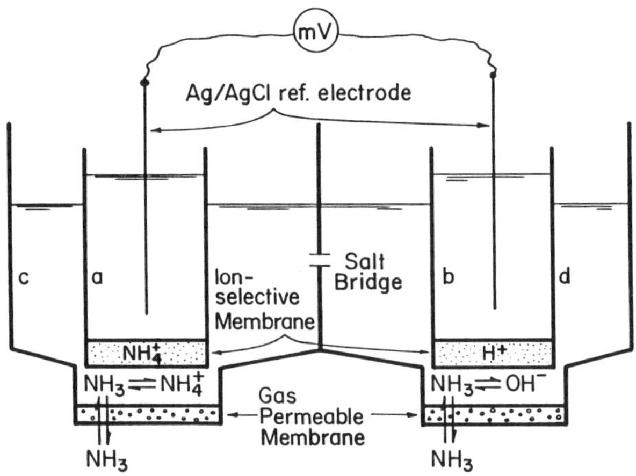

Figure 9. Schematic of differential ammonia gas sensor fabricated with two different polymer ion-selective membranes: (a) 0.1 M NH_4Cl; (b) 0.2 M phosphate buffer, pH 7.0, containing 0.1 M NaCl; (c) 0.1 M Tris-HCl buffer, pH 7.8; (d) 0.05 M NH_4Cl.

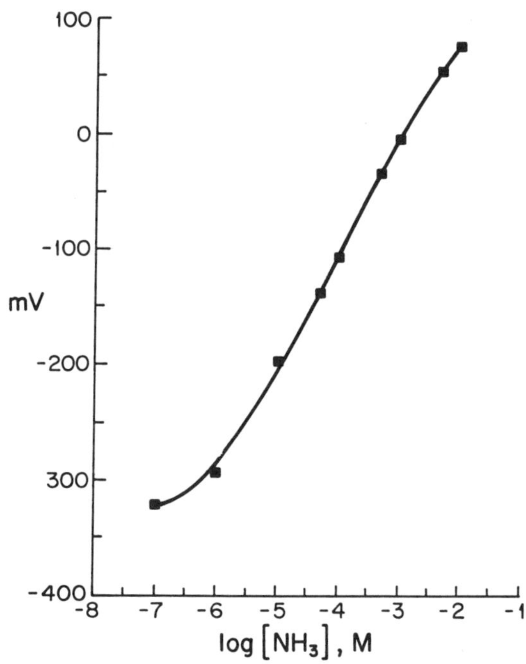

Figure 10. Typical calibration curve for differential ammonia sensor shown in Figure 9, illustrating enhanced ammonia gas sensitivity.

studying the anion-coordinating properties of different metal-ligand complexes in polymer membranes, it is likely that sensors for other important anions can be devised. Furthermore, as demonstrated with the new sulfite sensor, the development of anion responsive membranes also directly impacts advances which can be made in the area of gas selective sensing systems. Indeed, reconfiguring the new sulfite polymeric membrane electrode as a detector in a flow-through gas phase sniffing arrangement (8) should enable the continuous and selective detection of trace levels of ambient SO_2. The sensitivity of this and other gas sensing electrode systems may be enhanced by taking advantage of the novel differential cell concept, introduced here for the detection of dissolved ammonia.

Acknowledgments. The authors gratefully acknowledge financial support from the National Institutes of Health (GM-28882) and Mallinckrodt Sensor Systems. The authors also wish to thank Professor John T. Groves, Department of Chemistry, Princeton University, for providing several of the metalloporphyrin structures used in these studies.

Literature Cited

1. Morf, W.E. The Principles of Ion-Selective Electrodes and of Membrane Transport; Elsevier: Amsterdam, 1981.
2. Ammann, D.; Morf, W.E.; Anker, P.; Meier, P.C.; Pretsch, E.; Simon, W. Ion-Selective Electrode Rev. 1983, 5, 3-92.
3. Oesch, U.; Ammann, D.; Simon, W. Clin. Chem. 1986, 32, 1448-1459.
4. Solsky, R.L. Anal. Chem. 1988, 60, 106R-113R.
5. Koryta, J. Anal. Chim. Acta 1988, 206, 1-48.
6. Meyerhoff, M.E.; Robins, R.H. Anal. Chem. 1980, 52, 2383-2388.
7. Fraticelli, Y.M.; Meyerhoff, M.E. Anal. Chem. 1981, 53, 992-997
8. Pranitis, D.M.; Meyerhoff, M.E. Anal. Chem. 1987, 59, 2345-2350.
9. Opdycke, W.N.; Parks, S.K.; Meyerhoff, M.E. Anal. Chim. Acta 1983, 155, 11-20.
10. Opdycke, W.N.; Meyerhoff, M.E. Anal. Chem. 1986, 54, 950-956.
11. Yu, R.Q. Ion-Selective Electrode Rev. 1986, 8, 153-171.
12 Arnold, M.A.; Solsky, R.C. Anal. Chem. 1986, 58, 84R-101R.
13. Meyerhoff, M.E.; Pretsch, E.; Welti, D.H.; Simon, W. Anal. Chem. 1987, 59, 144-150.
14. Wuthier, U.; Pham, H.V.; Zund., R.; Welti, D.; Funk, R.J.J.; Bezegh, A.; Ammann, D.; Pretsch, E.; Simon, W. Anal. Chem. 1984, 56, 535-538.

15. Oesch, U.; Ammann, D.; Pham, H.V.; Wuthier, U.; Zund, R.; Simon, W. J. Chem. Soc. Faraday Trans. 1986, 82, 1179-1186.
16. Arnold, M.A.; Glazier, S.A., U.S. Patent 4, 735, 692, 1988.
17. Schulthess, P.; Ammann, D.; Krautler, B.; Caderas, C.; Stepanek, R.; Simon, W. Anal. Chem. 1985, 57, 1397-1401.
18. Chaniotakis, N.A.; Chasser. A.M.; Meyerhoff, M.E.; Groves, J.T. Anal. Chem. 1988, 60, 185-188.
19. Ammann, D.; Huser, M.; Krautler, B.; Rusterholtz, B.; Schulthess, P.; Lindermann, B.; Holder, E.; Simon, W. Helv. Chim. Acta 1986, 69, 849-854.
20. Haley, N.J.; Axelrad, C.M.; Tilton, K.A. J. Public Health 1983, 73, 1204-1207.
21. Trinder, P. Biochem. J. 1954, 57, 301-303.
22. Stewart, M.S.; Watson, I.D. Ann. Clin. Biochem. 1987, 24, 552-565.
23. Smith, M.J.H.; Smith, P.K. The Salicylates, A Critical Bibliographic Review; Interscience: New York, 1966; Chapter 1.
24. Levy, G. Drug Metabs. Rev. 1979, 9, 3-19.
25. Ma, S.C.; Chaniotakis, N.A.; Meyerhoff, M.E. Anal. Chem. 1988, 60, 2293-2299.
26. Severinghaus, J.W.; Bradley, A.F. J. Appl. Physiol. 1958, 13, 515-520.
27. Bailey, P.L.; Riley, M. The Analyst 1975, 100, 145-156.
28. Parczewski, A.; Stepak, R.. Fresenius' Z. Anal Chem. 1983, 316, 29-31.
29. Stepak, R. Fresenius' Z. Anal. Chem. 1983, 315, 629-630.
30. Ross, J.W.; Riseman, J.A.; Krueger, J.A. Pure Appl. Chem. 1973, 35, 473-487.
31. Meyerhoff, M.E.; Fraticelli, Y.M.; Opdycke, W.N.; Bachas, L.G.; Gordus, A.D. Anal. Chim. Acta 1983, 154, 17-31.

RECEIVED March 9, 1989

Chapter 3

Electrical Properties of Phospholipid Bilayer Membranes Measured with a Light Addressable Potentiometric Sensor

George B. Sigal, Dean G. Hafeman, J. Wallace Parce, and Harden M. McConnell

Molecular Devices Corporation, 4700 Bohannon Drive, Menlo Park, CA 94025

Techniques for using a silicon-based light addressable potentiometric sensor (LAPS) to measure the electrical properties of phospholipid bilayer membranes were developed. Membrane conductance, capacitance, and potential could all be measured when the membrane was painted on an aperture between the silicon surface and a controlling electrode. The sensor was tested by observing changes in membrane properties on the addition of simple ion carriers and channels.

In previous work we described a light addressable potentiometric sensor that can be used for a variety of analytical purposes (1). Included among these are measurements of pH, redox potential and trans-membrane potential, all at a multiplicity of sites on a single dielectric-coated silicon surface. In this earlier work it was shown that potassium ion concentrations can be measured with this device when a valinomycin-containing polyvinyl chloride membrane is deposited on the dielectric-coated silicon surface. For many biochemical and biophysical experiments it is desirable that this sensor system be adapted for use with phospholipid bilayers. This is not a trivial problem, given the electrical characteristics of silicon, and the silicon oxynitride dielectric coating that is necessary to make this system compatible with aqueous solutions. The long term objective of this project is to provide a system where (a) the electrical properties of lipid bilayers can be determined and (b) where these bilayers can be stabilized by placing them in close proximity to a solid surface. The present paper illustrates the techniques we have developed to accomplish the first part of this objective, using the LAPS configuration. In the present work no effort was made to provide stable, or microscopic sensor - bilayer configurations. Studies on the latter part of the objective will be provided elsewhere.

0097–6156/89/0403–0046$06.00/0

Theory

The LAPS. The LAPS is diagrammed in Figure 1a. The sensor uses photoexcitation of the semiconductor to probe the surface potential at the insulator - liquid interface (1). Typically this is used for potentiometric pH or redox measurements in enzyme linked immunoassays. In the absence of illumination, the LAPS behaves like a metal-insulator-semiconductor (MIS) diode (see Reference 2 for a general description of the MIS diode). For n-type silicon, where the majority charge carriers are electrons, a positive bias potential (Ψ) applied across the system will drive the semiconductor into accumulation. Majority carriers will distribute evenly through the bulk semiconductor except at the semiconductor - insulator interface where negative charges accumulate to balance the charge along the solution - insulator interface. In accumulation, therefore, the system is analogous to a capacitor made up of two metal plates separated by the insulator. The capacitance C_i is simply a function of the width of the insulator, the area, and its permittivity (Equation 1).

$$C_i = \varepsilon \frac{A}{w} \tag{1}$$

If a negative potential is applied across the sensor, electrons are repelled from the semiconductor - insulator interface forming a layer depleted in majority carriers. The depletion layer can be modeled as an insulator with a width (and therefore capacitance, C_d) that is a function of the bias potential. As the bias potential becomes more negative, the depletion layer width increases, and C_d decreases, until the maximum depletion layer width is attained.

Illumination of silicon with infrared light results in the generation of hole-electron pairs (semiconductor photoresponses are reviewed in Reference 3). Photogenerated hole-electron pairs in the semiconductor may diffuse, recombine, or be seperated in an electric field. When the semiconductor is biased into depletion, hole-electron pairs that have diffused into the depletion layer, or that are formed in the depletion layer, are seperated in the electric field on a time scale short with respect to that for recombination. Upon illumination of the semiconductor with a constant intensity light source, charge seperation of photogenerated hole-electron pairs in the depletion region leads to a transient current which acts to collapse the depletion layer and charge the insulator. This current decays as the depletion layer width reaches a new steady state value. Turning off the illumination leads to a transient current of opposite polarity as the insulator discharges and the depletion layer returns to its original width. If the intensity of the light source is modulated

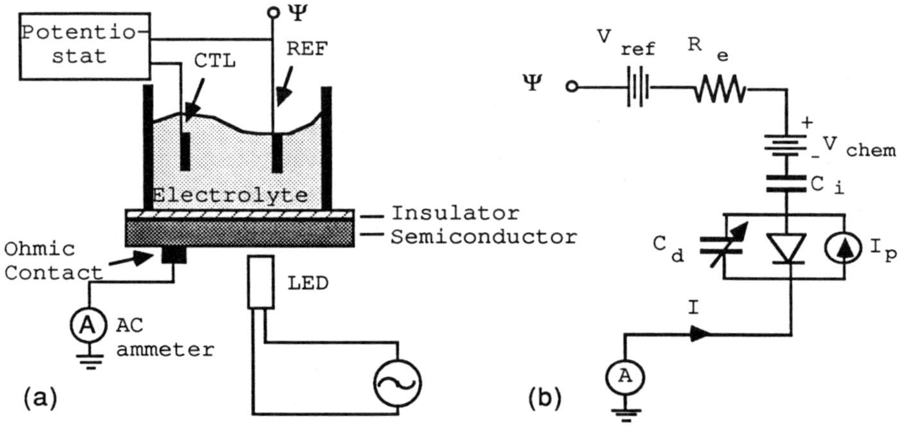

Figure 1. A schematic diagram (a) and a partial equivalent circuit (b) are given for the LAPS. The components Ψ, C_i, C_d, R_e, V_{ref} and V_{chem}, respectively represent the applied bias potential, the insulator and depletion layer capacitances, the electrolyte resistance, the potential across the reference electrode, and a chemically sensitive surface potential. I_p represents the photogeneration of hole-electron pairs, and I the measured alternating photocurrent. Solution potential is maintained by a potentiostat using a Pt controlling electrode (CTL) and Ag/AgCl reference electrode (REF). The potential Ψ is defined as the potential from the output of the reference electrode to ground.

with a period short compared to the decay time constants of the transient currents, an alternating current is produced because there is never enough flow of charges in either direction to significantly modulate the depletion layer width.

Figure 1b represents an approximate equivalent electronic circuit for the LAPS under photoexcitation by an intensity-modulated LED. Modulating the light intensity at frequency (f) results in alternating charging currents through the insulator and depletion layer capacitances at the same frequency. The amplitude, $|I|$, of the insulator charging current through the external circuit is the measured quantity. Under strong depletion, and with a wide range of modulation frequencies, $|I|$ is a function of the rate at which hole-electron pairs form in or diffuse into the depletion layer (I_p), and of the capacitances of the illuminated area as given by Equation 2 where $|I_p|$ is the alternating componant of I_p.

$$|I| = \frac{C_i}{C_i + C_d} \times |I_p| \tag{2}$$

Under conditions of accumulation, although charge seperation occurs, little current in the external circuit results due to the large value of C_d, and the low resistance to charge recombination across C_d (represented by the diode in Figure 1b).

Figure 2 shows typical plots of $|I|$ vs. Ψ. At the most positive bias potential, the semiconducter is in accumulation and no photocurrent is measured. As the semiconductor is biased into depletion, $|I|$ increases as C_d decreases. When the maximum depletion width is attained $|I|$ reaches a maximum value.

The LAPS may be used to monitor chemically-sensitive surface potentials on the insulator surface (V_{chem} in Figure 1b). For example, the potential of a surface with proton binding capacity will increase on addition of protons (4). Therefore, on dropping the pH, the bias potential must be decreased to maintain the same electrical field in the semiconductor and hence the same photocurrent. The result is a shift of the $|I|$ vs Ψ curves along the Ψ axis due to changing the pH as shown in Figure 2. The potential at which the inflection point in the curves occurs (Ψ_{pip}), is used to follow V_{chem}. For a silicon nitride surface Ψ_{pip} vs. pH is linear from pH 0 to 12 with a slope close to the Nernstian 59 mV/pH (1). Similarly, coating the insulator with a metal surface makes the sensor instead sensitive to redox potential (1).

An additional attribute of the LAPS is that of light addressability. From a measurement, V_{chem} is determined only near the illuminated region of the sensor. Therefore by using more than one light source, or by directing a light source to different regions of the sensor, values of

V_{chem} can be determined for discreet sites on the sensor surface. Although this feature is not exploited in the present study, it should prove useful in an extension of this work, where lipid bilayers will be placed directly on the sensor surface.

Phospholipid Bilayer Membranes. A lipid bilayer supported on an aperture between the LAPS and the controlling electrode can be represented by adding the equivalent circuit for a lipid bilayer (5) in series with the LAPS, as in Figure 3. As a thin insulating film, the membrane has a resistance (R_m), and a capacitance (C_m) associated with it. In addition, ion gradients across the membrane may cause trans-membrane potentials. These potentials are represented by the voltage source V_m in the equivalent circuit. It is assumed that no low resistance pathways around the membrane are present. We would like to be able to measure R_m, C_m, and V_m.

Measurement of Trans-membrane Potential. For a constant bias potential, negligible current flows through the insulator on the semiconductor chip, so that there is no steady state potential across R_m. The potential across the chip is $\Psi - V_m$. If in the absence of the membrane an photocurrent inflection point potential was measured at $\Psi = \Psi_{pip}$, then in the presence of the membrane it will shift positive by the value of the trans-membrane potential as indicated by Equation 3. Measuring V_m is clearly analogous to using the sensor to make pH or redox measurements.

$$V_m = \Psi_{pip} - \Psi_{pip}(\text{no membrane}) \tag{3}$$

Measurement of Membrane Capacitance. A sufficiently high frequency of light modulation is chosen so that R_m is large compared to the AC impedance (i.e. capacitive reactance) of the membrane. Therefore, in depletion, $|I|$ can still be described by the relative magnitudes of the internal and external capacitances (Equation 4). As long as $C_m \ll C_i$ (which can be controlled by the size of the illuminated region on the insulated semiconductor), Equation 4 can be transformed into a linear form (Equation 4a).

$$|I| = \frac{\dfrac{C_m C_i}{C_m + C_i}}{\dfrac{C_m C_i}{C_m + C_i} + C_d} \times |I| \tag{4}$$

$$\frac{|I|}{C_m} = \frac{|I_p|}{C_d} - \frac{|I|}{C_d} \tag{4a}$$

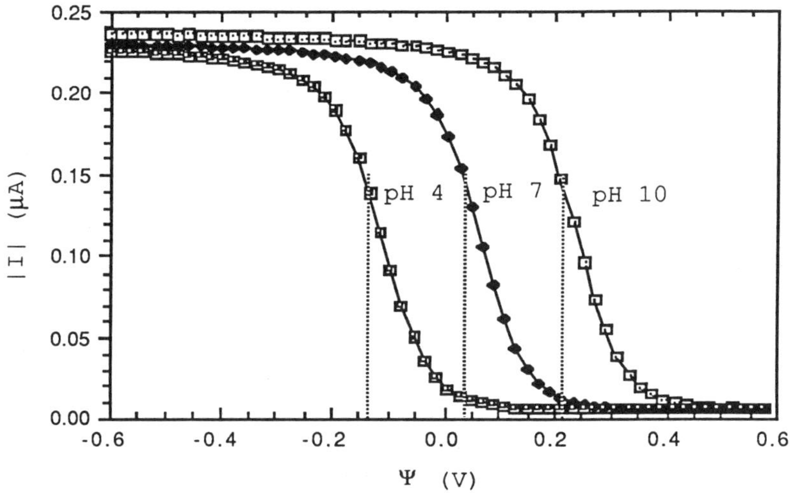

Figure 2. Alternating photocurrent amplitude (|I|) as a function of bias potential (Ψ) for the LAPS illuminated with 10 kHz intensity-modulated light. The three curves shown were obtained with 50mM potassium biphthalate buffer, pH 4.0; 50 mM potassium phosphate buffer, pH 7.0; and 50 mM potassium borate buffer, pH 10.0. The photocurrent inflection potentials (Ψ_{pip}), marked by dashed lines, were -138 mV, 39 mV, and 215 mV respectively. Photocurrent is given in RMS amplitude.

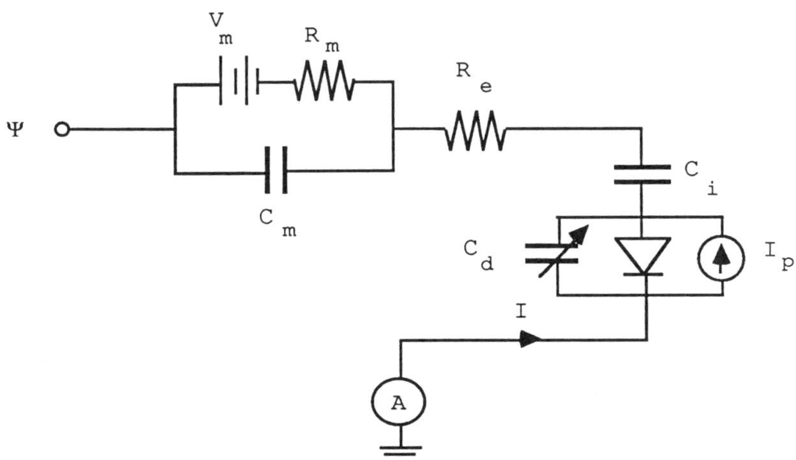

Figure 3. Equivalent electrical circuit for LAPS with lipid bilayer suspended between controlling electrode and sensor surface. R_m, C_m, V_m are the membrane resistance, capacitance, and trans-membrane potential.

If the bias potential and $|I_p|$, and therefore C_d, are kept constant, then changes in C_m can be calculated from $|I|$ once C_d is determined. In practice, calibrated capacitors (to model C_m) can be inserted into the circuit prior to formation of the membrane. Calibration curves of the modelled C_m vs. $|I|$ and vs. ϕ, where ϕ is the phase angle between $|I|$ and the LED current (light intensity), can then be used to determine C_m once the membrane is formed.

Measurement of Membrane Resistance. For a given value of C_m, $|I|$ is a function of the potential across the semiconductor and insulator (V_t). The equivalent circuit can therefore be simplified as shown in Figure 4. For bias potentials near Ψ_{pip}, $|I|$ follows Ψ approximately linearly. Because no significant direct current can flow through the insulator, R_m doesn't enter into the DC characteristics of this system, but only affects transient responses to changes in bias potential. Let us consider the case where the bias potential is stepped with time. Immediately after the potential step, the new bias potential will be distributed across the capacitances with the time constant τ_1 (Equation 5). C_I is the total capcitance of the insulator in contact with electrolyte and C_D is the total capacitance of the depletion region under the electrolyte (therefore C_t is calculated for the area of the semiconductor in contact with solution and not just the illuminated area as in Equation 4). R_e is due to the resistance of the electrolyte.

$$\tau_1 = R_e\left(\frac{C_m C_t}{C_m + C_t}\right) \text{ where } C_t \text{ (the total chip capacitance)} = \left(\frac{C_I C_D}{C_I + C_D}\right) \tag{5}$$

Because R_m is very large (typically > 1 MΩ) it passes very little current in this initial transition compared to the capacitors and doesn't enter into this time constant. On a longer time scale than this initial charging, the membrane capacitance discharges through R_m with a time constant τ_2 approximated by Equation 6 for a small electrolyte resistance R_e.

$$\tau_2 = R_m(C_m + C_t) \tag{6}$$

The change in potential across the insulated semiconductor in response to a step in Ψ from V_0 to V_f may be approximated by Equation 7 when τ_1 and τ_2 are sufficiently different.

$$V_t(t) = V_f + (V_0 - V_f)\left(\frac{C_m}{C_m + C_t}\right)e^{-t/\tau_1} + (V_0 - V_f)\left(\frac{C_t}{C_m + C_t}\right)e^{-t/\tau_2} \tag{7}$$

When R_m is at least three orders of magnitude larger than R_e, the fast transient can be considered as almost instantaneous and that term can be eliminated from the expression. Near Ψ_{pip}, $|I|$ responds with the same time constant as V_t to a step in bias potential (Equation 8). I_0 is the measured alternating photocurrent amplitude immediately after distribution of the step in potential across the capacitances. I_f is the measured alternating photocurrent amplitude after complete dissipation of the potential across the membrane capacitance. When $\ln(I_f - |I|)$ is plotted vs. time, the slope is equal to $-1/\tau_2$ (Equation 9). R_m can then be calculated from Equation 6.

$$|I|(t) = I_f + (I_0 - I_f)\left(\frac{C_t}{C_m + C_t}\right)e^{-t/\tau_2} \tag{8}$$

$$\ln(I_f - |I|) = \ln(I_f - I_0) - \frac{t}{\tau_2} \tag{9}$$

Experimental

Materials. Dioleoyl phophatidyl choline (DOPC) was purchased from Avanti Polar Lipids Inc. (Pelham, AL) and was stored at -78°C under nitrogen as a 20 mg/ml solution in chloroform. All other chemicals were purchased from Sigma Chemical Co. (St. Louis, MO) and used without further purification. Solutions of DOPC in decane were prepared fresh daily by evaporating the chloroform from an aliquot of the DOPC stock and replacing the solvent with an equal volume of decane. Alamethicin, gramicidin D, and valinomycin were stored at 4°C as 1 mM solutions in ethanol.

The semiconductor chip used in the lipid bilayer membrane apparatus was n-type silicon with a 1000 A° insulating layer of silicon nitride deposited over 300 A° of thermally grown oxide. A grid of 1% arsenic in gold was deposited on the opposite side and alloyed to provide ohmic contact.

Lipid Bilayer Membrane Apparatus. The device consists of two liquid filled chambers separated by a small aperture. An exploded cross sectional view of the device is provided in Figure 5. The front chamber was machined out of Delrin. It is open at the top and has circular openings cut in the front and back which fit a Lucite window and a Teflon septum respectively. The Teflon septum has a narrow aperture (0.5 mm I.D. x 0.5 mm length) leading into a small cylindrical back chamber which is sealed against the insulating surface of the silicon chip by an o-ring. The device is supported above, but vibrationally isolated from, a magnetic stirring plate.

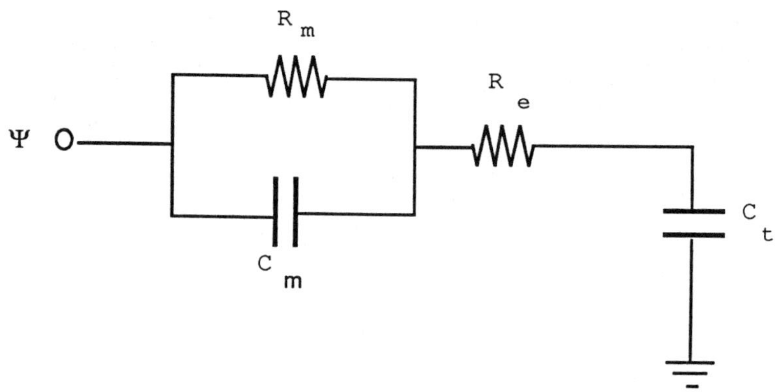

Figure 4. Simplified equivalent circuit for examining the response of the lipid bilayer - sensor configuration to a step in the applied bias potential. C_t is the combined series capacitances of C_I (the capacitance of the total insulator area in contact with electrolyte) and C_D (the capacitance of the depletion region under C_I).

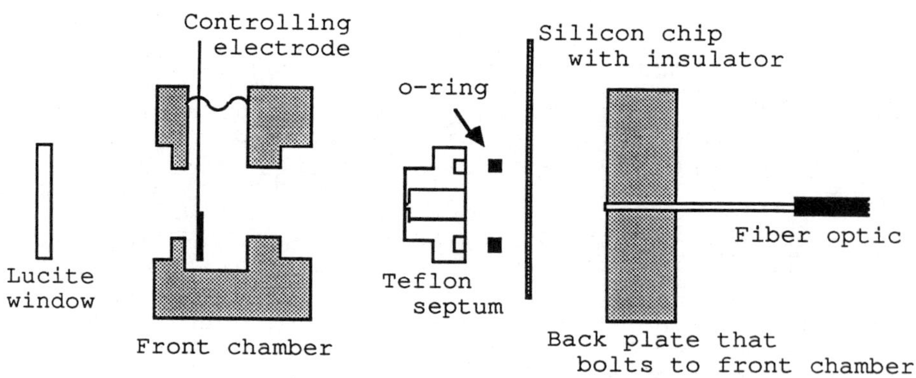

Figure 5. A cross sectional exploded side view of the lipid bilayer membrane apparatus.

Measurement of Photocurrent with Lipid Bilayer Membrane Apparatus. For simplicity, a reference electrode was not used. The solution potential in the front chamber was controlled by a platinum electrode connected directly to a function generator. The changes made to the electrolyte composition during the course of the experiments were small enough not to induce significant drifts in the platinum electrode potential as confirmed by the absence of significant drift during measurements of photocurrent inflection potential (see below). The circuit was completed by tying the semiconductor to virtual ground at the I to V input of a Stanford Research Systems SR510 lock-in amplifier. A fiber optic coupled to an infrared LED provided 10 kHz intensity-modulated illumination of the semiconductor. The magnitude of the 10 kHz photocurrent in the external circuit and its phase relative to the LED were measured by the lock-in amplifier.

The system could be run in three modes: To measure the alternating photocurrent inflection point potential (Ψ_{pip}), the bias potential applied to the platinum electrode was ramped from positive to negative while the photocurrent amplitude ($|I|$) was measured . The potential which gives an inflection point in $|I|$ was determined by finding the zero crossing of the second derivative ($d^2|I|/d\Psi^2 = 0$). In the second mode, changes in photocurrent amplitude and phase were monitored as a function of time at constant bias potential. In the third mode, the response of $|I|$ to a step in bias potential was followed as a function of time. For some experiments and for calibrations, a lipid bilayer was modeled by inserting a resistor and capacitor in parallel between the function generator and the controlling electrode.

Formation of Phospholipid Bilayer Membranes. Membranes were formed by the classic painting technique (6,7). After pretreating the aperture in the Teflon septum with 5 µl of 20 mg/ml DOPC in decane and allowing it to dry, the back chamber was filled with degassed phosphate-buffered saline (PBS) and the device assembled. Care was taken not to trap air bubbles in the back chamber. The front chamber was then immediately filled with PBS. Ψ_{pip} was monitored continuously until a stable value was reached. The bias potential was then adjusted to Ψ_{pip} - 50 mV (i.e. the high photocurrent side of Ψ_{pip}) and a membrane painted over the aperture with a quick stroke of a camel's hair brush onto which 20 µl of 20 mg/ml DOPC in decane had been pipetted. The thinning of the membrane to a bilayer could be followed either visually or electrically and was complete within an hour. After use the phospholipid bilayer membrane apparatus was taken apart and cleaned by soaking in 1N NaOH and 1N HCl, rinsed with water and methanol and air dried.

Ionophores and Ion Channels. After the formation of a phospholipid bilayer membrane, ethanolic solutions of valinomycin, gramicidin D, or alamethicin were added with stirring to the front chamber. The membrane was allowed to equilibrate for 15 min. after each addition before data were taken. For the valinomycin experiments, the buffer used was PBS + 1 mM KCl. To measure the effect of a potassium gradient across the membrane, the potassium concentration in the front chamber was adjusted by addition of 2 M KCl.

Results and Discussion

To model the effects of a phospholipid bilayer membrane supported on an aperture between the controlling electrode and the semiconductor surface of the LAPS, a resistor and capacitor in parallel were connected between the controlling electrode and the function generator driving the electrode. Figure 6a shows a plot of the amplitude of the alternating photocurrent ($|I|$) and its phase (ϕ) relative to the modulated light source vs. the value of the capacitor (C_m). Both $|I|$ and ϕ show a strong dependance on the membrane capacitance thereby demonstrating the capability of the system to measure membrane capacitances between 0.1 and 10 nF. A plot of $|I|/C_m$ vs. $|I|$ (Figure 6b) is linear for large values of C_m as predicted by Equation 4a. For values of $C_m<C_d$, however, a deviation from linearity not predicted from the equation is observed. This is an indication that the impedance through the membrane is sufficiently high that the 10 kHz modulation of I_p results in a 10 kHz variation in potential across the depletion layer that is significant with respect to the steady state depletion layer potential. The phase shift, ϕ, as opposed to $|I|$ is a preferable value to measure because it is less affected by small changes in light intensity. Unknown values of C_m may be determined by comparing measured phase values to the calibration plot of phase vs. C_m.

A plot of the time constant for the photocurrent response to a step in the bias potential vs. resistance for the model system is given in Figure 7. The potential steps were kept relatively small, from Ψ_{pip} - 50mV to Ψ_{pip} + 50 mV, to stay in a region where photocurrent changes approximately linearly with potential across the semiconductor. The modelled C_m was held constant at 1.5 nF. As predicted by Equation 6, the time constant changes linearly with resistance. The slope of the plot is 28.6 nF which is equal to the sum of the membrane and sensor capacitances ($C_m + C_t$). The sensor capacitance is therefore 27.1 nF. Using this value, R_m can be determined from the measured values for C_m and τ_2 (Equation 7). Membrane conductance is simply the reciprocal of R_m.

Because C_t changed slightly each time the sensor was assembled, it was necessary to recalibrate before the formation of each membrane.

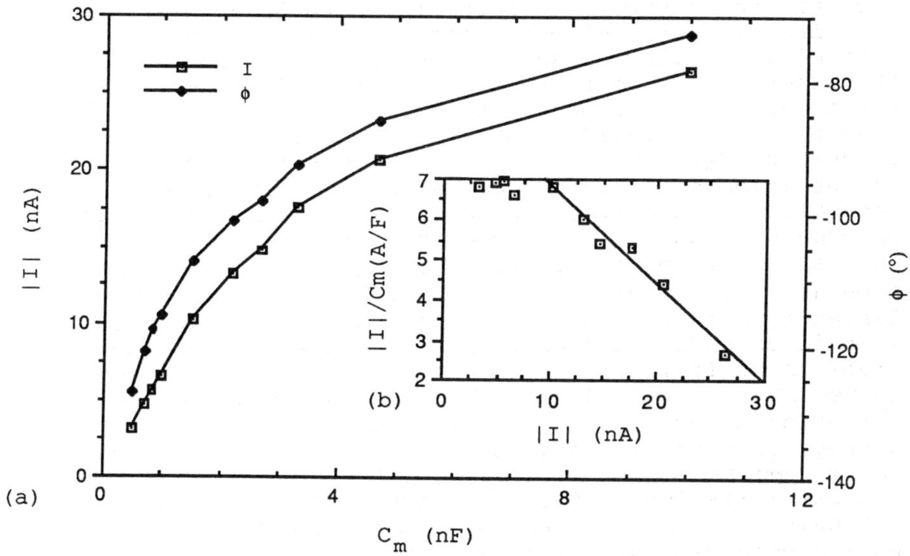

Figure 6. (a) Sensor photocurrent amplitude (|I|) and phase (f) vs. capacitance when the lipid bilayer is replaced with standard electrical components. R_m=200MΩ. (b) Plot of photocurrent divided by capacitance vs. photocurrent according to Equation 4a. Photocurrent is given as RMS amplitude.

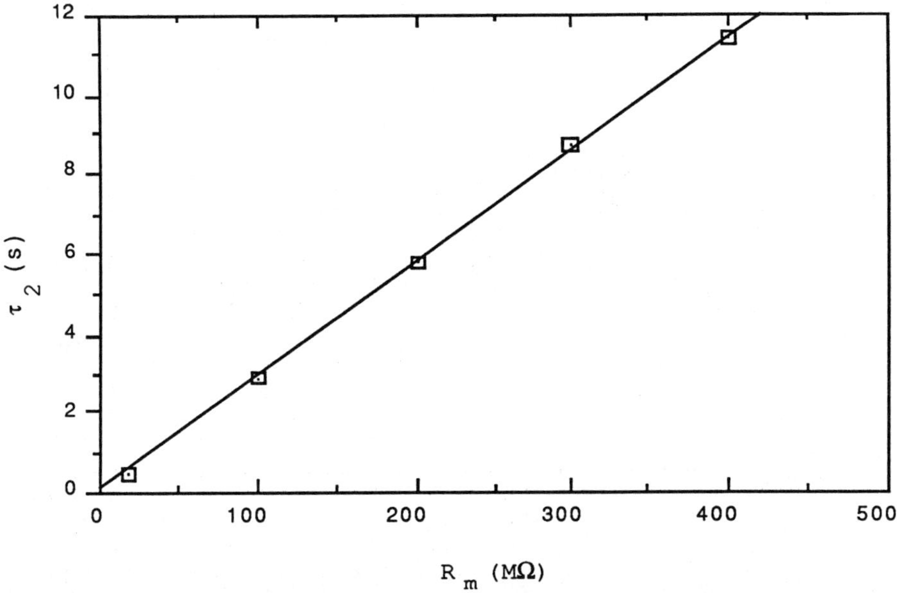

Figure 7. Time constant of sensor photocurrent response to step in bias potential vs. resistance when the lipid bilayer is replaced with standard electrical components. C_m = 1.5 nF.

Electrical Measurements on Phospholipid Bilayer Membranes. Phospholipid bilayer membranes were prepared by painting solutions of DOPC in decane across a Teflon aperture between an open chamber and a small closed chamber (6,7). Closing one chamber is reported to add considerably to the stability of the membrane (8). Painting resulted in stable membranes on at least 80% of attempts. Immediately after painting, the membrane capacitance would drop to less than 50 pF as determined from changes in ϕ, but would slowly begin to rise over approximately an hour to a stable value as the lipid-solvent film thinned to a bilayer. The membranes were usually stable for over 5 hours, although they could be broken by strong vibrations or large voltage spikes. The membrane capacitances ranged from 1 - 2 nF. For an aperture of approximately 0.002 cm^2, this corresponds to a specific capacitance of between 0.5 and 1 $\mu F/cm^2$. The generally accepted value for bilayer capacitance is 0.9 $\mu F/cm^2 \pm 0.1\ \mu F/cm^2$, although the presence of a torus of solvent around painted phospholipid bilayer membranes has been known to drop the measured capacitance to 0.5 $\mu F/cm^2$ and below (9,10).

Figure 8 shows the response in photocurrent to a step in bias potential for a typical membrane. The plot shows the expected fast jump in the alternating photocurrent as the potential step is divided between all the sensor capacitances and then the slow exponential rise as the potential across the membrane discharges through its resistance R_m. Figure 8 (inset) is a plot of the same data after subtracting the the photocurrent $|I|$ from the final equilibrium value I_f and then taking the logarithm. The transformed plot is linear as predicted by Equation 9, indicating that the photocurrent response was indeed exponential. The time constant (τ_2) is the negative reciprocal of the slope or 59.9 sec. The phase of the photocurrent relative to the illumination intensity was -102.3° which corresponds to a C_m equal to 2.0 nF or 1.0 $\mu F/cm^2$. From a capacitance calibration curve similar to Figure 6b and Equation 6, a value of 15.4 nF was obtained for C_t when C_m was zero. Membrane resistances in the absence of ionophores ranged from 300 MΩ to 20 GΩ as determined from τ_2. These values varied considerably because they are probably measures more of the quality of the seal of the membrane to the aperture than of a real property of the membrane.

Valinomycin. To observe the effect of the potassium binding polypeptide valinomycin on lipid bilayer resistance, a phospholipid bilayer membrane was formed under PBS containing 1 mM KCl, and valinomycin was added with stirring to the front chamber. While lipid bilayers are normally highly impermeable to charged species, K^+ ions bound to the lipophilic valinomycin diffuse freely through the hydrophobic core of a bilayer (11,12). Valinomycin therefore increases membrane conductance in the presence of potassium ions. A plot of membrane conductance ($1/R_m$ in Siemens) vs. valinomycin concentration is given in Figure 9a. All data were taken from the same membrane. The plot is linear in agreement with the literature. From the slope of the plot a specific conductance of $4x10^5$ S cm^{-2} per M valinomycin per M K^+ is obtained. This agrees reasonably well with the value of 10^5 S cm^{-2} M^{-2} measured by Stark and Benz (12) considering the differences in lipids and apparatus.

Trans-membrane potentials develop across valinomycin containing membranes in the presence of potassium gradients. A phospholipid bilayer membrane was prepared under PBS containing 1mM KCl and valinomycin added to the front chamber to a concentration of 1 µM. KCl was then added to the front chamber with stirring, and the resulting trans-membrane potential monitored by measuring shifts in the photocurrent inflection point potential. Figure 9b is a plot of Ψ_{pip} vs. the log of the ratio of K^+ concentrations on either side of the membrane. The plot is linear with a slope of 54 mV/log unit. The Goldman equation predicts a slope of 59 mV/log unit for the trans-membrane potential across a membrane permeable to only one ionic species (5). The difference is at least partially explained by the non negligible permeability of sodium ion in the presence of valinomycin. The permeability is about 1000 times less for sodium than for potassium (11), however, sodium was present at 150 mM compared to 1mM potassium behind the membrane.

Gramicidin D. As opposed to valinomycin, which is an ion carrier, the polypeptide gramicidin D transports cations across phospholipid bilayers by forming channels through the membrane (5,12). This makes it more suitable as a model for protein ion channels. Also unlike valinomycin, it has a broad selectivity for cations. The specific conductance of a phospholipid bilayer membrane in the presence of increasing concentrations of gramicidin D was measured in a manner similar to that for valinomycin. Specific conductance vs. gramicidin concentration is plotted in Figure 10. Specific conductance increases in a nonlinear fashion with gramicidin concentration. This behavior which has been previously reported in other systems, occurs because gramicidin D must dimerize in the membrane in order to form a channel (12).

Alamethicin. The polypeptide alamethicin adds a level of complexity over gramicidin D. Like gramicidin D it forms cation channels through phospholipid bilayers. Unlike gramicidin D, alamethicin exhibits the voltage dependant gating which is characteristic of many protein ion channels . The alamethicin channel only transports cations when the membrane potential is greater than a threshold value (13). Conductance measurements of a phospholipid

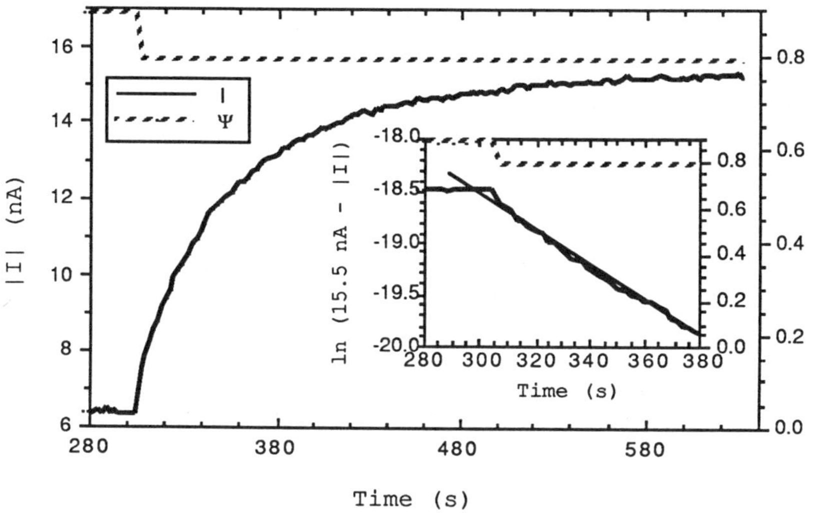

Figure 8. Response of photocurrent amplitude (|I|) to step in bias potential (Ψ) across sensor with lipid bilayer membrane. The inset shows the same data linearized according to Equation 9. Photocurrent is given as RMS amplitude.

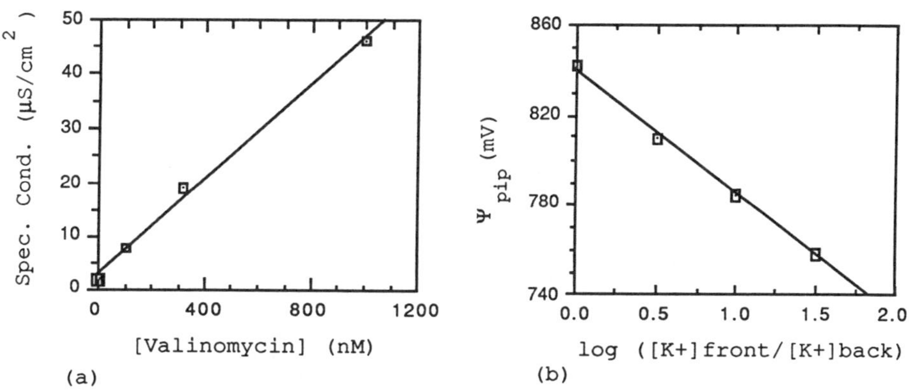

Figure 9. (a) Specific conductance of lipid bilayer membrane as a function of valinomycin concentration in the surrounding solution. (b) Photocurrent inflection potential as a function of the log of the ratio of potassium on the two sides of a lipid bilayer in the presence of 1 µM valinomycin.

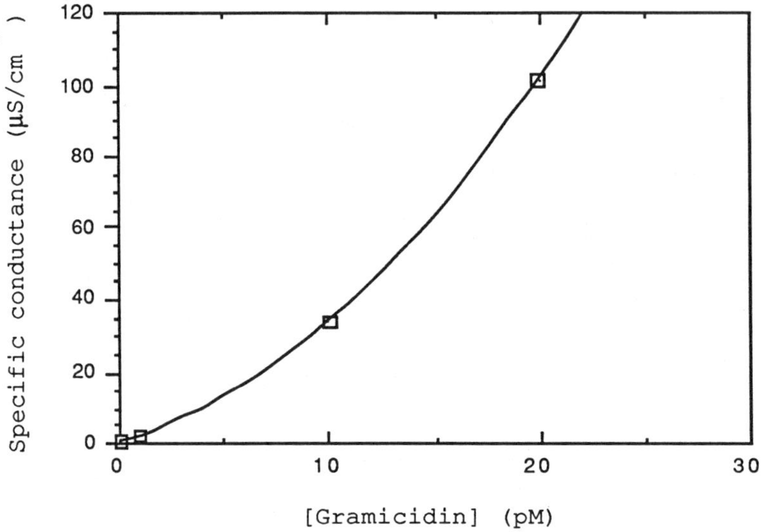

Figure 10. Specific conductance of a lipid bilayer membrane as a function of gramicidin concentration in the surrounding solution.

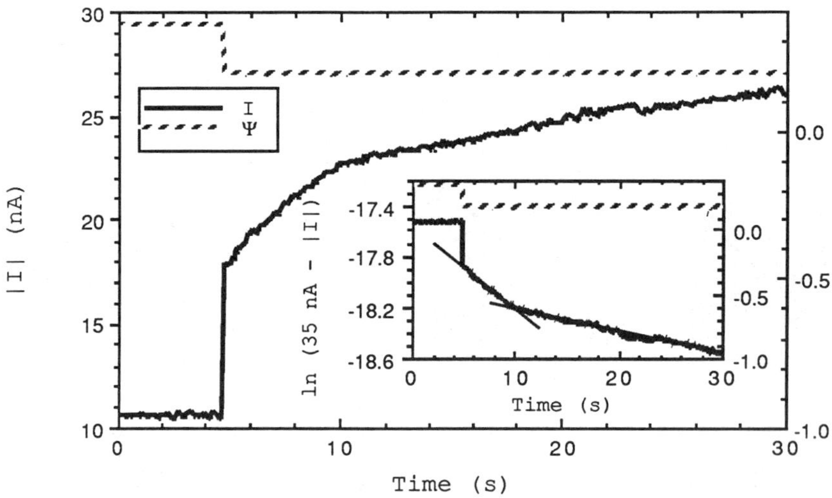

Figure 11. Response of photocurrent amplitude (|I|) to step in bias potential (Ψ) across sensor with lipid bilayer membrane in the presence of 100 pM alamethicin. The inset shows the same data, linearized according to Equation 9. Photocurrent is given as RMS amplitude.

bilayer membrane in the presence of alamethicin were made. The alternating photocurrent response to a step in bias potential was found to be biphasic (Figure 11). After stepping the bias potential, |I| initially changes with a short time constant which abruptly increases to a much longer time constant. The two time constants correspond to conductances of 2.1 nS (480MΩ) and 59 pS (17 GΩ) respectively. The initial high conductance was due to cation transport through alamethicin channels opened by the large potential applied across the membrane. When the membrane potential decayed to the threshold value, the channels shut off and only the conductance of the membrane is observed.

Conclusions

The methods described in the present paper provide a means to measure phospholipid bilayer conductance, capacitance, and trans-membrane potentials in conjunction with the LAPS. It is quite plausible that the stability of this bilayer system can be significantly improved for many purposes by decreasing the distance between the bilayer and the silicon by micromachining the silicon. Studies in this direction will be reported elsewhere.

Acknowledgments

We thank Joe Kercso for his technical assistance. This work was supported in part by the Army Research Office and Defense Advanced Research Projects Agency under ARO contract DAAL03-86-C-0009.

Literature Cited

1. Hafeman, D. G.; Parce, J. W.; McConnell, H. M. Science 1988, 240, 1182.
2. Sze, S. N. Physics of Semiconductor Devices; Wiley: New York, 1981; Chapter 7.
3. Lewis, N. S. Ann. Rev. Mater. Sci.; 1984, 14, 95.
4. Bousse, L.; Meindl, J. D. ACS Symp. Ser. 1986, 323, 79.
5. Hille, B. Ionic Channels of Excitable Membranes; Sinauer Associates: Sunderland, MA, 1984; Chapter 1.
6. Mueller, P.; Rudin, D. O.; Tien, H. Y.; Wescott, W. C.; J. Phys. Chem. 1963, 67, 534.
7. Finklestein, A. Methods in Enzymology 1974, 32, 545.
8. Vodyanoy, V.; Murphy, R. B. Biochim. Biophys. Acta 1982, 687,189.
9. Montal, M.; Mueller, P. Proc. Natl. Acad. Sci. 1972, 69, 3561.

10. Niles, W. D.; Levis, R. A.; Cohen, F. S. Biophys.J. 1988, 53, 327.
11. Lauger, P. Science 1972, 178, 24.
12. Stark, G.; Benz, R. J. Memb. Biol. 1971, 274, 294.
13. Hladky, S. B.; Haydon, D. A. Biochim. Biophys. Acta 1972, 274, 294.
14. Mueller, P.; Rudin, D. O. Nature 1968, 217, 1713.

RECEIVED March 9, 1989

Chapter 4

Electropolymerized Films in the Construction of Biosensors

Robert J. Geise and Alexander M. Yacynych

Department of Chemistry, Rutgers, The State University of New Jersey, New Brunswick, NJ 08903

A glucose sensor was constructed for the determination of glucose in blood serum using flow injection analysis. A platinized, reticulated vitreous carbon (RVC) electrode was used as a combination support, enzyme reactor, and detector. Glucose oxidase was immobilized on the surface of the RVC electrode with glutaraldehyde. Various electropolymerized films were tested for their effect in eliminating interferences and electrode fouling. The use of an electropolymerized film has the additional advantage of forming a basis for an all-chemical method of construction, which can be used for any shape or size of biosensor.

Biosensors have recently become an area of great interest, especially in clinical chemistry. These sensors can be used to determine analyte concentrations in clinical samples, such as blood serum and urine. Electrochemical biosensors are constructed by incorporating a biochemical system, such as an enzyme, with an electrode. Amperometric biosensors often use oxidase enzymes, such as glucose oxidase, as the sensing enzyme. A major product of these types of enzymatic reactions is hydrogen peroxide, which can be oxidized at an electrode surface. The current produced by the oxidation of hydrogen peroxide is directly proportional to the enzymatic substrate concentration.

A major difficulty arises because there are many other species present in physiological samples that are oxidized at the same potential necessary to oxidize hydrogen peroxide. This results in substantial interference, and compromises the excellent selectivity of the enzymatic system. For this reason, it becomes mandatory to discriminate between the hydrogen peroxide produced in the enzymatic reaction, and easily oxidizable species commonly present in blood serum and urine. Unfortunately, for complex samples such as serum, there are many interferents (1). Several attempts have been made to circumvent this problem. Masoom and Townshend (2) have incorporated

0097–6156/89/0403–0065$06.00/0

either a dialyzer or a column of copper(II)diethyldithiocarbamate on controlled porosity glass in front of an enzyme column. Yao et al. (3) place an electrolytic column before the electrode.

Conventional enzyme electrodes employ discrete-macroscopic membranes to overcome problems associated with interferences, enzyme immobilization, and electrode fouling. While these types of enzyme electrodes have been commercially developed, there are some limitations with this approach. Some sensors use three relatively thick membranes, resulting in a slow and complex diffusion path for reactants reaching the enzyme and hydrogen peroxide reaching the electrode. Slow diffusion in this type of system adversely affects the response and recovery time, decreasing sampling rate. Each sensor must be individually constructed, and this construction technique is limited to two-dimensional surfaces. In addition, for sensors that have complex and slow diffusion paths, rates of diffusion must remain constant, otherwise calibration of the biosensor, and more important the maintenance of calibration, are difficult. A variety of factors can influence rates of diffusion, and consequently the performance of the enzyme layer and the performance of the sensor. These complicated, and most often uncharacterizable, properties have made the development of most biosensors difficult.

Mell and Maloy report that the ideal biosensor configuration, based on a digital simulation, would use the thinnest possible biochemical layer and associated membranes, and that the biochemical layer have the highest possible activity (4). With this in mind, it was our goal to greatly minimize the diffusional limitations associated with conventional biosensors, while still protecting the electrode against fouling and interferences. A direct benefit could then be realized in a wide range of performance characteristics. A biosensor constructed using electropolymerized films can have significantly improved diffusional properties due to the thinness of the film. By engineering the components and properties of a biosensor on a microscopic scale, rather than using "bulk-technology" and physically assembling discrete macroscopic components, as is the conventional practice, an all-chemical method of construction can be achieved. All-chemical methods of construction would be important for miniaturized sensors, lithographically made sensors (i.e. "sensors on a chip"), and for ease of manufacture in general. For this type of sensor, the sensitization layer thickness (this includes everything between the electrode surface and the solution, i.e. immobilized enzyme, polymer film, etc.) is approximately 10 nm, while for conventional sensors this thickness is about 10 micrometers. This is an improvement of three orders of magnitude, which results in better diffusional properties and operating characteristics.

Polymer-film modified electrodes are used for a variety of applications, such as electrocatalysis (5-8), analysis (9,10), and more recently for their permselectivity characteristics (11-14). These polymer films are formed by casting the film on an electrode surface (12), using radio-frequency plasma (15), or by electropolymerization (13,16,17). Alternatively, a polymer film is formed as a discrete membrane and subsequently applied to an electrode (18). The use of both cast and discrete membrane films is essentially limited to two-dimensional electrode surfaces, as it is nearly impossible to control the reproducibility, uniformity, and thickness of the polymer film on an intricately complex surface, such as reticulated vitreous carbon (RVC), which is useful as a flow-through electrode (19,20). Electropolymerization, however, permits controlled film thickness, homogeneity, and reproducibility.

Dubois et al. (21,22) have showed that the electrochemical oxidation of phenol and its derivatives, on metal surfaces, produced hydrophobic, adherent, and insulating polymer films of uniform thickness. Both Yacynych and Mark (17), and Heineman et al. (13) showed the oxidation of 1,2-diaminobenzene to be irreversible, and with successive cyclic voltammetric scans formed an insulating polymer film completely covering the electrode surface. Heineman et al. (13) further showed that 1,2-diaminobenzene forms a polymeric film over a pH range of 4 to 10, and that platinum electrodes coated with the poly(1,2-diaminobenzene) provided a nearly Nernstian response to pH. Cheek et al. (16) studied the pH response of platinum and vitreous carbon with polymer films of either 1,2-diaminobenzene or phenol. These polymer films are selective enough to allow the permeation of protons, while limiting access to larger molecules, which could be potential interferents.

Electropolymerization offers the advantages of producing a very thin and self-insulating film which can be coated on any conducting, three-dimensional surface. The formation of a self-insulating film prevents electropolymerization on any part of the electrode that has already been coated, thus providing inherently uniform, reproducible coverage of any surface, regardless of its geometry.

EXPERIMENTAL

APPARATUS. An ECO (ECO instruments, Newton, MA) model 549 potentiostat/galvanostat was used for platinization. All other experiments were done using an EG&G Princeton Applied Research (Princeton, NJ) model 264A potentiostat. For flow-injection analysis (FIA), a Varian (Walnut Creek, CA) 8500 HPLC pump and a model 7125 Rheodyne (Cotati, CA) injection valve fitted with a 5 uL injection loop were used. A Houston Instrument division of Bausch and Lomb (Houston, TX) Omnigraphic 2000 X-Y recorder was employed. A Rainin (Woburn, MA) Rabbit peristaltic pump was used for platinization, enzyme attachment, and electropolymerization procedures. A saturated calomel reference electrode (SCE) was used for all experiments.

MATERIALS. Phosphate buffer (0.1M, pH 6.5) was prepared with distilled-deionized water using ACS certified (Fisher, Springfield, NJ) phosphate salts, and the pH was adjusted to 6.5 with concentrated phosphoric acid or potassium hydroxide. Other chemicals used were L-ascorbic acid (Fisher, certified ACS), 4-acetamidophenol (a.k.a. acetaminophen) 98% (Aldrich, Milwaukee, WI), and B-D(+)-glucose (Sigma, St. Louis, MO). Hydrogen peroxide solutions were prepared in phosphate buffer by making appropriate dilutions of a 3% solution (York Pharmacal, Brookfield, MO). Hydrogen hexachloroplatinate (IV) hydrate salt (Aldrich) was used for platinization.

1,2- and 1,3-diaminobenzene (DAB), 98% (Aldrich) were purified by recrystallization with dichloromethane three times using ordinary techniques. 1,4-DAB, 99.5% (Pfaltz-Bauer, Waterbury, CT) was used as obtained. Catechol (99+%), resorcinol (98%), and hydroquinone (99%) (all from Aldrich) were recrystallized by ordinary techniques. Glucose oxidase was from Sigma (Type II from Aspergillus Niger), and glutaraldehyde, 25% (wt.%) was from Aldrich. Reticulated vitreous carbon (RVC - 80S) was obtained from The Electrosynthesis Co., Inc., East Amherst, NY.

PROCEDURE. The current was monitored at a constant potential of +0.60V vs SCE. The electrochemical set-up including a description of the electrode cell has been previously described (23). Electrochemical deposition of platinum on RVC electrodes was done using an ECO model 549 potentiostat/galvanostat in the galvanostatic mode at a constant current density setting of -3.03 mA. The actual current density was $-0.39 mA/cm^2$. A 0.025M solution of hexachloroplatinate salt in phosphate buffer was circulated using a peristaltic pump at a rate of 5 mL/min for 6 h. The deposition was characterized by testing its response to 1mM and 50mM hydrogen peroxide, 1mM acetaminophen, and 1mM ascorbic acid at +0.60V vs SCE.

Glutaraldehyde cross-linking of glucose oxidase (GOX) onto the RVC electrodes was done following a procedure previously described (23). Electropolymerization of diaminobenzene and dihydroxybenzene isomers was performed using a 3mM solution of the isomer in potassium phosphate buffer (0.1M, pH 6.5). The buffer was deaerated with high purity nitrogen for 0.5 h, and a nitrogen atmosphere was maintained over the solution throughout the procedure. For mixed polymers, the total concentration of the components was 3mM. A peristaltic pump circulated the solution during electropolymerization at a flowrate of 0.80 mL/min. Cyclic voltammetry was used for electropolymerization of the film, and to follow its progress, on the RVC surface. The electrode potential was cycled continuously from 0.00V to +0.80V vs SCE at 2 mV/sec until the current decreased to a minimum. This process took 18-24 h (depending on the monomer(s) used), and subsequent electropolymerizations of specific films were consistent from electrode to electrode, as evident from monitoring the process with cyclic voltammetry. Once the electropolymerization was complete, the sensor was tested for response to glucose, hydrogen peroxide, and various selected interferents using the FIA system.

RESULTS AND DISCUSSION

A flow-injection analysis system, shown in Figure 1, was used for all determinations. Reticulated vitreous carbon (RVC) was selected as the electrode material for its flow-through properties and large surface area (19). The RVC was partially platinized to permit a lower working potential for the oxidation of hydrogen peroxide (+0.60V vs SCE). A lower working potential results in less noise and a lower background current. Glucose oxidase (GOX) was immobilized by cross-linking with glutaraldehyde. A schematic of the biosensor surface (our best guess) and the reactions taking place are shown in Figure 2.

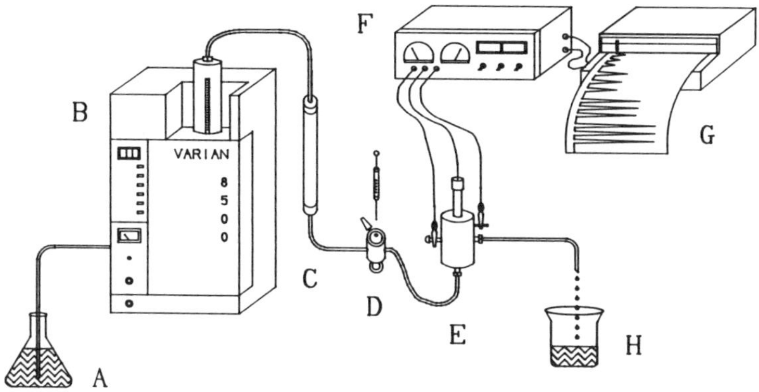

Figure 1 Schematic of the Flow-injection analysis (FIA) system.

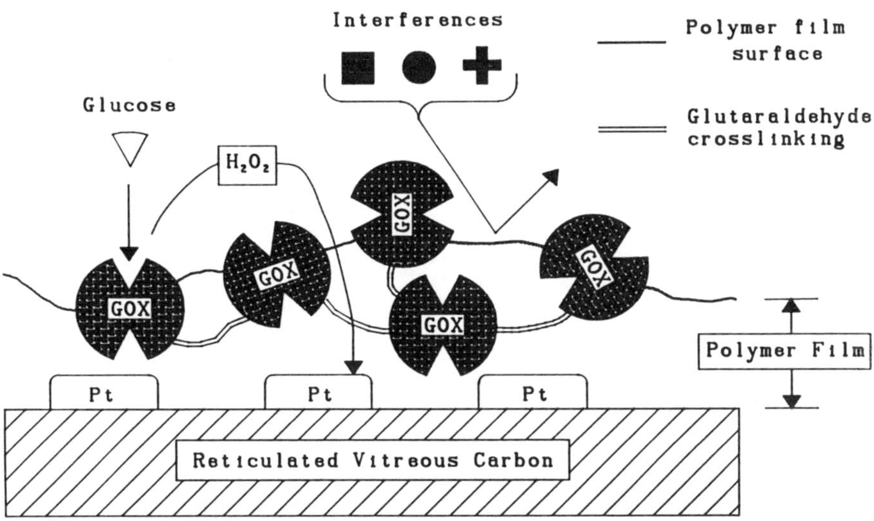

Figure 2 Schematic representation of a possible surface morphology of the biosensor.

The electropolymerization was done with the enzyme already immobilized on the electrode surface, resulting in a complete glucose biosensor. The limiting factor is now the electropolymerized film and the length of time that common interferents can be screened from the electrode surface. The cross-linked glucose oxidase enzyme is stable for months cross-linked on RVC, with or without the polymer film. Although the polymer film does provide additional enzyme stability.

It is well-known that diaminobenzene (DAB) and dihydroxybenzene isomers form polymer films on an electrode surface when oxidized. Six monomers were used, either by themselves, or in combination to form the electropolymerized films: 1,2-;1,3-; and 1,4-Diaminobenzene; 1,2-dihydroxybenzene (Catechol); 1,3-dihydroxybenzene (Resorcinol); and 1,4-dihydroxybenzene (hydroquinone). For all polymer films made from mixtures of monomers, the compounds were used in 1:1 ratios by weight.

Sasso et al. used electropolymerized 1,2-DAB in the construction of glucose biosensors (<u>23</u>). However, it was necessary to repolymerize the 1,2-DAB on the electrode every week to prevent interferences. It was the goal of this work to find an electropolymerized film which was more stable than that formed with 1,2-DAB, and which lasted longer than the enzyme.

Ascorbic acid and acetaminophen are common electroactive interferents found in blood serum samples. Both are effectively screened out by different polymer films to varying degrees. However, ascorbic acid is difficult to use due to air oxidation. Therefore, we used acetaminophen as a test interferent for its stability, and because of its small size it is one of the more difficult interferents to eliminate. It is a very popular non-aspirin pain reliever whose normal blood concentration does not exceed 0.2mM (<u>25</u>).

It should be noted that the test concentrations of acetaminophen are at least five times greater than the physiological concentration. When choosing a polymer film, its ability to screen out acetaminophen and other electroactive interferents must be considered (<u>23,24</u>). Sasso, et al. have shown that 1,2-DAB films effectively screen out ascorbic acid, uric acid and L-cysteine (<u>23</u>). Ideally, the glucose sensor should not respond to acetaminophen but have a good response to 5mM glucose. In reality, a response to 1mM acetaminophen which is 5% or less of the 5mM glucose response, for at least 2 months is acceptable. Electropolymerized 1,2-DAB does not satisfy this requirement, as can be seen in Figure 3, the response to 1mM acetaminophen increases steadily over a period of 60 days.

<u>LONG TERM RESPONSE TO ACETAMINOPHEN</u>. Table I lists various electropolymerized films studied for their ability to screen out interferents. Hydroquinone does not appear to sufficiently electropolymerize because the current from the cyclic voltammogram did not decrease to a minimum, but remained at a steady-state after an initial decrease. Mixtures of the six compounds were chosen by pairing structural isomers, i.e. 1,2-DAB with Catechol (1,2-dihydroxybenzene) and 1,3-DAB with Resorcinol (1,3-dihydroxybenzene). The combination of 1,2- and 1,3-DAB was also studied.

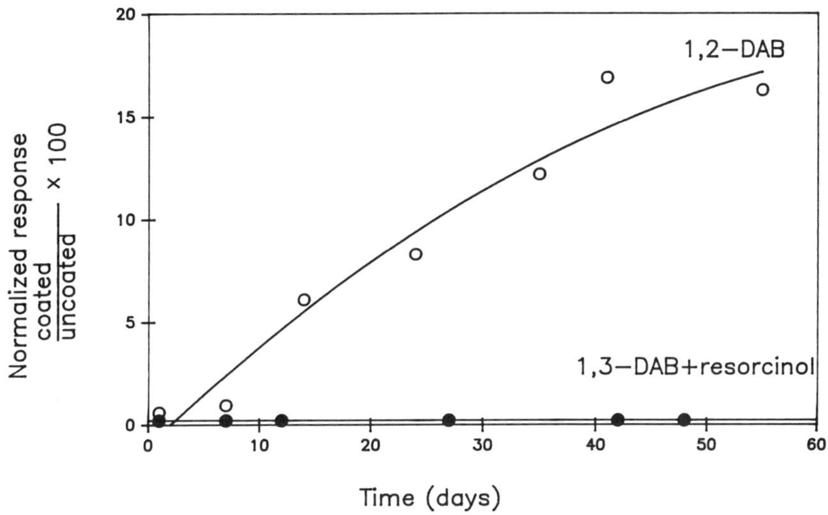

Figure 3 Response to 1mM acetaminophen for 1,2-DAB and 1,3-DAB/Resorcinol.

TABLE I. Electropolymerized Films for Glucose Sensors: Long term slope of response to 1mM acetaminophen

Polymer Film	Slope x 100	Days Studied
1,2-DAB	2.6 - 25	80 - 110
1,3-DAB	0.5 - 2	64 - 118
1,4-DAB	0.6 - 4.3	76 - 114
1,2- + 1,3-DAB	1.2 - 7.7	64 - 123
1,2-DAB + catechol	8 - 50	40 - 166
Catechol	0.7 - 1.5	111 - 115
Resorcinol	8.5 - 9.0	80 - 82
1,3-DAB + resorcinol	0 - 0.054	48 - 131
Hydroquinone	Does not sufficiently electropolymerize on the electrode surface.	

The long term slope of the response to 1mM acetaminophen was used to evaluate the various electropolymerized films formed on platinized RVC electrodes. The parameter of slope x 100 (m X 100) is used for convenience because some of the slopes were so small. The number of days that each film was studied is indicated by a minimum and maximum time for each type of polymer film electrode. At least three electrodes with each polymer film were studied, except for resorcinol (2 electrodes). A value of 2.0 or less was arbitrarily assigned to be an acceptable value for m X 100 because this is a clear improvement over the 1,2-DAB polymer film. Also, a narrow range for these values is desirable as this indicates reproducibility of the polymer film. Resorcinol, despite the narrow range of slope is not a suitable polymer film (m X 100 values of 8.5 and 9.0) because the slopes are much greater than 2.

The 1,2-DAB + catechol polymer film can be eliminated because it fails in both areas, the slopes are too high and the range is too wide (8 to 50). Table I shows that these two polymer films, both of which incorporate 1,2-DAB, have by far the widest ranges of slope. The third polymer film that incorporates 1,2-DAB does not have nearly as wide a range as the previous two 1,2-DAB film electrodes, but the combination of 1,2- and 1,3-DAB gives a wider range of slopes than any other polymer film in which 1,2-DAB is not a component (1.2-7.7).

1,4-DAB has a range of 0.6 to 4.3 for its slope X 100 values, and is not acceptable. The 1,3-DAB polymer film appears to be better, with a range of slopes from 0.5 to 2.0 (5 electrodes studied) it has both a small slope and a narrow range. This suggests that 1,3-DAB is a clear improvement over 1,2-DAB in screening out acetaminophen and also in reproducibility (as seen from the narrow range of slope values). The same argument can be made for catechol as a polymer film with a m X 100 value range of 0.7 to 1.5. However, none of these approaches the effectiveness of the polymer film made from 1,3-DAB and resorcinol. Table I shows that the slope X 100 range is 0 to 0.054, at least an order of magnitude less than 1,3-DAB, 1,4-DAB and catechol, and 2-3 orders of magnitude better than 1,2-DAB or 1,2-DAB and catechol.

CALIBRATION CURVES. Figures 4 and 5 show the calibration curves for glucose sensors with polymer films of 1,2-DAB and 1,3-DAB/Resorcinol respectively. Both curves are similar; they are linear from 2-20mM glucose, and level off at concentrations above 20mM due to limiting oxygen concentrations. This has been shown by using an electron mediator (benzoquinone) in the flowstream which increases the linear range to 100mM glucose (24). Other polymer films in Table I also have similar calibration curves. This suggests that the various polymer films do not have an observable effect on the calibration curve. In fact, the calibration curves obtained for the enzyme electrodes after they are covered with the electropolymerized film are similar to those before electropolymerization, except for a lower current response due to diffusion limitations caused by the polymer film.

RESPONSE/RECOVERY TIME. For glucose determinations, a flow rate of 2 mL/min results in an analysis time of less than 30 seconds from

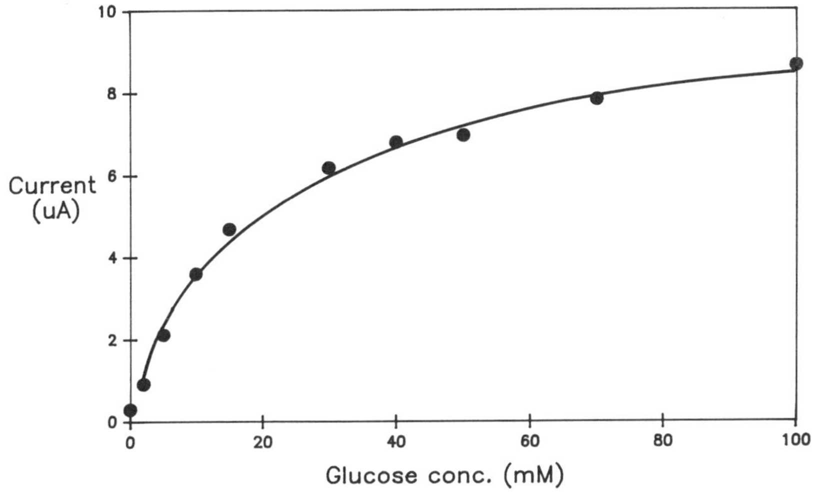

Figure 4 Calibration curve of a glucose sensor with 1,2-DAB as the polymer film. Range: 2-100mM glucose.

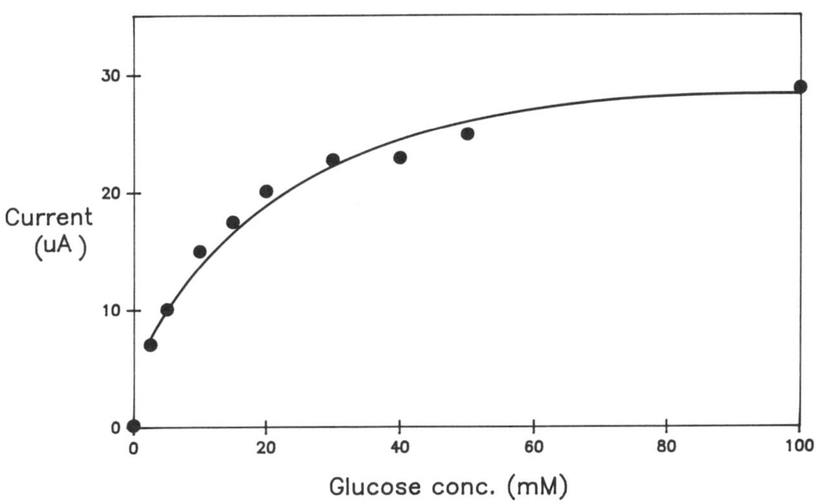

Figure 5 Calibration curve of a glucose sensor using 1,3-DAB/Resorcinol as the polymer film. Range: 2.5-100mM glucose.

injection until the reestablishment of baseline. This allows the construction of a five-point calibration curve with 10 repetitions for each concentration to be completed in about 0.5 h. A response/recovery study conducted using a flow stream of 5mM glucose instead of separate injections had a response of 60 sec, with or without polymer film, and a recovery time of about 2 min without the film, 3 min with the film. It is believed that these response/recovery times more accurately reflect the hydrodynamics of the flow cell rather than the intrinsic response of the biosensor.

COMPARISON OF POLYMER FILMS: SIGNAL TO INTERFERENT PLOTS. Figure 6 shows the response over time to 5mM glucose and 1mM acetaminophen for an RVC/GOX glucose sensor with 1,2-DAB as the polymer film. While the response to 5mM glucose remains relatively constant (5-6uA over 3 months), the current response to 1mM acetaminophen initially increases sharply, and after only 30 days is more than half the glucose response. After 100 days the acetaminophen response exceeds the glucose response. It should be noted that at this point the response to acetaminophen is less than 10% of the original response of the electrode without the 1,2-DAB film. Although the electrode would no longer function as an effective biosensor, the polymer film is still present and Figure 6 is not meant to imply that the film completely opens up or is no longer present on the electrode. An acetaminophen response of 5% or less of the glucose response is considered acceptable.

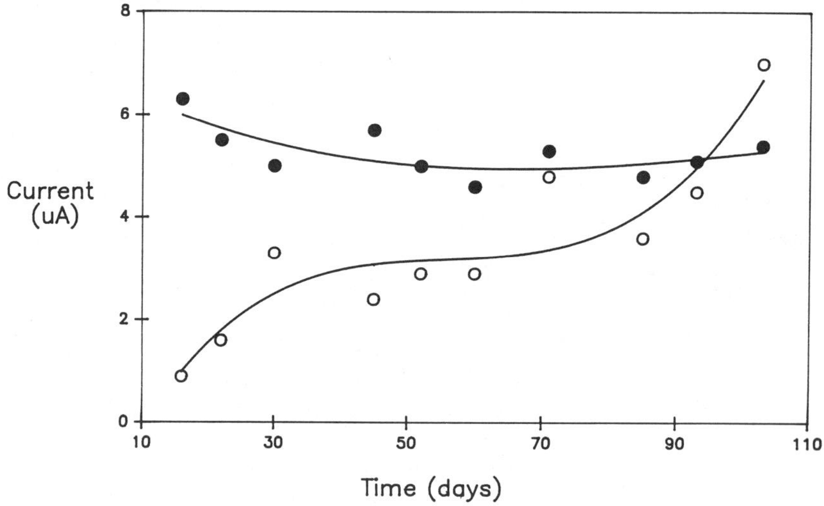

Figure 6 Response of glucose sensor with 1,2-DAB polymer film. Open circles: 1mM acetaminophen; Filled circles: 5mM glucose.

Figure 7 is similar to Figure 6 except that the polymer film used is catechol, which is intermediate in effectiveness between 1,2-DAB and 1,3-DAB/resorcinol. This graph indicates that catechol does not approach 1,3-DAB/resorcinol in its ability to screen out acetaminophen over time, but it does appear better than 1,2-DAB. Similar conclusions can be drawn from plots for 1,3-DAB, 1,4-DAB and 1,2-,1,3-DAB electrodes.

Figure 8 shows the same studies as for Figures 6 and 7, except the polymer film is 1,3-DAB/resorcinol. For over 5 months (160 days) the ratio of the glucose to acetaminophen response is constant. There is significant drift in the current response to 5mM glucose within the first 40 days, these readings level off at about 5uA. Except for an unusually high response to 1mM acetaminophen on day 40, the response does not exceed 0.3uA over a 5 month period (it should be remembered that the test concentration is at least five times greater than that found in normal blood serum). Clearly the combination of 1,3-DAB and resorcinol is by far the best polymer film for screening out acetaminophen over time. The use of electropolymerized films from 1,3-DAB/resorcinol provides for a biosensor whose lifetime is limited by the enzyme and not the polymer film, as was the case for 1,2-DAB polymer films.

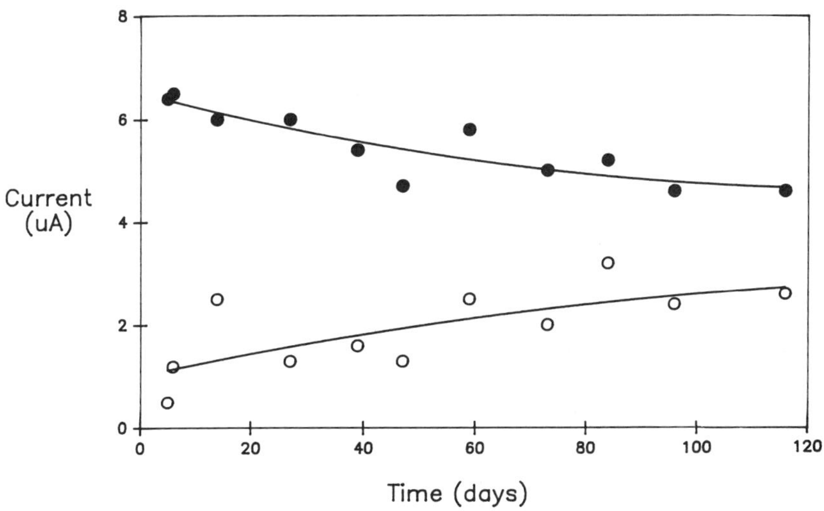

Figure 7 Response of glucose sensor with catechol polymer film. Open circles: 1mM acetaminophen; Filled circles: 5mM glucose.

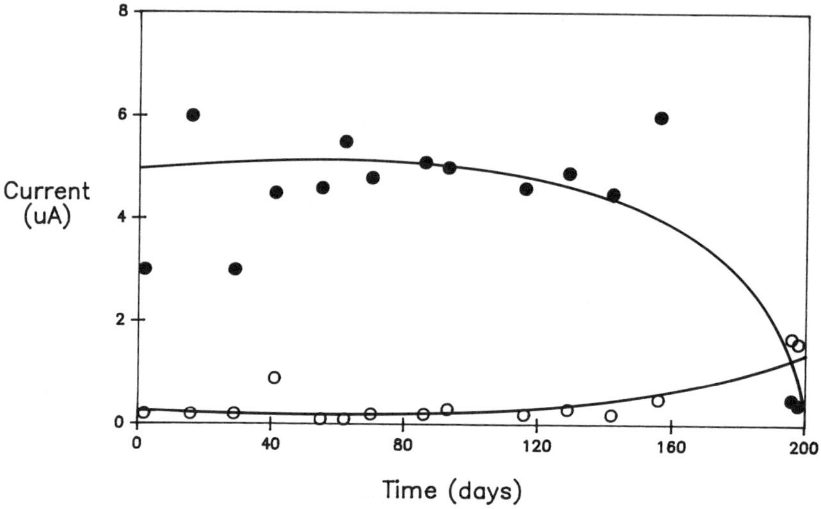

Figure 8 Response of glucose sensor with 1,3-DAB/resorcinol polymer film. Open circles: 1mM acetaminophen; Filled circles: 5mM glucose.

LITERATURE CITED

1. Palleschi, G.; Rahni, M.A.N.; Lubrano, G.J.; Ngwainbi, J. N.; Guilbault, G.G. Anal. Biochem. **1986**, 159, 114.
2. Masoom, M.; Townshend, A. Anal. Chim. Acta **1984**, 166, 111-118.
3. Yao, T.; Kobayashi, Y.; Sato, M. Anal. Chim. Acta **1983**, 153, 337.
4. Mell, L.D.; Maloy, J.T. Anal. Chem. **1975**, 47, 299.
5. Oyama, N.; Anson, F.C. J. Am. Chem. Soc. **1979**, 101, 739.
6. Oyama, N.; Anson, F.C. Anal. Chem. **1980**, 52, 1192.
7. Bull, R.A.; Fan, F.R.; Bard, A.J. J. Electrochem. Soc. **1983**, 130, 1636-1638.
8. Volkov, A.; Tourillon, G.; Lacaze, P.C.; Dubois, J.E. J. Electroanal. Chem. **1980**, 115, 279-291.
9. Wier, L.M.; Guadalupe, A.R.; Abruna, H.D. Anal. Chem. **1985**, 57, 2009-2011.
10. Guadalupe, A.; Abruna, H.D. Anal. Chem. **1985**, 57, 142-149.
11. Ohnuki, Y.; Matsuda, H.; Ohsaka, T.; Oyama, N. J. Electroanal. Chem. **1983**, 158, 55-67.
12. Sittampalam, G.; Wilson, G.S. Anal. Chem. **1983**, 55, 1608-1610.
13. Heineman, W.R.; Wieck, H.J.; Yacynych, A.M. Anal. Chem. **1980**, 52, 345-346.
14. Inzelt, G.; Chambers, J.R.; Kensite, J.F.; Day, R.W.; Lange, M.A. Anal. Chem. **1984**, 56, 301-302.
15. Nowak, R., Schultz, F.A.; Umana, M.; Abruna, H., Murray, R.W. J. Electroanal. Chem. **1979**, 94, 219.
16. Cheek, G.; Wales, C.P.; Nowak, R.J. Anal. Chem. **1983**, 55, 380-381.

17. Yacynych, A.M.; Mark, H.B. J. Electrochem. Soc. **1976**, 123, 1346-1351.
18. Schichiri, M.; Kawamoni, R.; Yamasaki, Y.; Hahui, N.; Abe, H. Lancet **1982**, 1129-1131.
19. Wang, J. Electrochimica Acta **1981**, 26, 1721-1726.
20. Strohl, A.; Curran, D. Anal. Chem. **1979**, 51, 353.
21. Bruno, F.; Pham, M.C.; Dubois, J.E. Electrochimica Acta **1977**, 22, 451-457.
22. Pham, M.C.; Lacaze, P.C.; Dubois, J.E. J. Electroanal. Chem. **1978**, 86, 147-157.
23. Yacynych, A.M.; Sasso, S.V.; Heider, G.M.; Wieck, H.J. Proc. Sensor Science and Technology, The Electrochemical Society, Inc., Pennington, NJ, **1987**, p. 85.
24. Yacynych, A.M.; Sasso, S.V.; Reynolds, E.R.; Geise, R.J. Proc. Biosensors International Workshop, VCH publishers, New York, NY, **1987**, p.69.
25. Nimmo, W.S.; Prescott, L.F. British Journal of Clinical Pharmacology **1978**, 5, 348.

RECEIVED March 13, 1989

Chapter 5

Electrochemical Sensors Based on Polymer Films Immobilized by Gamma Irradiation

Louis A. Coury, Jr., Christos Galiatsatos, Edward W. Huber, Sandra Sponaugle, and William R. Heineman[1]

Edison Sensor Technology Center, Biomedical Chemistry Research Center, and Department of Chemistry, University of Cincinnati, Cincinnati, OH 45221-0172

Gamma irradiation is used as a means of immobilizing water-soluble polymers on spectroscopic graphite electrodes for the purpose of imparting certain selectivity characteristics to electrochemical sensors. Poly(N-vinylpyrrolidone) enables catecholamine neurotransmitters to be determined in the presence of a 10-fold excess of ascorbate by concentrating the catecholamine at the electrode surface while discriminating against ascorbate. This polymer also imparts resistance to electrode fouling from oxidation products generated during the detection of various phenols. Glucose oxidase can be immobilized in poly(vinyl alcohol) by irradiation to form a long-lived glucose sensor with rapid response time.

The development of electrochemical sensors for use in biological systems has been of interest for many years (1). Since samples of biological origin are complex, electrochemical sensors must demonstrate a high degree of selectivity for successful monitoring of the analyte of interest without interference. A typical strategy in sensor development to gain this selectivity is electrode modification, which involves the immobilization of a species at the surface of the electrode. Electrode modification can alter the microenvironment of the electron-transfer reactions that occur at the electrode surface, thus leading to the fabrication of electrodes with unique analytical properties.

The modification of electrode surfaces with polymer films has received considerable attention because of many advantageous properties of polymers (2,3). Polymer films are chemically stable, provide diffusional barriers that can lead to selectivity based on size or charge exclusion properties, provide a means of preconcentrating analytes by ionic or other complexation interactions, and are a convenient matrix for the immobilization of other reagents, such as enzymes. Coating electrode surfaces with polymer films takes advantage of these properties. Complexation of a specific

[1]Address correspondence to this author.

0097-6156/89/0403-0078$06.00/0

analyte(s) by a polymer can increase the effective concentration and hence the analytical signal due to the species of interest at the electrode surface while the signal due to interferents that do not interact with the polymer is attenuated by a combination of decreased concentration near the electrode surface and hindered mass transport through the polymer. The coupling of these properties with the very sensitive instrumental methods available in electrochemistry results in sensors with high selectivity and sensitivity. Electrochemical biosensors based on enzymes or enzymatic substrates immobilized within the polymer film at the electrode surface gain their selectivity and sensitivity from very specific enzyme-substrate reactions that can be monitored electrochemically. Fundamental to progress in expanding the applications and reliability of electrochemical-based sensors is the availability and understanding of modification strategies for imparting the requisite selectivity and sensitivity to these devices.

Strategies for the polymer modification of electrodes have been based on dip coating or spin casting polymers onto the electrode from volatile solvents, adsorption or precipitation from solution, chemical coupling to reactive surface groups, insolubilization by degradation or the formation of crosslinks, and polymerization. The polymer adheres to the electrode surface by some combination of adsorption, covalent bonding, physical entrapment, and low solubility in the contacting solvent. Enzymes or other reagents/catalysts are immobilized into the polymer films by physical entrapment or by bond formation as a result of the modification technique used.

Once prepared, polymer-modified electrodes are transferred to an electrolyte/solvent medium for electrochemical study. The solvent permeates and swells the polymer, decreasing ohmic resistance and allowing permeation of solution species. For some polymers this swelling has led to unsatisfactory results because of film dissolution. Procedures to overcome this instability in swelling solvents often involve the formation of cross-links. As an example, polymer films have been stabilized by in situ cross-linking using organosilane condensation (4), divinylbenzene (5), triallyltrimellitate (6), and heat curing (7).

We have investigated the immobilization of water soluble polymers by exposure to gamma radiation (8-10). This chapter describes some of our work in this area as it relates to the development of electrochemical sensors, especially for use in biological systems. The article focuses on two polymers: poly(N-vinylpyrrolidone) and poly(vinyl alcohol).

Gamma Radiation Immobilization of Polymer Films

Gamma rays are highly energetic photons that are efficient in the generation of radicals in polymer solutions and films (11). Radical sites on different polymer chains can react to form new bonds, or cross-links, and thereby drastically increase the molecular weight of the macromolecular species. Under certain conditions, all of the polymer chains in a given sample may become interconnected, forming a polymer network that is insoluble in a given solvent, but still retains the ability to swell when immersed in that solvent (12). Gamma irradiation may also result in further polymerization of the polymer with monomer species present, or in main chain degradation of the polymer (13), either of which may result in insolubilization of the polymer.

Gamma irradiation has several advantageous features as a method of immobilizing polymers on electrode surfaces. It is a simple, relatively inexpensive procedure. The resulting insoluble gels that are formed swell in solution permitting access for solvent, supporting electrolyte and analyte to the electrode surface (8,14). Varying the radiation dosage or monomer content of the film cast on the electrode can be used to control the degree of swelling of the polymer when it is placed in solvent post-irradiation (15). This allows for control of the relative permeability of the resulting swollen film (8,16). Gamma rays are highly penetrating compared with electron beam or ultraviolet (UV) radiation, hence they can be used to cause gel formation in polymer solutions (12) and thick polymer films (8). No additives such as initiators or sensitizers need to be added to the polymer before gamma irradiation (17), unlike most UV irradiation protocols. Radiation induced grafting of dissimilar polymers has been used to create novel blood-compatible hydrogels for biomedical applications (18). Because this type of polymer contains a high percentage by volume of water when swollen (and no residual initiators or cross-linking agents), such hydrogels would be expected to be well-suited for use in biosensor development. Furthermore, gamma radiation has been used to sterilize surgical instruments and pharmaceutical preparations (19), hence the subjection of polymer coated devices to gamma irradiation will also sterilize the device if the dosage is sufficiently high. This is potentially advantageous for the fabrication of sensors for medical use and fermentation monitoring. Some disadvantages of the gamma irradiation method are the relatively poor control over some free radical reactions, the need for access to a gamma source such as ^{60}Co, and potential radiation damage to components of the device.

The procedure for irradiation of electrodes prepared in our laboratory is as follows. An aliquot of polymer solution is placed onto the surface of an electrode and the solvent evaporated. The electrodes that we have used primarily are spectroscopic graphite, which has a slightly porous surface that aids in the anchoring of the polymer film (8). The approximate surface area of these electrodes is 0.166 cm^2. The electrode is then put in a vial that is purged with argon or nitrogen and then sealed. The vials are placed in a specially designed wooden holder for irradiation shown in Figure 1. The holder is placed inside an aluminum irradiation chamber (i.d. 6 in), which is sealed and lowered into the ^{60}Co gamma radiation source at the University of Cincinnati Nuclear Engineering Facility. The source consists of 18 ^{60}Co rods held vertically in a ring and submerged in a concrete pool filled with water. The chamber is placed in the center of the rods during irradiation. The radiation flux in the chamber varies as a function of position in the source. The dosage received by each electrode, which is the product of flux and irradiation time, can

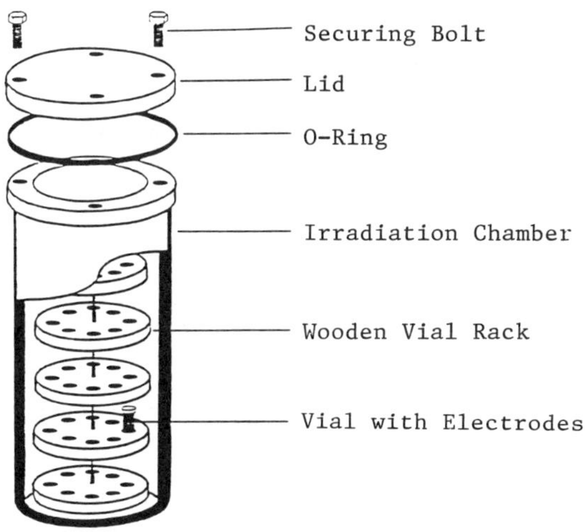

Figure 1. Irradiation apparatus.

be varied by placing the electrodes at different heights in the chamber or at different distances from the chamber center. Figure 2A shows a plot of the dosage from a 95-hour irradiation at a given height from the chamber base for various X,Y coordinates. Figure 2B shows the variation in dosage at a fixed radius as a function of height in the chamber.

Poly(N-vinylpyrrolidone) Modified Electrodes

Poly(N-vinyl-pyrrolidone), or PNVP, is a biocompatible, water-soluble (11) polymer that has previously been used as a synthetic blood plasma extender, as a component of soft contact lenses, and in a complex with iodine as the commercially available topical antiseptic Betadine (20). It also forms a remarkable variety of chemically specific complexes with drugs and hormones such as the tetracyclines, phenothiazine antipsychotics, the antiepileptic drug phenytoin, procaine local anesthetics, and methyldopa antiparkinsonian drugs (21). In its insoluble, chemically cross-linked form, PNVP has also been used as a concentrating agent for a variety of phenols and catechol compounds (22).

Our work with PNVP-modified electrodes has thus far focused on the interaction of the polymer with electroactive catecholamine neurotransmitters and their metabolites (23-26) and on the ability of the electrode coating to reduce fouling of the electrode surface from oxidatively electrogenerated species in the determination of phenols and methoxyphenols. The polymer has been immobilized through gamma irradiation of films cast onto electrode surfaces containing the polymer mixed with its monomer, N-vinylpyrrolidone. The results of NMR experiments have demonstrated that the irradiation process generates radicals in the film, causing polymerization of the vinyl monomer in the presence of the polymer, leading to insolubilization (25). The structure of the polymer formed is shown in Figure 3A.

Enhanced Response to Catechols. An important problem in bio-electroanalytical chemistry is the determination of catechol compounds in the presence of ascorbic acid (vitamin C) (27). The development of carbon fiber ultramicroelectrodes has allowed for the monitoring of these compounds *in vivo* (28). Because the oxidation potential of ascorbate is very similar to those of catechol compounds, because the concentration of ascorbate is at least an order of magnitude higher than that of the catechol species present in the brain (27), and because ascorbate may become involved in an electrocatalytic reaction sequence with oxidized catechol species leading to loss in voltammetric resolution (29,30), the ability to detect catechols in the presence of ascorbate is a non-trivial problem.

When a PNVP-coated electrode is placed in a solution containing a catechol compound, the concentration of the catechol analyte in the polymer film increases with time as the polymer begins to complex the analyte. Figure 4 compares the cyclic voltammetric response obtained in a 1.00 mM solution of dihydroxyphenylacetic acid (DOPAC) at a PNVP-coated electrode (top) and a bare graphite electrode (bottom). DOPAC is one of the

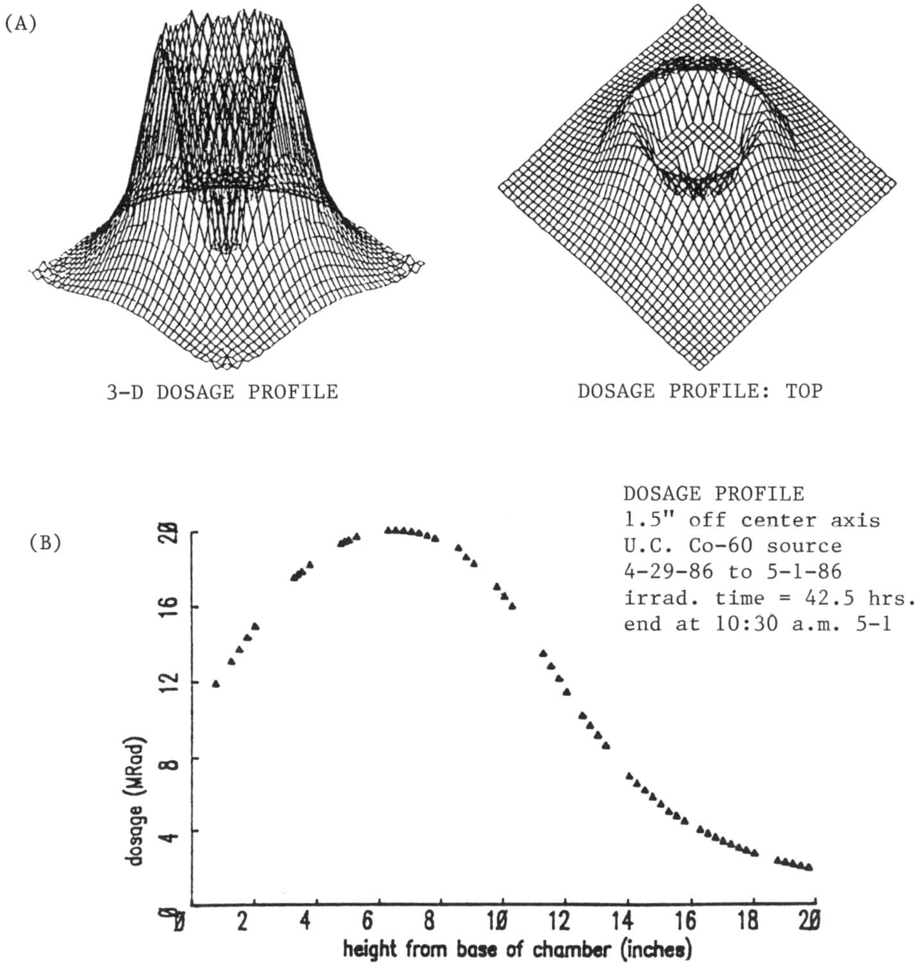

Figure 2. Dosage profile for irradiation time of 95 h (A) as a function of radius at fixed height, 6.25 in., and (B) as a function of height at fixed radius, 1.5 in.

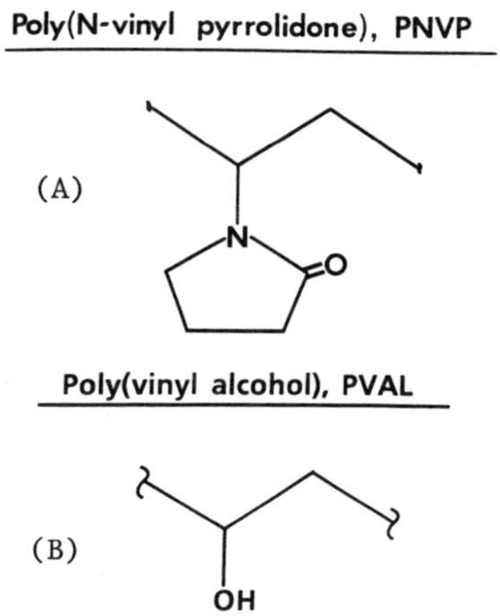

Figure 3. Structures of (A) poly(N-vinylpyrrolidone), PNVP, and (B) poly(vinyl alcohol), PVAL.

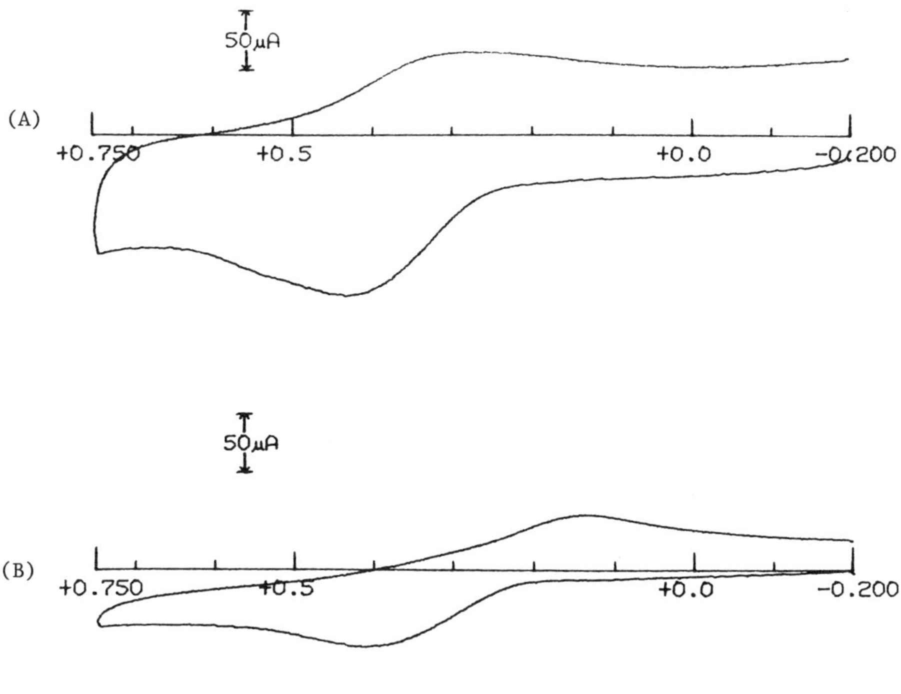

Figure 4. Cyclic voltammogram after exposure of electrode to 1.0 mM DOPAC for (A) PNVP-coated electrode and (B) bare graphite electrode. pH 7 phosphate buffer, scan rate = 0.1 V/s, Ag/AgCl reference.

primary metabolites of the adrenergic neurotransmitter dopamine. The cyclic voltammetric peak current densities were compared for a series of catechol compounds at bare and PNVP-coated graphite electrodes. It was found that, after a 30-min equilibration time, the response at the coated electrode was 1.38 times higher for dopamine at the PNVP-coated electrode, 1.40 times higher for DOPAC, but only 0.98 times as high for noradrenaline (23). The lesser extent of complexation of adrenaline compounds has been reported elsewhere (22), and was attributed to the steric interference of the hydroxyl group *alpha* to the catechol ring, which prevents proper alignment of the adrenaline species with the pyrrolidone ring.

Response to Catechols in the Presence of Ascorbic Acid. In addition to the enhanced response for most catechol compounds, the voltammetric signals due to species in solution that are not complexed by the polymer are often diminished. Because the solvent-swollen polymer occupies space near the electrode surface, it effectively decreases the concentration of uncomplexed solution species. Furthermore, the polymer hinders diffusion of all species to the electrode surface. In the case of catechols, the increase in concentration in the film offsets this effect, but for species that do not bind with the polymer (e.g. ascorbic acid), the rate of mass transport (and subsequently the oxidation current monitored) is attenuated. This effect can be very useful when determining catechol in biological samples.

When a voltammetric scan is initiated at a bare carbon electrode placed in a solution of 10 mM ascorbate/1 mM DOPAC, the peaks due to the oxidation of the two compounds are poorly resolved (24-26). If, however, a PNVP-coated electrode is placed in the same solution, the two peaks are partially resolved, as illustrated in the upper square wave voltammogram shown in Figure 5. Because the concentrations of the two species are more nearly equal close to the electrode surface, the oxidative signals are partially resolved. If the electrode potential is held sufficiently positive to oxidize both species for several minutes, and another scan is acquired, the lower voltammogram shown in Figure 5 is obtained. In this case, ascorbate was oxidized with subsequent rapid hydrolysis to an electroinactive form whereas DOPAC was oxidized to its ortho-quinone form. When the potential was stepped back to a more negative value, the DOPAC quinone was reduced back to the dihydroxy- form. Thus the lower scan shown exhibits a peak due primarily to oxidation of DOPAC since the ascorbic acid present in the film was oxidatively depleted. We have recently reported similar results obtained at gamma-irradiated, PNVP-coated, carbon fiber ultramicroelectrodes (24,25).

Other researchers have demonstrated the utility of Nafion-coated electrodes for the exclusion of ascorbate from the electrode surface during the determination of protonated (cationic) catecholamines (31,32). Although the rejection of ascorbate by these coatings is substantially superior to that demonstrated by PNVP coatings, Nafion films function on the basis of ion-exchange. For this reason, negatively-charged catechol compounds are rejected with anionic interferents such as ascorbic and uric acids. PNVP

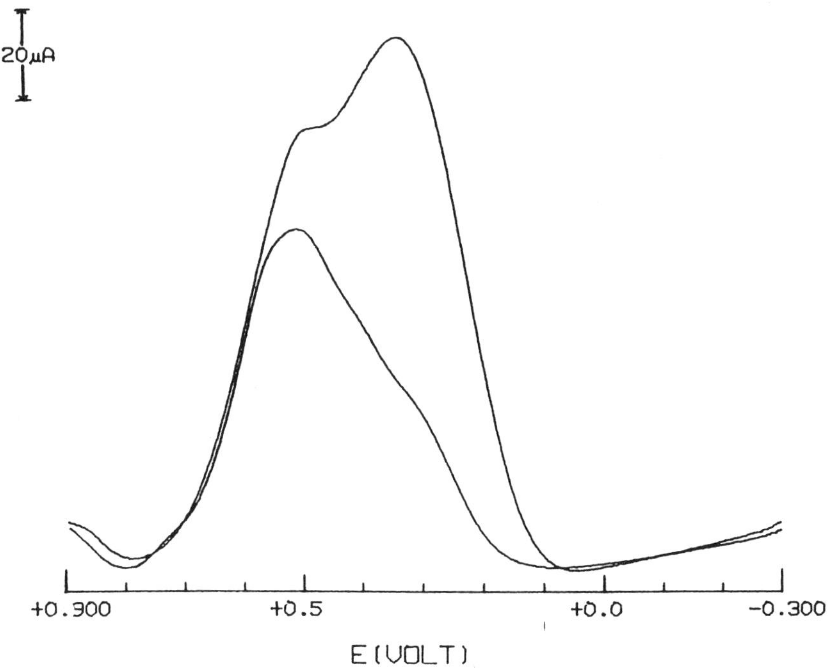

Figure 5. Square wave voltammograms obtained in 10 mM ascorbate/1mM DOPAC, pH 7 phosphate buffer, at a PNVP-coated electrode. Top voltammogram is first scan after equilibration with solution; bottom scan is after oxidative depletion of ascorbate in PNVP film.

electrode coatings allow for the observation of cationic, anionic and neutral catechol compounds in the presence of ascorbate.

Resistance to Electrode Fouling. A potential problem with voltammetric analyses, especially in biological systems, is fouling or passivation of the electrode surface (33). This can be due either to adsorption of compounds from solution or adsorption of products of the electron transfer reaction at the electrode surface. The oxidation of phenols and methoxyphenols, for example, is particularly well-known for causing this latter problem (34). Because of the biological importance of these compounds, there is an increased interest in the development of fouling-resistant electrodes for their detection (35). Phenol has anesthetic and antiseptic properties, and it has been used as the enzymatic product monitored in electrochemical enzyme immunoassays (36). Tyrosine is a phenolic amino acid that has been previously investigated electrochemically (37). Methoxyphenols including 4-hydroxy-3-methoxyphenylacetic acid (homovanillic acid or HVA) are metabolites of neurotransmitters (27) in the brain and are often determined electrochemically in the clinical laboratory. The phenolic compound 4-allyl-2-methoxyphenol (eugenol) is used as a dental analgesic (38) and has been used in the electrocatalytic determination of NADH (39). 4-Hydroxyanisole (Para-methoxyphenol) has been used in the treatment of malignant melanomas (40).

Figure 6 shows the oxidation of tyrosine by square wave voltammetry as a function of time after the electrode is immersed in solution. The oxidation mechanism of this compound has been characterized as involving both radical formation and anodic hydroxylation (41). Voltammograms recorded at the bare electrode (part B) show a loss in current signal due to fouling caused by adsorption of oxidation products of the electrode reaction. By comparison, the current signal at the PNVP-modified electrode (Part A) is greater after 30 min due to partitioning of tyrosine into the polymer film.

Figure 7 is a plot of cyclic voltammetric peak current density as a function of time for the oxidation of eugenol and is representative of the manner in which this analyte may be incorporated into the polymer film. At the bare electrode surface the current signal falls with each voltammogram to ca. 50% of its original value after 60 min due to the fouling process. The response of the PNVP-modified electrode increases as a result of complexation of eugenol into the polymer film. Thus, the response of the bare electrode is attenuated while the response of the modified electrode increases and reaches a maximum after 45 min. Similar behavior has been observed for para-methoxyphenol, homovanillic acid, tyrosine, phenol and vanillin (4-hydroxy-3-methoxybenzaldehyde).

Reliable methods for preventing electrode fouling are needed for long-lived amperometric biosensors. In some cases gamma irradiated polymer coated electrodes such as PNVP could provide a simple and effective way to accomplish this goal. The increased response due to the incorporation of the analyte into the polymer film observed for some compounds makes this approach particularly attractive.

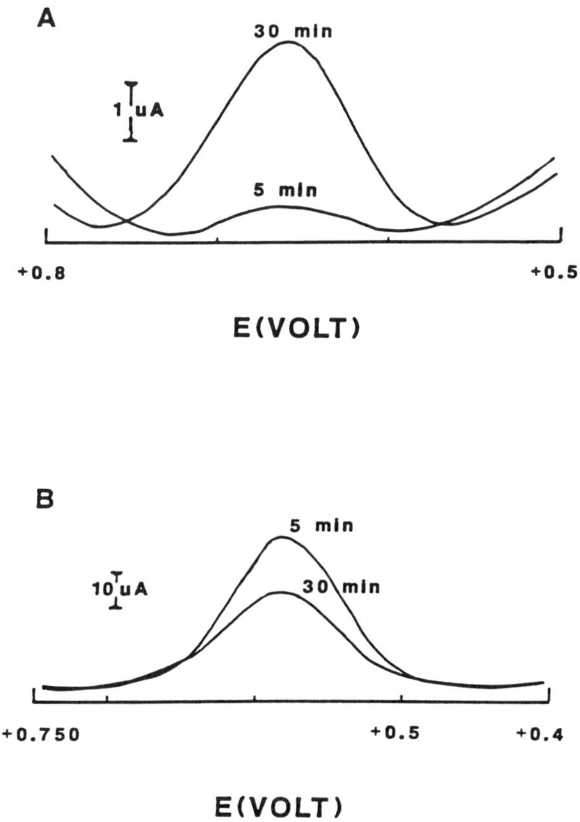

Figure 6. Square-wave voltammograms of 0.1 mM tyrosine, pH 7.45 at (A) PNVP-modified graphite electrode and (B) bare graphite electrode (19.5 mrad).

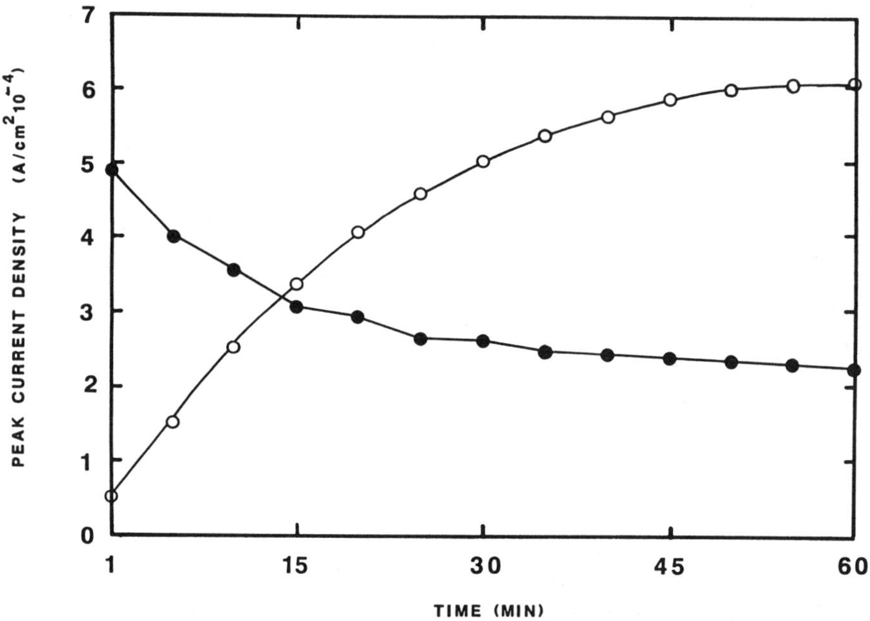

Figure 7. Plot of cyclic voltammetry anodic peak current density vs. time for the oxidation 0.58 mM eugenol, pH 7 phosphate buffer at ● nonmodified graphite and ○ PNVP-modified graphite (19.5 mrad).

Immobilization of Glucose Oxidase in Poly(vinyl alcohol) Modified Electrodes

Immobilized enzymes have been used for the formation of biosensors that are selective for numerous enzyme substrates (42). We have evaluated the potential of gamma irradiation as a means of immobilizing an enzyme in a polymer matrix on an electrode surface. The previously mentioned advantageous features of the gamma radiation method are potentially useful in the formation of biosensors based on enzymes immobilized in polymer films.

Poly(vinyl alcohol), or PVAL, is the polymer that was used for this purpose. It is a neutral, partially water-soluble polymer (43,44) that is generally considered to be biocompatible and non-toxic (45-47). Its structure is shown in Figure 3B. PVAL has been used as a matrix for entrapment of several enzymes in gels formed by gamma irradiation of aqueous solutions of polymer/enzyme mixtures (45,48-50). In some of our experiments, allyl methacrylate (AM) was added to the polymer to facilitate cross-linking at lower doses.

Glucose oxidase (GO) was used as a model enzyme for the evaluation of the immobilization technique. It is commonly used for the measurement of glucose by means of the following reaction

$$\beta\text{-D-glucose} + O_2 + H_2O \xrightarrow{GO} \text{D-gluconic acid} + H_2O_2$$

in which H_2O_2 is detected by oxidation. The PVAL/GO films were prepared on spectroscopic graphite electrodes onto which a layer of platinum had been deposited to enable detection of H_2O_2 at 300 mV vs. Ag/AgCl (14).

Response to Glucose. The electrochemical response of GO immobilized on a platinized graphite electrode to glucose has been observed by dipping it into air-saturated solutions of β-D-glucose and monitoring the H_2O_2 generated. Typical hydrodynamic amperometric responses of a 5 Mrad irradiated PVAL/AM/GO electrode to different concentrations of β-D-glucose injected into stirred pH 7 phosphate buffer are shown in Figure 8. The response times of several electrodes varied between 20-60 s depending on the film thickness and the age of the electrode. Older electrodes and electrodes with thicker films responded more slowly. The linear response range for this compound is 0.05 to 3.0 mM with a sensitivity of 1.39 μA/mM. The detection limit as determined by signal equal to twice the average noise is *ca.* 0.05 mM.

Effect of Gamma Irradiation Dose. A potential problem with the use of gamma irradiation as a means of immobilizing enzymes in polymer films is denaturation of the enzyme. The response of PVAL/GO on platinized electrodes to β-D-glucose as a function of dosage is shown in Figure 9. The electrodes give maximum response when irradiated in the 3 to 5 Mrad range. Unirradiated electrodes gave an initial response that is 20-40% of the maximum response obtained for irradiated electrodes, which indicates that some enzyme is retained on the electrode surface at zero dose. However, this response diminishes over a period of several hours as the uncross-

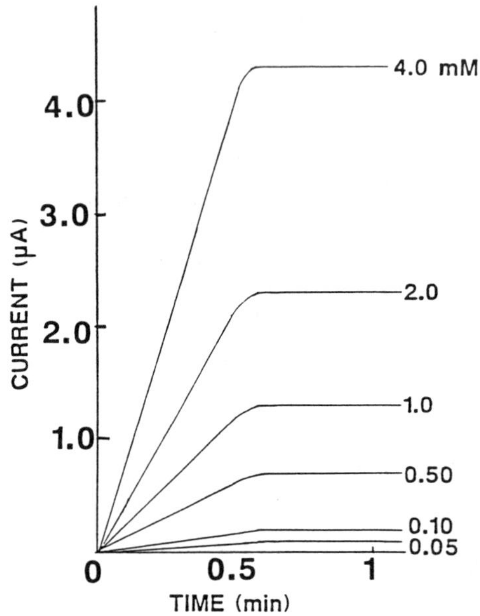

Figure 8. Hydrodynamic amperograms of a PVAL/AM/GO electrode (5 Mrad irradiation) to addition of different concentrations of β-D-glucose in pH 7.0, 0.1 M phosphate buffer.

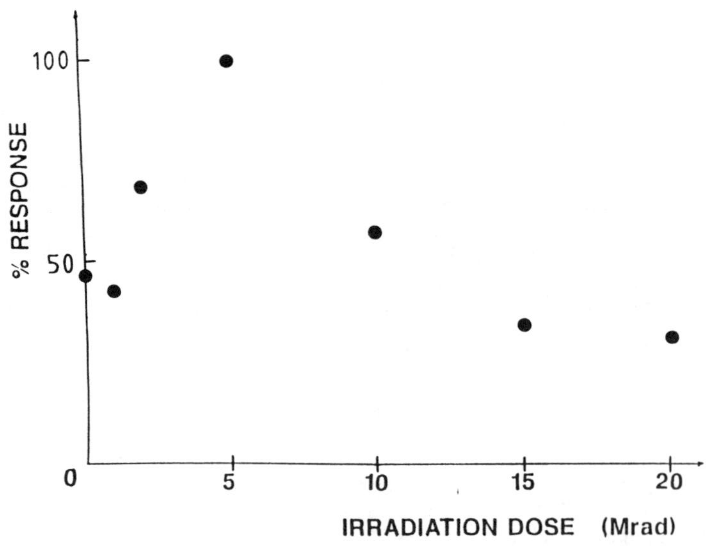

Figure 9. Effect of gamma-irradiation dose on the PVAL/GO electrode response on platinized graphite at 300 mV to 1.00 mM β-D-glucose in pH 7.00, 0.1M phosphate buffer.

linked polymer film slowly releases the enzyme into solution. By comparison, irradiated electrodes gave a stable response for up to two months. Above *ca.* 5 Mrad, electrodes exhibited a progressive loss of response with increasing dosage. This is attributed to radiation damage to the enzyme.

Conclusions

Gamma irradiation has been demonstrated as a means of immobilizing water soluble polymer films on electrode surfaces for the purpose of imparting selectivity properties to the modified electrode. The polymer can impart selectivity for an analyte as a result of a specific chemical interaction between the polymer and the analyte as illustrated by PNVP and catechols. Alternatively, selectivity can be achieved by immobilization of an enzyme in the polymer films such as the case of glucose oxidase immobilized in PVAL. A dosage range was found that is sufficiently high to form a stable film of PVAL and to effectively immobilize the enzyme, yet that is not so high as to destroy enzyme activity. The successful extension of this methodology to other enzymes will depend in large part on the sensitivity of the biomolecule to gamma radiation. These results suggest that immobilization of polymers and biomolecules by this procedure may provide an effective avenue for the fabrication of chemical sensors for cases in which the advantages outlined in the Introduction are important.

Literature Cited

1. Lunte, C.E.; Heineman, W.R. "Electrochemical Techniques in Bioanalysis" in *Topics in Current Chemistry*; E. Steckhan, Ed.: Springer-Verlag: Berlin Heidelberg, 1988; Vol. 143, p 1.
2. R.W. Murray in "Electroanalytical Chemistry"; Bard, A.J., Ed.; Marcel Dekker: New York, 1984; Vol. 13, p. 191.
3. Murray, R.W. *Ann. Rev. Mater. Sci.* 1984, *14*, 145.
4. Nakahama, S.; Murray, R.W. *J. Electroanal. Chem.* 1983, *158*, 303.
5. Faulkner, L.R.; Shaw, B.R.; Haight, G.P. *J. Electroanal. Chem.* 1982, *140*, 147.
6. Blubaugh, E.A.; Bushong, W.C.; Shupack, S.I.; Durst, R.A. *Anal. Lett.* 1986, *19*, 1777.
7. Anson, F.C.; Ohsaka, T.C.; Saveant, J.M. *J. Phys. Chem.* 1983, *87*, 640.
8. De Castro, E.S.; Huber, E.W.; Villarroel, D.; Galiatsatos, C.; Mark, J.E.; Heineman, W.R.; Murray, P.T. *Anal. Chem.* 1987, *59*, 134.
9. De Castro, E.S.; Smith, D.A.; Mark, J.E.; Heineman, W.R. *J. Electroanal. Chem.* 1982, *138*, 197.
10. Kelly, M.J.; Heineman, W.R. *J. Electroanal. Chem.* 1987, *222*, 243.
11. Chapiro, A. *Radiation Chemistry of Polymeric Systems*; Interscience: New York, 1962.
12. Chapiro, A.; Legris, C. *Radiat. Phys. Chem.* 1986, *28*, 143.
13. Kiran, E.; Rodriguez, F. *J. Macromol. Sci.-Phys.* 1973, *B7(2)*, 209.

14. Galiatsatos, C.; Ikariyama, Y.; Mark, J.E.; Heineman, W.R. Biosensors, in press.
15. Chapiro, A.; Legris, C. Eur. Polym. J. 1985, 21, 49.
16. Huber, E.W.; Heineman, W.R. unpublished results.
17. Davis, J.E.; Senogles, E. Aust. J. Chem. 1981, 34, 1413.
18. Ratner, B.D.; Hoffman, A.S.; Hanson, S.R.; Harker, L.A.; Whiffen, J.D.; J. Polym. Sci.: Polym. Symp. 1979, 66, 363.
19. Jacobs, G.P. Radiat. Phys. Chem. 1985, 26, 133.
20. Blecher, L.; Lorenz, D.H.; Lowd, H.L.; Wood, A.S.; Wyman, D.P. In Handbook of Water-Soluble Gums and Resins; Davidson, R.L., Ed.; McGraw-Hill: New York, NY, 1980, pp 21.1-21.21.
21. Molyneux, P. Water-Soluble Synthetic Polymers: Properties and Behavior, CRC Press: Boca Raton, FL, 1983; Vol. 2, pp. 93-96.
22. Carpenter, A.; Siggia, S.; Carter, S. Anal. Chem. 1976, 48, 225.
23. Coury, Jr., L.A.; Birch, E.M.; Heineman, W.R. Anal. Chem. 1988, 60, 553.
24. Coury, Jr., L.A.; Birch, E.M.; Heineman, W.R. Proc. Electrochem. Soc., 1988, 87-15 (Proc. Symp. Sensor Sci. Tech.), 104.
25. Coury, Jr., L.A.; Huber, E.W.; Birch, E.M.; Heineman, W.R, J. Electrochem. Soc, in press.
26. Coury, Jr., L.A.; Huber, E.W.; Heineman, W.R. In Applied Biosensors; Wise, D.L., Ed.; Butterworths: Stoneham, MA, in press.
27. Adams, R.N. Anal. Chem. 1976 48, 1126A.
28. Wightman, R.M. Science 1988, 240, 415.
29. Stamford, J.A. Anal. Chem. 1986, 58, 1033.
30. Stamford, J.A. J. Neurosci. Meth. 1986, 17, 1.
31. Whitely, L.D.; Martin, C.R. Anal. Chem. 1987, 59, 1746.
32. Nagy, G.; Gerhardt, G.A.; Oke, A.F.; Rice, M.E.; Adams, R.N.; Moore, R.B.; Szentirmay, M.N.; Martin, C.R. J. Electroanal. Chem. 1985, 188, 85.
33. Johnson, D.C.; Polta, T.Z.; Neuburger, G.G.; Johnson, J.; Tang, A.P.-C.; Yeo, I.-H.; Baur, J. J. Chem. Soc., Faraday Trans 1 1986, 82, 1081.
34. Koile, R.C.; Johnson, D.C. Anal. Chem. 1979, 51, 741.
35. "Proceedings of the Symposium on Sensor Science and Technology"; Schumm, Jr., B.; Liu, C.-C.; Powers, R.A.; Yeager, E.B., Eds.; The Electrochemical Society: Pennington, NJ, 1987.
36. Heineman, W.R.; Halsall, H.B. Anal. Chem. 1985, 57, 1321A.
37. Bennett, G.W.; Brazell, M.P.; Marsden, C.A. Life Sci. 1981, 29, 1001.
38. The Merck Index, Tenth Edition, M. Windholz (Ed.), Rahway, NJ, 1983, p 563.
39. Ueda, C.; Tse, D.C.-S.; Kuwana, T. Anal. Chem. 1982, 54, 850.
40. Morgan, B.D.G.; O'Neill, T.; Dewey, D.L.; Galpine, A.R.; Riley, P.A. Clinical Oncology, 1981, 7, 227.
41. Papouchado, L.; Petrie, G.; Adams, R.N. J. Electroanal. Chem. 1972, 38, 389.
42. Turner, A.P.F.; Karube, E.; Wilson, G.S. Biosensors: Fundamentals and Applications, Oxford University Press: Oxford, 1987.

43. Pritchard, J.G. Poly(vinyl alcohol) Basic Properties and Uses, Gordon and Breach Science Publishers: New York, 1975.
44. Dieu, H.A. J. Polym. Sci. 1954, 12, 417.
45. Maeda, H.; Suzuki, H. Process Biochem., 1977, 12, 9.
46. Maeda, H. Biotech. & Bioeng., 1975, 17, 1571.
47. Imai, K; Shiomi, T.; Uchida, K.; Miya, M. Biotech. & Bioeng. 1986, 28, 198.
48. Mascini, M.; Mateescu, M.A.; Pilloton, R. Bioelectrochem. Bioeng. 1986, 16, 149.
49. Ichimura, K; Komatsu, T. J. Polym. Sci. 1987, 25, 1475.
50. Matsumoto, K.; Mizuguchi, H.; Ichimura, K. Kobunshi Ronbunshu, 1984, 41, 221; in Chem. Abstr., 101: 86528S.

RECEIVED March 17, 1989

Chapter 6

Pulse Voltammetric Biosensing Device for Real Time Determination of a Stationary Droplet Glucose Sample

Yoshihito Ikariyama[1], Naoki Shimada, and Shigeru Yamauchi

Research Institute, National Rehabilitation Center for the Disabled, Namiki 4–1, Tokorozawa, Saitama 359, Japan

A microbiosensing device fabricated by assembling an enzyme-embodied microelectrode with a counter electrode and a reference electrode was applied to the real time determination of a trace droplet of glucose sample. The microdevice demonstrated unique properties such as real time response, independence of sample volume, and no need for sample mixing by the combination of a microfabricated biosensing device with pulse voltammetry.

Microfabrication is one of the common trends in science and technology of our age, since microfabricated sensors will play indispensable roles for the construction of very small, sophisticated instruments as well as the microinstrumentation of analytical tools (1).

Being small has many advantages in electrochemical analysis of trace substances of biological importance with biosensors. In the early step of biosensor research many enzyme-immobilized membranes have been attached to Clark-type oxygen electrodes (2-6). In recent years further application of enzyme electrodes has been extended by the microfabrication of electrode-based biosensors, since multifuctional biosensors can be developed by taking advantages of silicon technology (7-9). Recent photolithographic techniques for deposition and patterning of thin noble metal films (50 - 300 nm) have been developed for the purposes of fabrication of planar integrated circuits (10). These techniques are also promising for the preparation of microtransducers for miniaturized biosensors. However, new favorable techniques for immobilizing enzyme onto microtransducers are required for the fabrication of microenzyme electrodes, sizes in the μm range, because the conventional enzyme immobilization techniques are not necessarily compatible with miniaturized transducers as long as enzymes are employed in polymer matrix-supported forms.

In previous papers we have reported fundamental properties of microenzyme electrodes with diameters in the range from 10 to 200 μm prepared by incorporating enzyme molecules onto a platinized surface of a microplatinum electrode (11,12). It was the platinization technique that enabled the microfabrication of the biosensor. It is a well-known fact that platinized platinum, porous microparticles of platinum, possesses a very large surface area and high catalytic activity for electrolytic processes. By employing a microelectrode as a base electrode for platinization and simultaneous enzyme immobilization, a micro-enzyme sensor can be easily fabricated (13). Enzyme molecules are stably retained in the porous platinum particles (14). In other words, the platinum particles play the dual roles of transducer of enzyme electrode and matrix for enzyme immobilization.

[1]Current address: Department of Bioengineering, Tokyo Institute of Technology, O-okayama 2–12–1, Meguro-ku, Tokyo 152, Japan

0097–6156/89/0403–0096$06.00/0

The fabricated microbiosensors for glucose demonstrate high performances such as low detection limit of 5 x 10^{-7} mol/L, wide dynamic range from 5 x 10^{-7} to 2 x 10^{-2} mol/L, rapid response of 3 s, satisfactory coefficient of variation [2%], and time stability of over one month (14,15).

Here we describe a unique microbiosensing device for the real time determination of droplet glucose sample. As the operation mode of the microdevice is voltammetry driven by the application of pulsed potential, a stationary trace glucose sample can be determined in 2 ms. The advantages of the microsensing device over the dry chemistry-based clinical test papers are also discussed.

Experimental Section

A microplatinum wire having a diameter from 10 μm to 200 μm, a platinum wire (200 μm diameter), and a silver wire (500 μm diameter) were sealed in a Teflon casing (2.5-mm inner diameter) with polyester resin, ground, and polished to leave exposed a circular surface. Electrodes were polished with Imperial™ wrapping films of alumina powders of different particle sizes. At the final stage of polishing, 0.1 μm Al_2O_3 was used as the finish and only the microplatinum electrode for enzyme immobilization was cycled between -0.25 and +1.28 V vs a silver-silver chloride electrode (reference electrode) in 0.5 M H_2SO_4 prior to platinization with a conventional three electrode system. The silver electrode was anodized in 1 M NaCl solution for the preparation of a silver-silver chloride electrode. The microplatinum electrode was platinized in a mixed solution of hexachloroplatinate and lead acetate by following the procedure described previously (14). After the platinization high purity glucose oxidase (EC 1.1.3.4) (from *A. niger*, 120 units/mg) was soaked into the porous surface of the microplatinum electrode, following which crosslinking of the enzyme molecules with glutaraldehyde was performed (14). Figure 1 shows a schematic illustration of the electrochemical microfabrication techniques. The measuring principle of the micro-biosensing device thus prepared is schematically shown in Figure 2.

A stock buffer solution of pH 6.8 was prepared from phosphate salts (50 mM) and NaCl (50 mM) as working electrolyte.

Experimental assays of droplet glucose sample were carried out with a potentiostat in a three-electrode configuation with the Ag/AgCl reference electrode of the microbiosensing device. The potentiostat was operated in an on-off potential application mode. A potential of +0.6 V vs Ag/AgCl was applied to the enzyme-embodied working electrode. The resulting amperometric signals were recorded by using an x-y-t recorder or a memory scope (MOS static RAM = 4 x 12 bit x 4096 word). The response time of the x-y-t recorder was 200 ms. In the later case each recorded datum was stored in a floppy disk, and then recored on an x-y-t recorder. A schematic diagram of the microbiosensing principle is shown in Figure 3. Glucose determination was performed by addition of an aliquot of glucose stock solution. Potential was applied at 10 s after the addition of a droplet glucose sample.

Results and Discussion

Typical Responses of Microbiosensing Device. The effectiveness of the microbiosensing device, pulse voltammetry combination was evaluated in glucose sample by loading an aliquot of standard glucose sample (10 mM). Every glucose sample (10 μL) was dropped on the three-electrode biosensing device, and a potential of +0.6 V was applied at 10 s after sample loading. Figure 4 shows typical data recorded on the x-y-t recorder. Also shown is the response curve of a blank sample (the phosphate-buffered NaCl solution). After the initial sharp peak, a gradually decaying curve was observed. As the response time of the x-y-t recorder is 200 ms, the initial peak is the sum of capacitive and faradaic currents. The difference between the peak height of glucose sample and that of blank sample was plotted against the concentration of glucose sample, which gives a linear relationship between the peak height current and the glucose concentration in the range from 1 mM to 20 mM. The calibration curve is a straight line with slope equal to ca. one-third. Since the slope of the curve is less than expected, we applied a

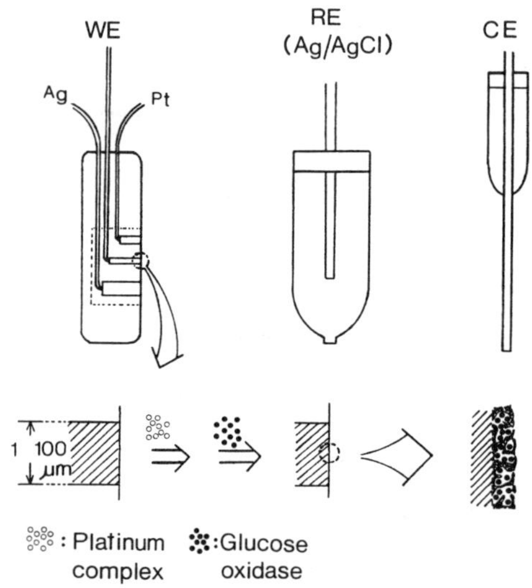

Figure 1. Schematic illustration of electrochemical fabrication technique.
After the preparation of the microplatinum electrode in a Teflon casing, platinization is performed by electrochemical reduction of hexachloroplatinate in the presence of lead acetate, following which enzyme incorporation is carried out. The electrode is immersed in an albumin solution and then treated with glutaraldehyde.

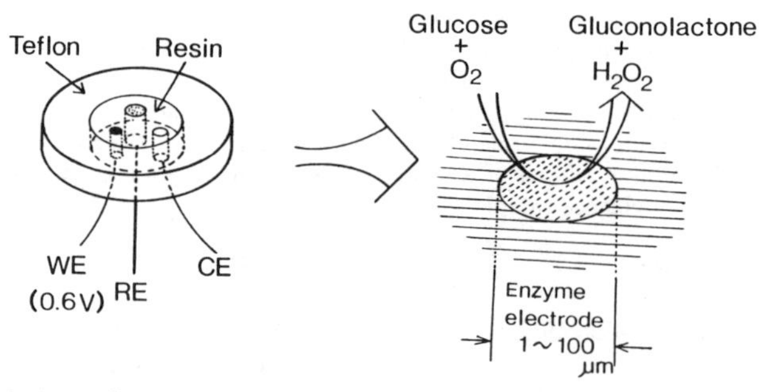

$$2\,H_2O_2 \xrightarrow{0.6\,V} 2\,H_2O + O_2 + 4e$$

Figure 2. Schematic drawing of microbiosensing device.
Glucose oxidase immobilized in the porous surface of the platinized microelectrode oxidizes glucose in a droplet sample, and the generated hydrogen peroxide is electrochemically oxidized to give anodic sensor signal.

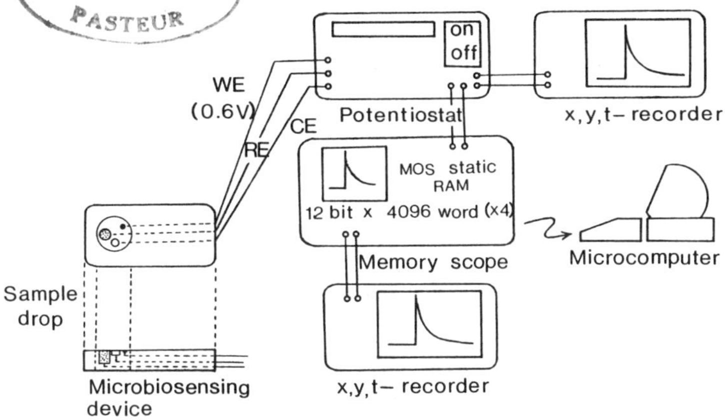

Figure 3. Schematic diagram of the sensing principle of microbiosensing device.
After the addition of a droplet glucose sample on the three-electrode biosensing device, a potential of +0.6 V is applied to the microdevice and the resulting transient response is recorded on an x-y-t recorder or on a memory scope.

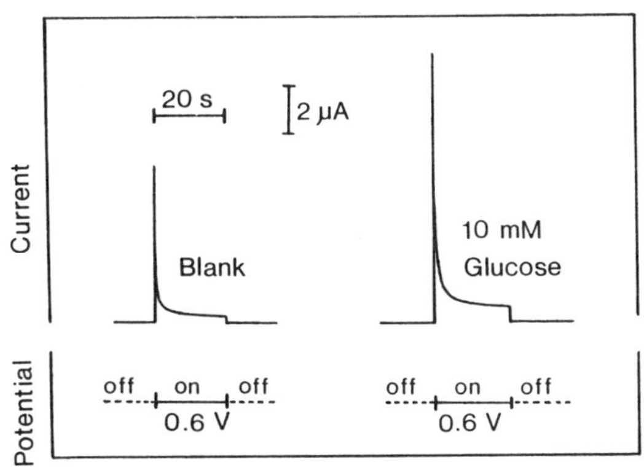

Figure 4. Typical responses of microbiosensing device.
The responses of the biosensing device to a blank sample (the phosphate-buffered NaCl solution) and a 10 mM glucose sample (in the same buffered solution) were recorded on an x-y-t recorder whose response time is 200 ms.

platinized microplatinum electrode to the pulse voltammetric determination of hydrogen peroxide, which gave a linear relation, with a slope of +1, between the initial peak height and the H_2O_2 concentration in the range of 1 - 100 mM hydrogen peroxide. These results explain the distinction between the enzyme-immobilized and the platinized (non-enzyme) electrodes.

The above results make it apparent that in order to measure faradaic current, a recording instrument capable of detecting fast electrodic process will improve the separation of faradaic from capacitive current.

A memoryscope was employed to investigate the transient response of the microdevice in further detail. In chronoamperometry current as a function of time is the system response. Figure 5 shows a detailed response of the microbiosensing device to a 10 mM droplet glucose sample (10 μL). In addition to a blank sample, we employed a fructose sample of the same volume and concentration to examine the specificity of the microdevice. All three samples demonstrated similar responses in the rapid decay of the initial peak; however, considerable difference among three samples was observed after the decay of charging current. The potential application initiated an instantaneous current as a result of the oxidation of hydrogen peroxide. The resulting current decreases as the electrodic reaction proceeds. In the faradaic process, therefore, the charge passed across the electrode interface is related to the amount of material that has been converted, and the generated current is related to the instantaneous rate at which the oxidation of hydrogen peroxide produced by the immobilized glucose oxidase occurs. In general, monitored response to the potential excitation signal for a solution containing supporting electrolyte only is the background response (capacitive current) which should decay rapidly when the electrode has been charged to the applied potential. However, the response to a blank sample shows the faradaic current even after the decay of capacitive current. It should be noted that when a platinum electrode is used as a transducer electrode, faradaic current is observed due to the formation of surface oxide of platinum (16,17). After the anodic polarization the faradaic current for the surface oxide formation decreases significantly (18). With respect to the response curve of the sensing device in Figure 5, the difference in the transient response between a glucose sample and a blank sample can be a reasonable sensor signal of the pulse voltammetric biosensing device. As the behavior of the device to a fructose sample was almost similar to that to a blank sample, the microdevice was regarded to have specificity to glucose.

Relation between Transient Response and Glucose Concentration. As the greater difference in the transient current between a glucose sample and a fructose sample was observed at 2 ms after the application of +0.6 V to the three-electrode device, the relation between the difference in sensor output and glucose concentration was investigated. Figure 6 shows a linear calibration curve for the real time determination of glucose in the range from 1 mM to 100 mM. We have also here a linear relation with a slope of one-third. So far we cannot find a resonable explanation for the slope of the calibration curve. In the case of glucose determination in a batch system (14) and a flow system (1)5 where samples are in a hydrodynamic state, we have obtained a linear calibration with a slope of +1. In the determination with the present pulse voltammetric device, droplet samples are in a hydrostatic condition where the flux of hydrogen peroxide to the microelectrode is diffusion controlled. It is noteworthy that the enzyme reaction is also a diffusion-controlled process in the stationary solution. With this in mind, one can easily understand that the transient response seemed to reflect the non-steady state of the enzyme reaction as well as the electrodic reaction. The reasonable explanation for this intriguing phenomenon still remains to be studied in further detail.

Relation between Sensor Output and Sample Volume. The responses of the microbiosensing device to 10 mM glucose sample of different sample volume were obtained to know the relation between sensor output and sample volume. Almost identical responses were observed in this series of measurements where sample volume was varied from 1 μL to 20 μL. When the sample volume was less than 1 μL, the three-electrode microbiosensing device was not covered with the sample droplet. On the other hand, a sample droplet of more than 20 μL spilled off of the device surface. Figure 7 illustrates the relation between sensor output and sample volume. It should be noted that the sensing principle is independent of sample volume, which is

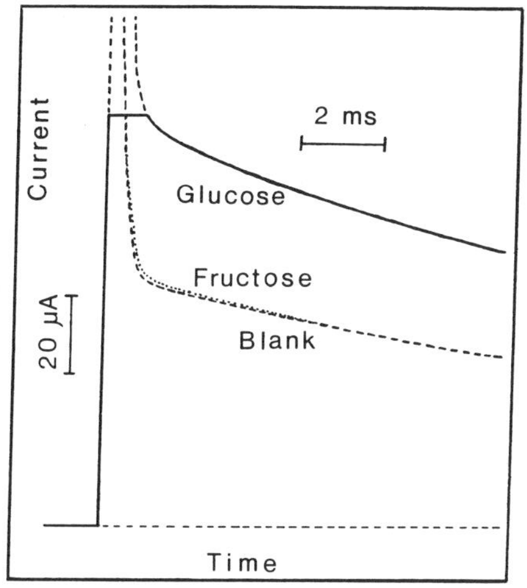

Figure 5. Transient responses of microbiosensing device.
The responses of the microdevice were obtained by applying a potential of +0.6 V after a glucose sample (20 mM), a fructose sample and a blank sample were separately loaded on the device surface.

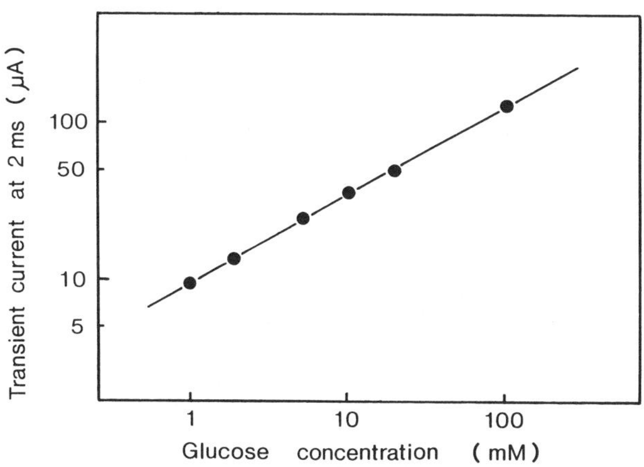

Figure 6. Calibration curve for real time determination of glucose.
The difference in transient response at 2 ms between glucose sample and blank sample is plotted against the glucose concentration.

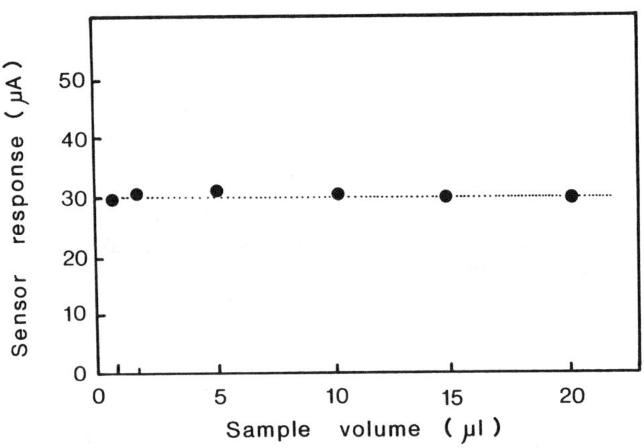

Figure 7. Relation between sensor output and sample volume.
The sensor output taken here is the difference in transient response at 2 ms between droplet glucose sample and droplet blank sample.

one of the advantages for the microinstrumentation in clinical situations, because the pulse voltammetric device does not require any attached stirring machines.

As the conventional voltammetry such as pulse voltammetry is independent of solution volume being tested, the results in Figure 6 are consistent with typical data obtained by pulse voltammetric techniques. It is also worth noting that almost all the measurements with chemical sensors are also independent of sample volume. To our knowledge, however, this is the first experimental demonstration that biosensors are applicable to hydrostatic samples without paying attention to the volume of the sample being tested. Although the present work is in the early stage of experimentation, we examined the precision of the pulse voltammetric biosensing device by using 10 mM glucose sample (10 μL). The resulting coefficient of variation was calculated to be + 4.4 % (N=10).

Conclusions

The microbiosensing device demonstrated several performance characteristics such as real time determination, no need for stirring equipment, and no need for a sampling tool. These advantages will allow several design criteria for microinstrumentation of clinical analyzers. The performance characteristics of the platinized platinum microelectrode-based biosensors in batch-system, flow system, and pulse voltammetric measurements are summarized in Table I.

By combining three micro-electrodes into a single sensing unit, we have succeeded in fabricating a miniaturized biosensing device. Therefore, a very small amount of sample can be applied to the microdevice, even if the droplet sample volume is only a few μL. The microsensing device described here has a couple of advantages over the dry chemistry-based clinical test papers. Microinstrumentation can be easily performed, because the device does not require additional equipment for the sampling of exact sample and for the stirring of sample solution. Furthermore, the microbiosensing device can be repeatedly used because of the stable, intact immobilization of enzyme molecules in the porous platinized electrode matrix. In addition to these advantages, another most important performance characteristic is that every sample can be determined instantaneously by taking advantage of be pulse voltammetric technique. Finally, while we have focused on the fabrication of a microbiosensing device which can be operated in the real time determination of analytes, the principle described here will also be a useful means for the microinstrumentation of portable-type biosensors for the daily surveillance of diabetic patients, especially of diabetic blind patients.

Table I. Performance Characteristics of Microbiosensing for Glucose

Measuring Principle	Electrode Diameter (μm)	Dynamic Range (M)	Processing Speed (s/sample)	C.V. (%)
Pulse voltammetry	10 - 100	10^{-3} - 10^{-1}	2×10^{-3}	4.4 (N=10)
Batch system	10 - 500	5×10^{-7} - 2×10^{-2}	3	2.0 (N=20)
Flow system	10 - 200	5×10^{-5} - 5×10^{-2}	5 - 10	1.0 (N=600)

Literature Cited

1. Haugen, G; Hieftze, H. *Anal. Chem.* 1988, **60**, 23A.
2. Clark, L. C. Jr.; Lyons, C. *Ann. N. Y. Acad.* 1962, **102**, 29.
3. Updike, S. J; Hicks, G. P. *Nature (London)* 1967, **214**, 986.
4. Aizawa, M.; Karube, I.; Suzuki, S. *Anal. Chim. Acta* 1974, **69**, 431.
5. Nanjo, M.; Guilbault, G. G. *Anal. Chim. Acta* 1974, **73**, 367.

6. Ikariyama, Y.; Furuki, M.; Aizawa, M. *Anal. Chem.* 1985, **57**, 496.
7. Caras, S.; Janata, J. *Anal. Chem.* 1980, **52**, 1935.
8. Miyahara, Y.; Moriizumi, T.; Ichimura, K. *Sensors and Actuators* 1985, **7**, 1.
9. Hanazato, Y.; Nakano, M.; Shiono, S. *IEEE Trans. Electron Devices* 1986,**ED-33**, 47.
10. Sze, S. M.in *VLSI Technology*; McGraw Hill Book Co.: New York.
11. Ikariyama, Y.; Yamauchi, S.; Yukiashi, T.; Ushioda, H. *Anal. Lett.* 1987, **20**, 1407.
12. Ikariyama, Y.; Yamauchi, S.; Yukiashi, T.; Ushioda, H. *Anal. Lett.* 1987, **20**, 1791.
13. Ikariyama, Y.; Yamauchi, S.; Yukiashi, T.; Ushioda, H. *J. Electrochem. Soc.* (in press).
14. Ikariyama, Y.; Yamauchi, S.; Aizawa, M.; Yukiashi, T.; Ushioda, H. *Bull. Chem. Soc. Jpn.* 1988, **61**, 3525.
15. Ikariyama, Y.; Yamauchi, S.; Yukiashi, T.; Ushioda, H. *Bull. Chem. Soc. Jpn.* (submitted for publication).
16. Polta, J. A.; Johnson, D. C. *Anal. Chem.* 1985, **57**, 1373.
17. Neuberger, G.; Johnson, D. C. *Anal. Chem.* 1987, **59**, 204.
18. Ikariyama, Y.; Yamauchi, S.; Yukiashi, T.; Ushioda, H. *J. Electroanal. Chem.* 1988, **251**, 267.

RECEIVED March 9, 1989

Chapter 7

Bioelectrochemistry at Microelectrodes

H. Allen O. Hill, Napthali P. Klein, A. Surya N. Murthy, and Ioanna S. M. Psalti

Inorganic Chemistry Laboratory and Oxford Centre for Molecular Sciences, South Parks Road, Oxford OX1 3QR, England

The successful detection of glucose, up to 15mM, employing disc microelectrodes (25μm to 60μm diameter) and 8mM, with dual microband gold electrodes (1μm thick), is described. The studies made use of the coupling of the glucose-glucose oxidase reaction to electro-generated ferricinium species. Detection of the analyte by the dual electrode is dependent upon interelectrode distance.

The electrochemistry of biological molecules has been studied thoroughly in recent years; that of simple redox proteins is now (1-2) well-understood. Two new features have recently come to prominence: the re-interpretation of some aspects of the electrochemistry of proteins in terms (Armstrong, F.A.; Bond, A.M.; Hill, H.A.O.; Oliver, B.N.; Psalti, I.S.M. J. Amer. Chem. Soc., 1989, in press) of the behaviour at multi-microelectrodes and the dynamics of movement both *of*, and *within*, protein-protein complexes when at the electrode surface (3-4). The latter came to the fore when the electrochemistry of protein-protein complexes was investigated. Reversible electron transfer of both proteins was observed at an electrode at which one component of the complex separately gave good electrochemistry and the other none. It was at first thought that some form of mediation of electron transfer was occuring, as indeed seems to be the case (Barker, P.D.; Hill, H.A.O.; Walton, N.J. J. Electroanal. Chem., in press) when a modified gold electrode is used. However, when a complex is prepared between zinc(II) cytochrome *c*, which is redox-inactive, and cytochrome b_5, the electrochemistry of the latter was observed. N.m.r. spectroscopic investigations have suggested (Driscoll, P.C.; Goodall, K.; Hill, H.A.O.; Redfield, C. University of Oxford, unpublished results) that the two proteins, *within the complex*, move with respect to one another. This implies that motion, both *within* the complex and *of* the complex at the electrode surface, is possible. The rate of electron transfer between electrode and protein, or protein-protein

0097–6156/89/0403–0105$06.00/0

complex, will obviously depend on both the rate of such dynamic motion and on the resultant distance from the electrode surface.

Larger proteins, especially enzymes, still cause many problems though there have been some encouraging reports (5-6) and recently (Guo, L.H.; Hill, H.A.O.; Hopper, D.J.; Lawrence, G.A.; Sanghera, G.S. J. Electroanal. Chem., submitted) excellent electrochemistry of the enzyme, *p*-cresolmethylhydroxylase, has been achieved. The problem over the years seems to have been that, even where modified electrodes were used, sufficient care was not taken to ensure that the enzyme was still *intact*, i.e., that the prosthetic group had not been released from the enzyme. In such cases, it was difficult to distinguish between the electrochemistry of the 'free' prosthetic group and the intact protein. This is particularly true when applied to flavoenzymes since the prosthetic group is often not covalently bound to the enzyme. At unmodified electrodes, the problem is more intense and *irreversible* adsorption with, or without, concomitant denaturation is often encountered. Additional problems concerned the presence of adventitious mediators or those released from the electrode material itself, e.g., in the use (7) of conducting organic salts as electrodes. Now, with well-defined electrode surfaces, a better appreciation of the necessity of using pure and stable enzymes and the correct criteria for assessing the successful attainment of electron transfer, we may expect an increasing number of reports in the near future.

ANALYTICAL BIOELECTROCHEMISTRY

There has long been an interest in the use of enzymatic methods, together with electrochemistry, as a way of combining the power and convenience of the latter with the elegance and efficiency of the former. Following the pioneering work of Clark (8), most of the methods employed made use of the electrochemistry of one of the enzyme's natural substrates or products, often dioxygen or hydrogen peroxide. Replacing either of these with alternative mediators, led to considerable problems with dioxygen interference. Ferrocene derivatives were introduced as mediators (9) since it was well known that they were only slowly oxidised. The existence of a huge variety of ferrocenes is another major advantage. A particular ferrocene can be chosen on account of its redox potential, solubility, charge, etc. One can readily make ferrocene derivatives of drugs (10) for immunoassays or even of proteins (11-12).

BIOELECTROCHEMISTRY AT MICROELECTRODES

Miniaturization of enzyme electrodes (13) has always been relevant for in vivo applications. A new generation of microelectrodes has already been employed (14) for in vivo electroanalysis. The diminished surface area of a microelectrode results in reduced capacitance, reduced ohmic losses as a consequence of the smaller current and increased rates of mass transport to and from the electrode resulting in a steady state response. This has led (15) to investigations of electrochemistry in highly-resistive solutions, studies of fast reactions with electron transfer rates greater than

50cm s^{-1} and, of course, implantation experiments. Furthermore redox proteins were recently shown to behave very well at a modified microelectrode (Bond, A.M.; Hill, H.A.O.; McCarthy, M.; Psalti, I.S.M.; Walton, N.J. J. Amer. Chem. Soc., submitted). We have made investigations of the properties of two types of microelectrode: gold or platinum discs (25µM to 60µm diameter) and dual microband gold electrodes.

DISC MICROELECTRODE. One drawback of microelectrodes is that information on reactions to which the electroactive species is coupled is usually lost due to the short time domain associated with these electrodes. We were thus interested in determining the limiting size of microelectrode at which one can detect the coupling of the glucose-glucose oxidase reaction to the electrochemical oxidation of ferrocene as shown in Figure 1a. It was encouraging to find that microelectrodes of 25µm in diameter (which is admittedly rather large) are sensitive to homogeneous coupled reactions as is shown by the catalytic enhancement of the current in Figure 1b. The studies performed employed a variety of ferrocene derivatives but most of the detailed work involved 1,1'-dimethyl-3-(1-hydroxy-2-aminoethyl)ferrocene (DMHAE-ferrocene) since interference by dioxygen is negligible. The dependence of the catalytic current (plotted as Δi) on the glucose concentration using a gold disc of diameter 60µm is shown in Figure 2. In all cases, the currents are almost independent of the glucose concentration above 15mM. Adsorption of the enzyme onto the electrode surface, possibly combined with kinetic factors, may be responsible for the absence of linearity of the response for concentrations greater than that 6mM. (Using a conventional electrode, a linear response, extending to 30mM of the analyte, was observed).

BAND MICROELECTRODE. These electrodes have an active surface with one dimension in the micrometer range and the other several millimetres long. They retain many of the features of the microelectrodes but have two additional advantages: they have a much higher current due to their large surface area and it is relatively easy to have two (or more) closely spaced. When the distance between the electrodes is in the micrometer range each electrode is within the diffusion layer of the other, and the reactions taking place at one influence the activity of the other and *vice versa* (16).

The glucose-glucose oxidase (GOD)-DMHAE ferrocene system was employed again. The chemical reaction of the ferricinium species generated at one electrode, during the time and space interval of their migration to the adjacent electrode (scheme in Figure 3), causes a change in the current at the second electrode (Hill, H.A.O.; Klein, N.A.; Psalti, I.S.M.; Walton, N.J. Anal. Chem, 1989, in press). The effect on the so-called collection currents depends on the gap (W_{gap}) between the electrodes. When the gap is large, most of the ferricinium ions react with GOD, giving a considerable decrease in the collection current compared to the system which lacks substrate and/or enzyme. When the gap is small, there is no appreciable decrease in the collection current but instead inversion of the normal hysteresis loop is observed (see direction

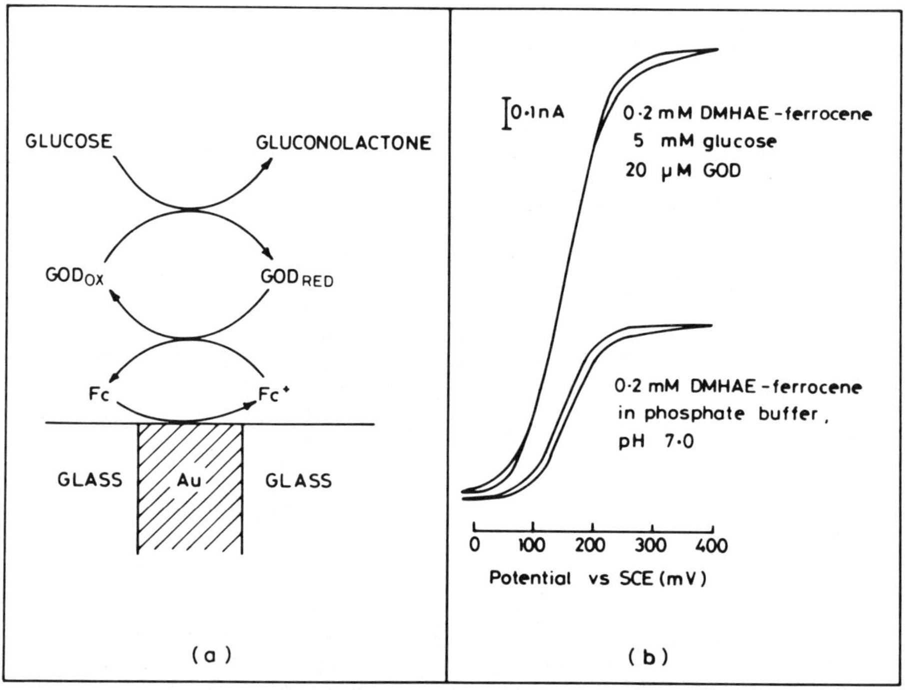

Figure 1. (a) Scheme for the reaction sequence of the electrochemically coupled enzymatic oxidation of glucose. (b) The effect of the enzymatic reaction on the electrochemistry of ferrocene using a gold disc of 25μm diameter at 5mV s^{-1}.

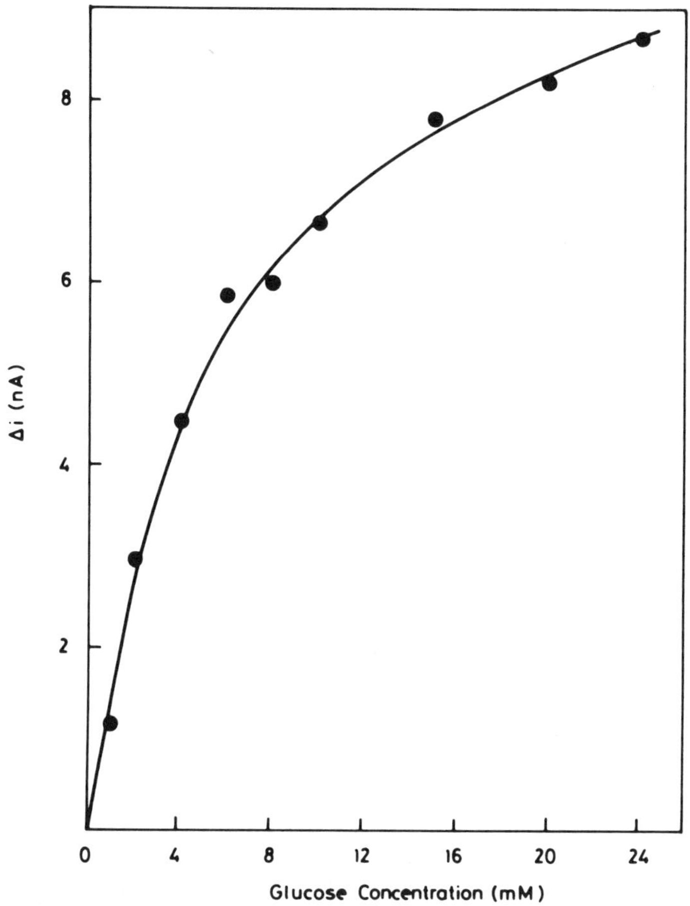

Figure 2. Dependence of the catalytic current on glucose concentration: gold electrode, 60μm diameter using 0.2mM DMHAE-ferrocene and 20μM GOD in phosphate buffer, pH 7.0, at 5mV s^{-1}.

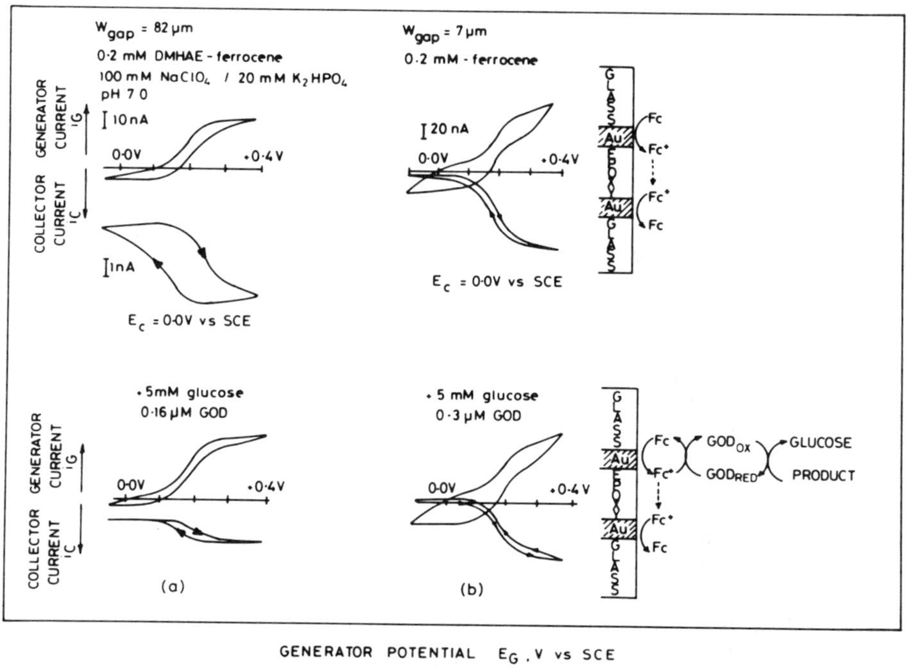

Figure 3. Effect of the enzymatic reaction with accompanying schemes on the collection current of a dual microband gold electrode (1μm thick) as a function of W_{gap}.

of arrows in Figure 3b). The large gap allows more time for the enzymatic reaction to take place and therefore to be monitored successfully. The minimum concentration of enzyme required for the catalytic reaction to be detected using this dual electrode configuration is about 0.05μM when used with 5mM glucose and 0.2mM DMHAE-ferrocene. Figure 4 shows the dependence of the collection

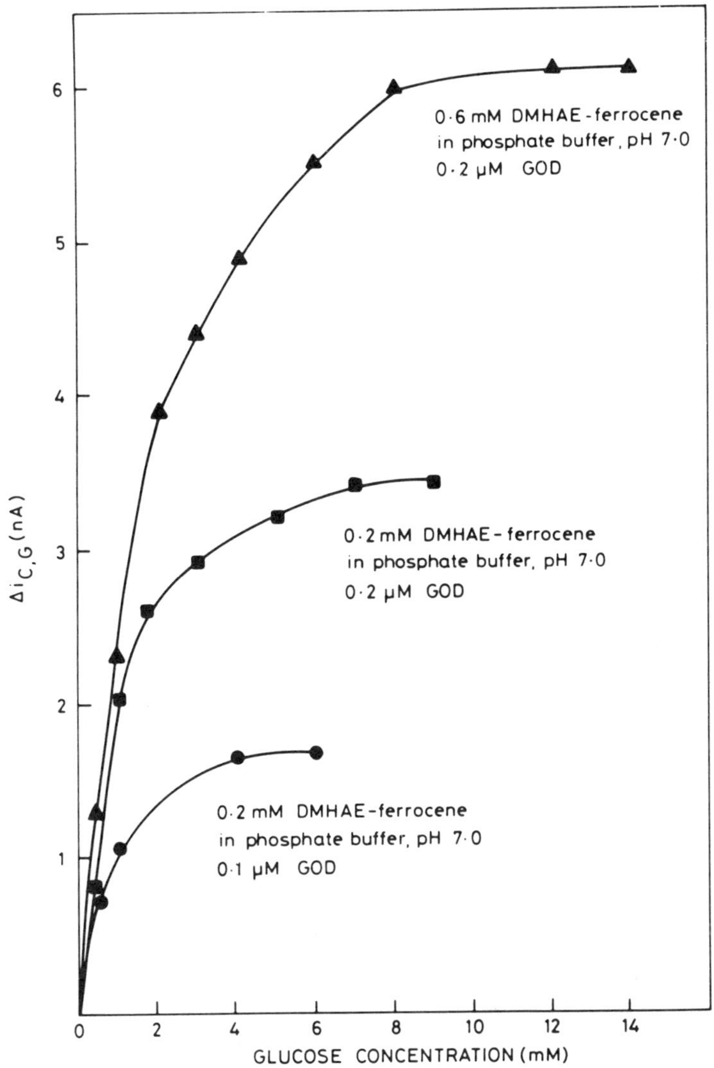

Figure 4. Dependence of the collecting current on glucose concentration for a dual microband gold electrode of W_{gap}=82μm at different DMHAE-ferrocene and GOD concentrations.

current ($\Delta i_{C,G}$) on the analyte concentration. Experiments in progress seek to find the dimensions of the dual electrodes which may result in significant extension of the range.

CONCLUSIONS

It will be apparent that the analytical use of bioelectrochemical methods depends on many factors including the use of the novel design of electrodes and the employment of the direct electrochemistry of enzymes, whether modified or not. Probably the defect of all the present methods is the modest sensitivity of bioelectrochemical methods. This is obviously important if these techniques are going to give rise to devices capable of sensing components of the immune system, to say nothing of DNA and RNA analyses. Obviously it is possible to provide considerable amplification from the electronic apparatus associated with the biochemical materials employed but the latter will have to be made more sensitive, perhaps, for example, by a series of coupled enzymatic reactions, before the full advantages of these techniques can be exploited.

ACKNOWLEDGMENTS

We thank MediSense Inc. for generous support.

LEGEND OF SYMBOLS

Δi calculated as

$$\Delta i = i_C - i_0$$

where i_C is the current at concentration C, and i_0 the current at zero concentration, of the analyte.

$\Delta i_{C,G}$ calculated as

$$\Delta i_{C,G} = i_{C,G} - i_0$$

where $i_{C,G}$ is the current detected by the collecting electrode at concentration C.

LITERATURE CITED

1. Frew, J.; Hill, H. A. O. Eu. J. Biochem. 1988, 172, 261.
2. Armstrong, F. A.; Hill, H. A. O.; Walton, N. J. Acc. Chem. Res. 1988, 21, 407, and references therein.
3. Barker, P. D.; Hill, H. A. O. In Oxidases and Related Redox Systems; King, T. E.; Mason, H. and Morrison, M., Eds.; M. Prog. Clinical Biological Research No. 274 Alan Liss Inc.: New York, N.Y., 1988; pp 419-432.
4. Barker, P. D.; Guo, L. H.; Hill, H. A. O.; Sanghera, G. S. Biochem. Soc. Trans. 1988, 16, 957-58.
5. Armstrong, F. A.; Lannon, A. M. J. Amer. Chem. Soc. 1987, 109, 7221.

6. Assefa, H.; Bowden, E. F. Biochem. Biophys. Res. Commun. 1986, 139, 1003.
7. Kulys, J. J. Biosensors 1986, 2, 3-13.
8. Clark, L. C.; Lyons, C. Ann. N.Y. Acad. Sci. 1962, 102, 29.
9. Cass, A. E. G.; Davis, G.; Francis, G. D.; Hill, H. A. O.; Aston, J.; Higgins, I. J.; Plotkin, E. V.; Scott, L. D. L; Turner, A. P. F. Anal. Chem. 1984, 56, 667.
10. Di Gleria, K.; Hill, H. A. O.; McNeil, C. J.; Green, M. J. Anal. Chem. 1986, 58, 1203.
11. Hill, H. A. O. European Patent Application No. 84303090.9, 1986.
12. Dagani, Y.; Heller, A. J. J. Phys. Chem. 1987, 91, 1285.
13. Carr, P. W.; Bowers, L. D. In Immobilised Enzymes in Analytical and Clinical Chemistry; Wiley: New York, 1980; p197. 14. Wightman, R. M. Anal. Chem. 1981, 53, 1125A-34A.
15. Pons, S.; Fleischmann, M. In Ultramicroelectrodes; Datatech Systems, Inc. 1987; pp 1-11.
16. Bard, A. J.; Crayston, J. A.; Kittleson, G. P.; Varco Shea, T.; Wrighton, M. S. Anal. Chem. 1986, 58, 2321.

RECEIVED March 17, 1989

Chapter 8

Microelectrodes To Probe Spatially Heterogeneous Concentrations

R. Mark Wightman, Leslie J. May, John Baur, David Leszczyszyn, and Eric Kristensen

Department of Chemistry, Indiana University, Bloomington, IN 47405

Electrodes with micrometer dimensions can be used to probe chemical concentrations which are spatially heterogeneous. With the use of a micromanipulator, the electrode can be rastered through solutions, and chemical information can be obtained with a resolution of a few micrometers. One example where such experiments provide unique information includes the examination of band broadening effects in flow injection analysis and liquid chromatography. In these experiments, the concentration of pulses of chemical substances is examined as a function of radial position in the transport tubing. Another type of experiment is the measurement of secretion of chemical substances from living cells. In this case, the measurements are made as a function of distance from the site of secretion.

In analytical measurements of the chemical composition of a sample, the values that are normally measured are the *average concentrations* of the constituents. Indeed, the sampling process is usually designed to ensure that the analyte reflects the overall composition of the material. Most samples are normally not homogeneous in their composition, and in some cases, it is of interest to examine these inhomogeneities. Chemical sensors with miniature dimensions provide a unique way to make measurements of chemical inhomogeneities. The sensor can be attached to a micromanipulator and rastered through solution regions where heterogeneous concentrations are expected to be found. In this way a concentration and composition map can be generated as a function of sensor position. This approach has been pioneered by Engstrom who has used sensors to define the chemical heterogeneities that exist in the diffusion layer at large size electrodes (1,2,3).

One sensor which we have developed, the carbon-fiber microelectrode, has been shown to be especially useful to obtain information on chemical heterogeneities (4). Carbon fibers are highly conductive and are commercially available in a variety of diameters ranging from 5 to 30 μm. The sensor is prepared by sealing a single fiber with epoxy into a glass pipette which is pulled to a tip diameter comparable to that of the fiber. The sensing area of the

0097–6156/89/0403–0114$06.00/0

sensor can have the geometry of a disk if the fiber is cut flush with the glass, a cylinder if the fiber is allowed to protrude from the glass (5), or an ellipse if the electrode is bevelled on a polishing wheel (6). Methods to coat the microelectrodes with polymer films have been developed to provide additional selectivity (7). When coupled with a sensitive potentiostat, these electrodes can sense submicromolar concentrations with a subsecond response time.

In this review we will describe several applications where we have explored different spatial chemical heterogeneities with this sensor. These applications include the concentration profile at the outlet of a chromatographic loop injector and a chromatographic column, and chemical heterogeneity of neurotransmitter storage and release in the rat brain.

Dispersion at the Outlet of a Loop Injection.

Ideally the output of a loop injector used in liquid chromatography or flow injection analysis would be a sharp concentration pulse. However, this is unlikely to be the case because of various dispersive forces which act on the concentration plug. Convective flow under laminar conditions in a circular tube, such as the outlet tube of a loop injector, tends to be much slower near the walls of the tube than in the center and this will distort the initial shape of the injected materials (8,9). In addition, radial and axial diffusion of material in the tube can alter its initial shape. The degree of dispersion can be evaluated (9) by the Peclet number (P_c) and the reduced time (τ). These values are defined as

$$P_c = au_0/D \quad ; \quad \tau = Dt/a^2$$

Where a is the radius of the tube, u_0 is the linear velocity of the liquid at the center of the tube, D is the diffusion coefficient of the analyte, and t is the time. For large values of P_c (>1000) and small values of time ($\tau < 0.02$) the dispersion should be primarily a result of convection (9). These conditions are those expected at a loop injector outlet under normal operating conditions.

To experimentally test these predictions (10), a disk shaped microvoltammetric electrode was placed in the exit tube, Figure 1, of a pneumatically controlled loop injector (Type 50, Rheodyne, Inc., Cotati, CA). The radial position of the electrode was controlled by a micromanipulator with 0.01 cm resolution, and measurements were made across the 0.08 cm diameter exit tube. The distance between the rotary position of the loop injector and the electrode was 7.5 cm. A pH 7.4 buffer was pumped through the system at a flow rate of 0.93 mL min^{-1}. Under these conditions $P_c > 4x10^4$ and τ~0.02. Dopamine was used as the substance injected through the loop injector at a concentration of 20 μM, and was detected at a fixed applied potential.

Current-time traces after introduction of dopamine with the loop injector are shown in Figure 2 as a function of radial position. When the sensor is placed near the center of the exit tube, the observed response approaches a square pulse. However, as the microelectrode is moved closer to the tube wall, the time for the sample bolus to reach the injector is increased and an increase in

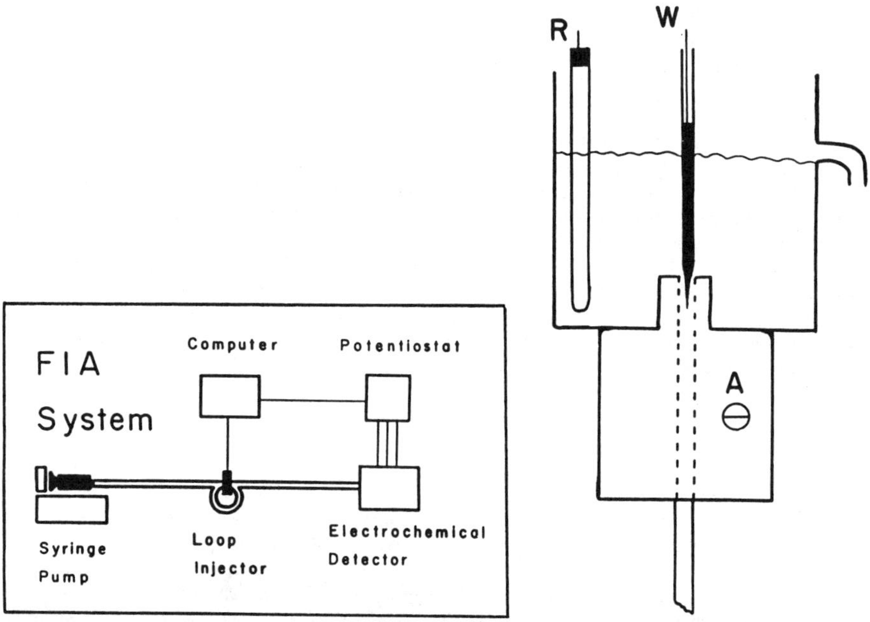

Figure 1. Stainless steel tube used with the microvoltammetric electrode: insert, block diagram of the flow injection system. (Reproduced from ref. 10. Copyright 1986 American Chemical Society.)

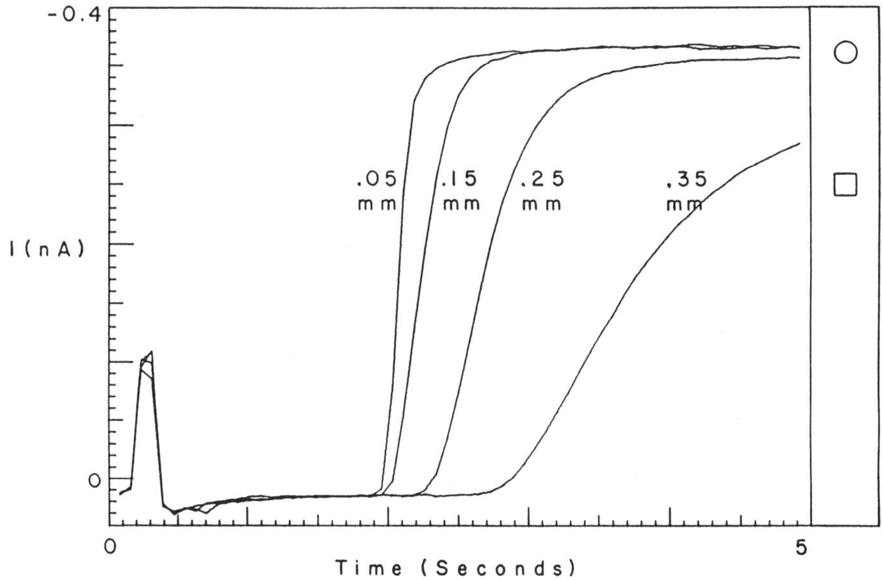

Figure 2. Average current response to a 30-s 20 μM dopamine pulse measured at different positions from the center of the tube. The average steady-state current for all the positions was measured at 20 s at the specified flow rate (circle) and a zero flow rate (square). (Reproduced from ref. 10. Copyright 1986 American Chemical Society.)

dispersion is observed. At each electrode position the current reaches a limiting, steady-state value by 20 s.

The radial velocity of the carrier stream in the tube (V_r) is given by:

$$V_r = 2V_0(1 - r^2/a^2)$$

where V_0 is the average velocity (3 cm s^{-1}) and r is the radial position. Thus, the transit time t_r from the loop injector to the sensor is $t_r = l/V_r + t_{offset}$ where l is the distance from the rotary portion of the loop injector to the electrode tip. The offset term is the time it takes the pneumatic activator to turn the loop injector (0.8 s). This curve is shown in Figure 3 and is superimposed on the time that the microelectrode response reached half of its limiting value. As can be seen, a good fit is obtained indicating that convection is the predominant cause of the observed radial dispersion.

To obtain the total dispersion in the tube, the individual responses were weighted by the area of the concentration ring in which each electrode was located, and these results were summed (10). Very good agreement is obtained between this result and the response measured with a channel type amperometric electrode connected directly to the loop injector (10).

These measurements clearly show that convective flow is the major origin of the dispersion observed at the output of a loop injector. However, these measurements were made in a time domain which is close to that where dispersion by diffusion should start to become apparent according to hydrodynamic theory. This is apparent in the data obtained closest to the wall where the dispersion is greatest. Solution linear velocity is lower in this region giving greater time for diffusion. The small size of the detector allows the radial concentrations to be probed with little perturbation of the stream. This is evidenced by the ability to reconstruct the bulk behavior from the individual radial measurements.

Dispersion in Chromatographic Columns.

Microelectrodes have also been used to examine the radial dispersion of bands eluting from commercial HPLC columns. Conventional chromatographic detectors detect the entire volume of a band eluting from a chromatographic column. For this reason, the primary figure of merit of a chromatographic column is the number of chromatographic plates measured in the axial direction. However, it has been shown that radial dispersion can also occur, and this may affect the axial dispersion measured with a bulk detector. Previous studies have modelled the chromatographic column with columns packed with spherical beads to characterize this phenomenon (11,12). These fundamental studies have shown that radial dispersion does not exert a great effect on column efficiency unless dispersion is sufficiently great that the injected species reaches the wall of the column. This "wall effect" causes a large degree of dispersion because of the lower efficiency of packing near the wall of the column. Thus, it is of interest to experimentally characterize the radial dispersion present with a commercial loop injector combined with a commercial chromatographic column (13).

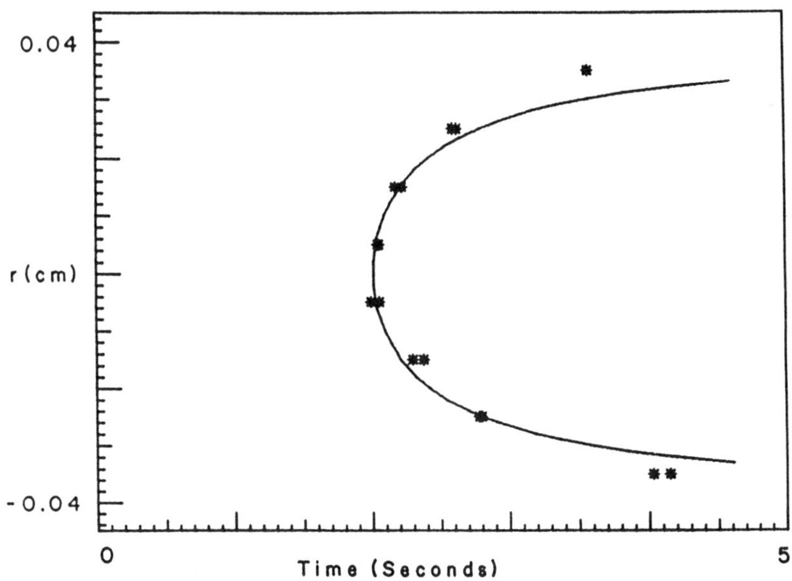

Figure 3. The time required to reach half of the steady-state current is plotted against the position of the electrode from the center of the tube. The asterisk symbol represents a single measurement whereas the solid line is that predicted by pure convection.

In this case the electrode had a cylindrical geometry (to improve sensitivity) and was placed at the exit frit of a 10-cm long reverse-phase column packed with 3-μm particles (Brownlee Labs, Santa Clara, CA). The column had a diameter of 3.2 mm, and the exit frit was crimped into place so that a column end fitting was not required. The geometry of the electrochemical cell is shown in Figure 4. The chromatographic system included a pneumatically operated loop injector (Model 3XL, Scientific Systems, Inc., State College, PA). The injection loop had a 10 μL volume, and mixtures of catecholamines were used as test compounds. The electrode could be positioned with 100 μm resolution in three dimensions with a micromanipulator (Narishige Co., Ltd., Tokyo, Japan). The vertical resolution was enhanced with a piezoelectric positioner with 0.5-μm resolution (Burleigh Instruments, Inc., Fishers, NY).

Representative chromatograms measured at two radial positions with the microelectrode are shown in Figure 5. The trace recorded at the center of the exit column exhibits sharper peaks than when the chromatograms are recorded at the edge. The chromatographic peaks recorded at the edge are broader, attenuated, significantly tailed, and have longer retention times.

Figure 6 shows the normalized concentration observed for one peak as a function of radial position. Superimposed on the curve is the dispersion expected for a centrally injected substance calculated for the column conditions and flow rate (1.0 mL min^{-1}) employed. It is clear that the dispersion is much greater for the experimental data. It is likely that the observed radial concentration profile is a consequence of the sample introduction processes as previously proposed by Kirkland (14).

The reduced plate height ($h_a = H_a/d_p$, where H_a is the axial plate height and d_p is the particle diameter) and retention time are plotted in Figure 7 as a function of radial position. In contradiction with that which is expected from the wall effect, the retention time is greater near the wall. This too may be a result of the sample introduction because a similar profile was observed at the outlet of the loop injector. However, the data clearly show that the separation efficiency decreases as the wall of the column is approached.

The separation efficiency was evaluated using the exponentially modified Gaussian model. With this procedure the peak is characterized by σ, the standard deviation of the peak, and τ, the exponential modifier. For measurements made in the center of the column, τ and σ both increased for longer retained compounds. The value of τ increased as the column wall was approached with little change in σ.

Comparison was made between the radial measurements and that made with a bulk detector, a commercial amperometric detector. When the individual radial measurements were weighted and summed as described earlier, the reconstructed chromatogram superimposed on that measured with the bulk detector. Separation efficiencies measured in the center of the column were at least 20% higher than those measured with the bulk detector. Most notable was the very high efficiency obtained with uric acid (k' = 0.33) where a reduced plate height of 1.5 was obtained at a flow rate of 0.56 mL min^{-1}. This represents a 90% improvement over the separation efficiency obtained with the bulk detector.

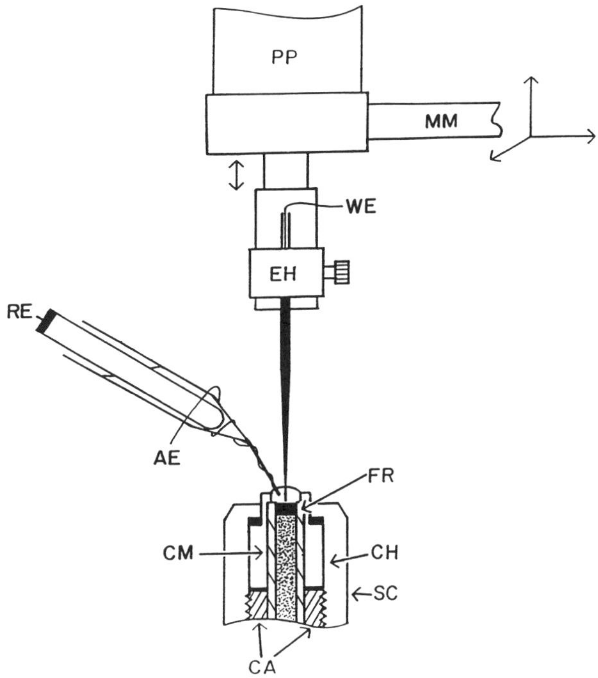

Figure 4. Column end-assembly configured for microvoltammetric electrochemical detector: AE, auxiliary electrode; CA, cartridge holder; CH, column holder; CM, Column; EH, electrode holder; FR, frit; MM, micromanipulator; PP, piezoelectric positioner; RE, reference electrode; SC screw cap; WE, working electrode. (Reproduced from ref. 13. Copyright 1988 American Chemical Society.)

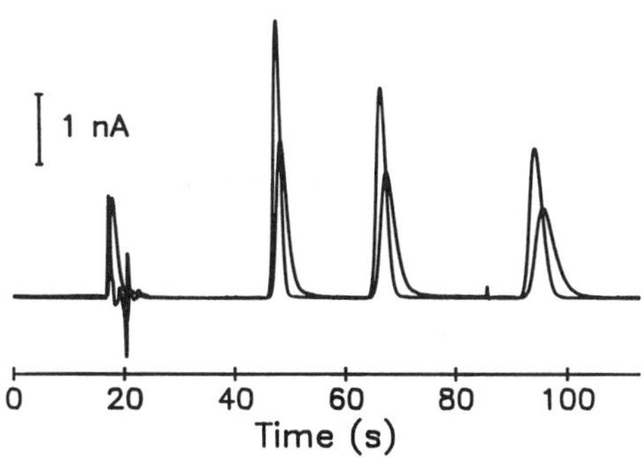

Figure 5. Chromatograms of 5 nmol each NE, E, and DHBA with microvoltammetric electrode near the center of the column and near the edge of the column. Flow rate = 1.0 mL min^{-1}. (Reproduced from ref. 13. Copyright 1988 American Chemical Society.)

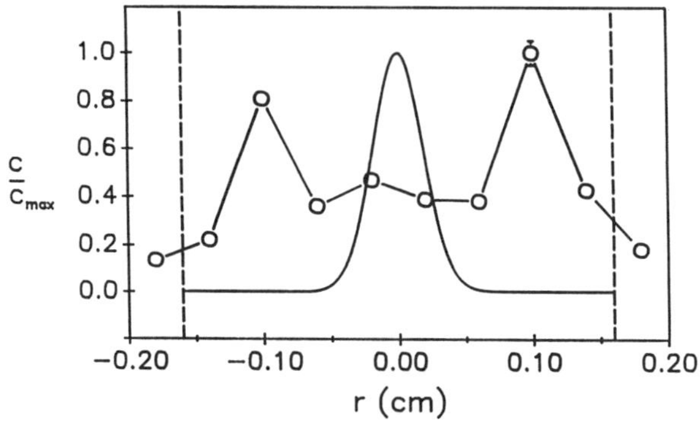

Figure 6. Normalized radial concentration distribution for NE. Solid line: theoretical for central injection with σ_R = 0.188 mm. Circles: experimental. The dashed lines are the estimated position of the column walls. (Reproduced from ref. 13. Copyright 1988 American Chemical Society.)

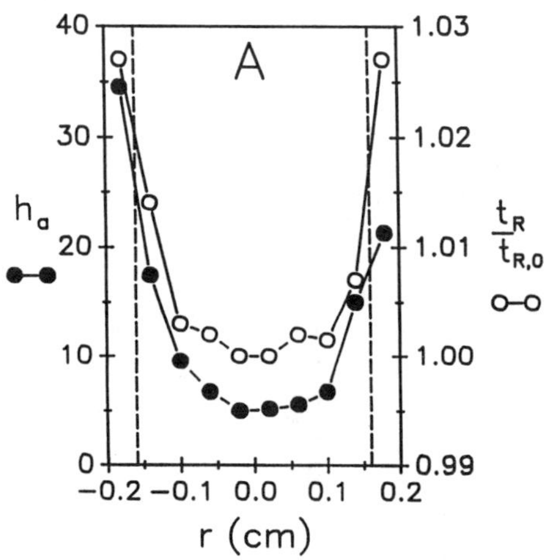

Figure 7. Normalized retention time and reduced axial plate height of 5 nmol NE as a function of radial position for a new column. Flow rate = 1.0 mL min^{-1}. The dashed lines are the estimated position of the column walls. (Reproduced from ref. 13. Copyright 1988 American Chemical Society.)

The microelectrode in this application is able to provide clear insight into radial dispersion processes in liquid chromatographic columns. The experiments show that dispersion is greater next to the column wall as predicted from fundamental studies. However, the observations that the retention time is slower near the walls, and that the injected bolus is distributed over the entire column differ from previous model studies. These differences are both likely to be due to the introduction of sample with the loop injector. These observations are significant because they indicate that a centrally placed detector will record chromatograms of very high efficiency in commercial columns. However, the increase in efficiency may be coupled with a decrease in sensitivity because a large portion of the eluting band is discarded. Preliminary data indicate that this is not a concern with amperometric detection. This is because the noise in amperometric detectors is proportional to the area, while the signal is proportional to the largest dimension. Thus, a cylindrical electrode exhibits superior signal-to-noise characteristics over a bulk amperometric detector until the fundamental limit of amplifier noise is reached.

Chemical Heterogeneity in the Brain.

The mammalian brain is an extremely heterogeneous structure on the cellular level. While only 10% of the total cell population is comprised of neurons, these are the all important cells which collect, integrate, and relay information (15). Neurons are arranged in a variety of networks, and, since there are over 10 million neurons in the human brain, the degree of heterogeneity can be seen to be extremely complex.

Most neurons communicate with one another by the secretion of chemical substances known as neurotransmitters. Neurotransmission occurs in a region known as the synapse which constitutes the input of one neuron and the output of the neuron secreting the transmitter. Since neurons form synapses throughout the brain, one would expect that the cellular heterogeneity would also be reflected in a chemical heterogeneity of neurotransmitter concentrations within the brain. This hypothesis can be tested for catecholamine neurotransmitters because carbon fiber microelectrodes implanted in the brain can detect them voltammetrically.

In vivo voltammetry has now become a reliable technique for the detection of dopamine dynamics in the brain (16). Dopamine, a catecholamine, is found in greatest concentration in a region of the brain known as the striatum. This region, which is present in all mammals, is important for normal coordinated movements. It has been shown that patients with Parkinson's disease have a severe lowering of their dopamine content in this brain region. This region is heterogeneous at the cellular level because many neuronal fibers pass through it in distinct tracts without forming synaptic contacts. In addition, it has been shown that the dopamine nerve terminals in the striatum are arranged in distinct patterns (17,18). Thus, heterogeneity of dopamine release from nerve terminals would be expected in this region as with other neurotransmitters in other brain regions.

The carbon fiber electrode used in this work is bevelled so that the active sensing area is an ellipse with a minor radius of 5 μm and a major axis of ~35 μm. The electrode is coated with a thin film of a perfluorinated cation exchange membrane which excludes dopamine metabolites and other interferences. In these experiments, in contrast to those described earlier in this work, the solution being sampled, the extracellular fluid of the brain, is static. Therefore, monitoring the current at a constant applied potential would alter the chemical composition surrounding the electrode. For this reason, the technique of fast cyclic voltammetry has been used. Voltammograms are repeated at 100 ms intervals at a scan rate of 300 V s^{-1} (19). Thus, the fluid is only sampled for 10 ms. The individual voltammograms are subtracted from the background signal and are used to identify dopamine as the detected substance (20). The integrated current at the potential for dopamine oxidation from each voltammogram provides a continuous readout of the concentration. The entire experiment is under computer control.

The experiments described here have been undertaken in the brain of anesthetized rats. The animal is placed in a frame which holds the head at a preset angle to ensure approximate reproducibility of the electrode placements (19). An incision is made on the scalp, the skull is exposed, and holes are drilled in the skull. The carbon fiber working electrode is lowered with a manipulator to the top of the striatum. To synchronize the release of dopamine with the measurement process, the dopamine containing neurons are stimulated electrically. The fibers of dopamine neurons which project to the striatum are closely packed in a region known as the medial forebrain bundle. Thus, a stimulating electrode placed in this region can be used to depolarize all of the dopamine neurons at once. The stimulation employed is a 60 Hz biphasic square wave with a duration of 2 s.

The results from a single stimulation are shown in Figure 8. The voltammogram obtained from the difference between that recorded before and during stimulation demonstrates that the detected substance is dopamine. The current from the peak oxidation potential for dopamine recorded from sequential voltammograms demonstrates that the concentration of dopamine increases during stimulation, and then rapidly decreases after the stimulation is over. The disappearance of dopamine is due to uptake back into dopamine cells (21). The appearance of dopamine during the stimulation is a combination of the synaptic release of this compound as well as the concurrent uptake. The combined effect of both of these factors, plus diffusion, can be modelled as indicated by the solid line in Figure 8.

The chemical heterogeneity of the brain is clearly evidenced when the experiment is repeated with the electrode placed at successively lower positions in the brain (22). In Figure 9 the maximum amplitude of dopamine concentration is recorded at 100 μm intervals. Results are shown from four different animals. As can be seen the pattern of chemical heterogeneity that is seen is different in each animal. However, the data are consistent with anatomical reports of the distribution of dopamine nerve terminals in the striatum (17,18). Studies are currently underway to establish whether an exact correlation exists.

The data in these experiments clearly show that the spatial resolution of carbon fiber electrodes is sufficient to measure chemical inhomogeneities in the intact brain. This type of information will be useful in establishing the role of different dopamine

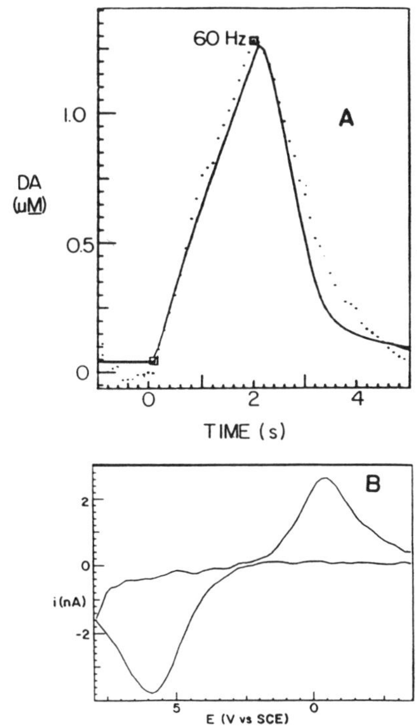

Figure 8. Data obtained in the caudate nucleus of an anesthetized rat from a Nafion-coated, carbon-fiber electrode with voltammetry (300 V s^{-1}) during stimulations of dopamine containing neurons. A: Temporal changes observed during electrical stimulations at 60 Hz. Each point represents the current from individual voltammograms integrated over the range 400-800 mV vs SCE. The response is converted to concentration by calibration with dopamine. The solid lines are the modelled response which involves the use of neurochemical kinetic parameters and diffusion from a distance of 10 μm. B: A subtracted cyclic voltammogram obtained during the 60-Hz stimulation. The shape is identical to that recorded in dopamine solutions.

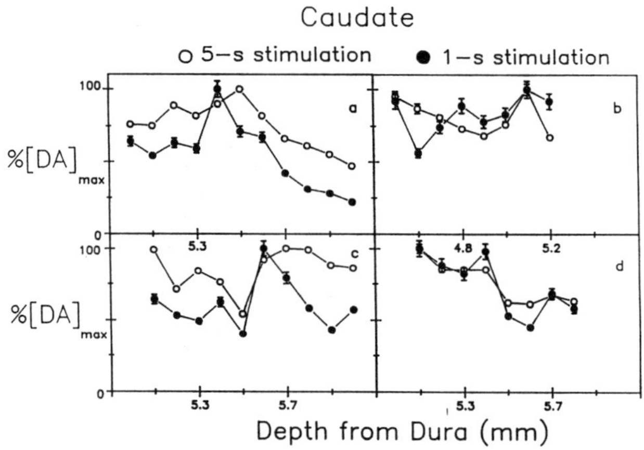

Figure 9. Heterogeneity of overflow in the caudate-putamen. Panels a, b, c, and d each represent responses from individual animals. The maximal [DA] observed is expressed as a percentage of the largest response in that animal and is plotted vs. vertical electrode position. Error bars represent the standard error of the mean for replicate measurements; when no error bars are indicated, the standard error is less than the symbol size. (Reproduced with permission from Ref. 22. Copyright 1989 Elsevier.)

neurons and the particular neuronal circuits in which they are involved.

Conclusions

The three examples shown in this paper indicate that microvoltammetric electrodes are useful tools to probe chemical heterogeneities in solution, and furthermore to characterize these phenomena under dynamic conditions. To achieve greater spatial resolution, smaller electrodes will need to be employed. Automation of this type of measurement would be desirable and can readily be accomplished with piezoelectric micropositioners and other such devices which can be remotely controlled. Such developments will lead to a form of dynamic chemical microscopy which would be useful to measure such events as secretion from single cells, corrosion processes in pits and cracks, or further studies of solution flow.

Acknowledgments

This research was supported by NIH and NSF.

Literature Cited

1. Engstrom, R. C.; Weber, M.; Wunder, D. J.; Burgess, R.; Winquist, S. Anal. Chem. 1986, 58, 844-848.
2. Engstrom, R. C.; Wightman, R. M.; Meaney, T.; Tople, R. Anal. Chem. 1987, 59, 2005-2010.
3. Engstrom, R. C.; Wightman, R. M.; Kristensen, E. W. Anal. Chem. 1988, 60, 652-656.
4. Dayton, M. A.; Brown, J. C.; Stutts, K. J.; Wightman, R. M. Anal. Chem. 1980, 52, 946-950.
5. Kovach, P. M.; Deakin, M. R.; Wightman, R. M. J. Phys. Chem. 1986, 90, 4612-4617.
6. Kelly, R.; Wightman, R. M. Anal. Chim. Acta 1986, 187, 79-87.
7. Baur, J. E.; Kristensen, E. W.; May, L. J.; Wiedemann, D. J.; Wightman, R. M. Anal. Chem. 1988, 60, 1268-1272.
8. Vanderslice, J. T.; Stewart, K. K.; Rosenfeld, A. G.; Higgs, D. J. Talanta 1981, 28, 11-18.
9. Ananthakrishnan, V.; Gill, W. N.; Barduhn, A. J. AIChE J. 1965, 11, 1063-1072.
10. Kristensen, E. W.; Wilson, R. L.; Wightman, R. M. Anal. Chem. 1986, 54, 986-988.
11. Knox, J. H.; Laird, G. R.; Raven, P. A. J. Chromatogr. 1976, 122, 129-145.
12. Eon, C. H. J. Chromatogr. 1978, 149, 29-42.
13. Baur, J. E.; Kristensen, E. W.; Wightman, R. M. Anal. Chem. 60, 2334, 1988.
14. Kirkland, J. J.; Yau, W. W.; Stoklosa, H. J.; Dilks, Jr., C. H. J. Chromatogr. Sci. 1977, 15, 303-316.
15. Cooper, J. R.; Bloom, F. E.; Roth, R. H. The Biochemical Basis of Neuropharmacology, Oxford University Press: Oxford, England, 1986.
16. Wightman, R. M.; May, L. J.; Michael, A. C. Anal. Chem. 1988, 60, 769A-779A.

17. Gerfen, C. R.; Herkenham, M.; Thibault, J. J. Neurosci. 1987, 7, 3915-3934.
18. Graybiel, A. M.; Radsdale, Jr., C. W. Proc. Natl. Acad. Sci. U.S.A. 1978, 75, 5723-5726.
19. Wightman, R. M.; Amatore, C.; Engstrom, R. C.; Hale, P. D.; Kristensen, E. W.; Kuhr, W. G.; May, L. J. Neuroscience 1988, 25, 513-523.
20. Howell, J. O.; Kuhr, W. G.; Ensman, R. E.; Wightman, R. M. J. Electroanal. Chem. 1986, 209, 77-90.
21. May, L. J.; Kuhr, W. G.; Wightman, R. M. J. Neurochem., in press.
22. May, L. J.; Wightman, R. M. Brain Res. 487, 311, 1989.

RECEIVED March 9, 1989

Chapter 9

A Fiber-Optic Electrode for Optoelectrochemical Biosensors

Masuo Aizawa, Masaru Tanaka, and Yoshihito Ikariyama

Department of Bioengineering, Tokyo Institute of Technology, O-okayama 2–12–1, Meguro-ku, Tokyo 152, Japan

A fiber-optic electrode was fabricated for the simultaneous generation and transmission of electrochemical luminescence by preparing a transparent electrode on the optical and surface of a fiber-optic. The opto-electrochemical properties of the micro-optical device were characterized in solutions containing the compounds required for luminol luminescence. The validity of sensitive measurement of electrochemiluminescence to be employed in a homogeneous immunoassay was evaluated by using potential step excitation of luminol in the presence and in the absence of hydrogen peroxide.

As the high flux of information density is one of the advantages of fiber-optic cables over other types of information carriers, fiber-optic communications have been providing a powerful means of data transmission with their extremely high bit rates. In the case of chemical sensors, several advantages of the fibers can additionally be exploited, namely such capabilities as remote sensing, miniaturization, electromagnetic immunity, and *in situ* measurement can be employed (1,2).

Optical biosensors are the devices that can transduce the information concerned with a determinant into optical signals such as the changes in absorbance, fluorescence, and luminescence (3-5). Generally these biosensors are the union of optoelectronic devices and biomaterials such as enzymes and antibodies in their solid matrix-bound forms (6).

Fiber-optic chemical sensors for substances such as protons, oxygen, and carbon dioxide have been developed. These sensors respond to specific substances in accordance with the fluorescent changes of fluorescent probes. An optical fiber-optic for pH can be coupled to enzyme sensors when the enzymes cause a change in pH, which results in a change in fluorescent property of the probe compounds. These optical sensors are named "Optrodes", and were extensively reviewed by Seitz (1).

An optoelectronic enzyme sensor, a combination of a photodiode and an enzyme membrane, has been constructed by using a luminescence-generating enzyme (7). The enzyme photodiode is applied to the determination of enzyme substrates such as hydrogen peroxide and glucose with the use of the luminescent reaction between luminol and hydrogen peroxide. On the other hand, enzyme-immobilized membranes are coupled with fiber-optics, when sensitive determinations of biological substances are required (8). We have also made a series of works on the luminescent immunoassay by using the luminescent reaction of luminol (9-13). Labeling agents such as hemin (10-12) and peroxidase (9,13) are employed for the competitive and sandwich assays of antigenic substances such as insulin, β_2-microglobulin and human albumin.

Recently the authors have proposed a new homogeneous immunoassay based on electrochemical luminescence, by employing aromatic hydrocarbons such as pyrene as an electrochemically active label (14). The principle of the proposed immunoassay is schematically

0097–6156/89/0403–0129$06.00/0

shown in Figure 1. The pyrene-labeled antigen emits luminescence by electrodic reaction, while the labeled antigen complexed with its antibody generates less luminescence depending on the antibody concentration, because antibody molecules prevent the labeled antigen from contacting the electrode surface. Although the generated photons are transmited to a photon counter through a fiber-optic, we have not succeeded in sensitizing the electrochemiluminescence-based homogeneous immunoassay due to the low quantum efficiency of the luminescent reaction of pyrene.

In order to improve the sensitivity of the homogeneous immunoassay we have developed a fiber-optic electrode for the efficient generation and transmission of photons, and we have exploited luminol which has a much higher quantum efficiency than pyrene as a candidate for an electrochemiluminescent probe.

Spectroelectrochemistry, a combination of optical spectroscopy and electrochemistry has provided a powerful means for elucidating complex redox processes near solution-electrode interfaces (15,16). A union of transparent electrodes with fiber-optics provides a new strategy for designing new biosensing devices, provided they can be assembled into one instrument (fiber-optic electrode) that will generate and transmit a luminescence signal in response to the substance to be determined.

We report here the performance characteristics of the fiber-optic electrode by applying the device to the electrochemiluminescence of luminol to clarify the feasibility of luminol as an electrochemiluminescent label for a sensitive homogeneous immunoassay.

Experimental Section

Fabrication of fiber-optic electrode. Platinum was sputtered on the flat end surface of a plastic fiber-optic (diameter : 2 mm, length : 150 cm) courteously supplied by Mitsubishi Rayon Co. Sputtering was performed by a Hitachi minisputter for 5 min at 15 mA. The platinum counter electrode was prepared as illustrated in Figure 2. The thin platinum layer for the working electrode was connected to a lead wire with silver paste, while that for the counter electrode was connected with solder. The contact points were then fixed in place and insulated by sealing with epoxy resin.

A photon counter of Hamamatsu Photonics consisted of a photon counter (Type C 767) and a power source (Type 752-01). Electrode potential was controlled by a function generator and a potentiostat (Hokuto Denko Co.) against a Ag/AgCl reference electrode.

Characterization of fiber-optic electrode. Luminol was prepared in a phosphate-buffered solution of pH 7.0 (0.1M). Luminol was dissolved in the buffer under sonication. Both the photon counter output and the current output from the fiber-optic electrode were digitized (8 bit) and monitored by a Hitachi digital memory scope (Type VC-6020). The digitized data were stored on a floppy disk by a NEC microcomputer (Type PC-9801 VX2) through a GP-IB interface. The data were processed in the microcomputer to reduce background luminescence. The experimental setup of measuring apparatus for electrochemiluminescence detection and the following data processing is illustrated in Figure 3.

Results and Discussion

Electrolytic luminescence of luminol was investigated with the fiber-optic electrode in a solution containing 1 mM luminol. Figure 4 shows a cyclic voltammogram of the solution used to study the electrochemiluminescence of luminol. The luminescence was observed from +0.4 V and reached a maximum at +0.65 V in the neutral solution when the potential was scanned from negative to positive potential. However, no luminescence was generated without the preceding negative potential application, the reason for which is described below. After repeated potential scanning, the fiber-optic electrode generated reproducible current and luminescence. From the transparency (20 %) of the electrode, the efficiency of the photon collection was estimated to be ca. 2 %. The estimated efficiency was calculated by considering only the photons emitted toward the fiber.

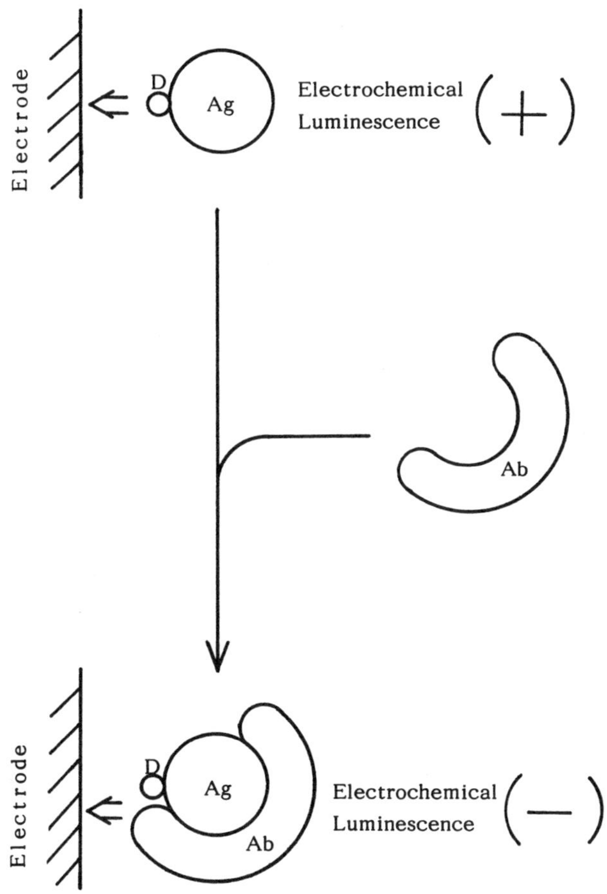

Figure 1. Principle of homogeneous immunoassay based on electrochemical immunoassay. In the presence of antibody the labeled antigen emits less photons due to the steric hindrance of antibody in electrodic reaction.

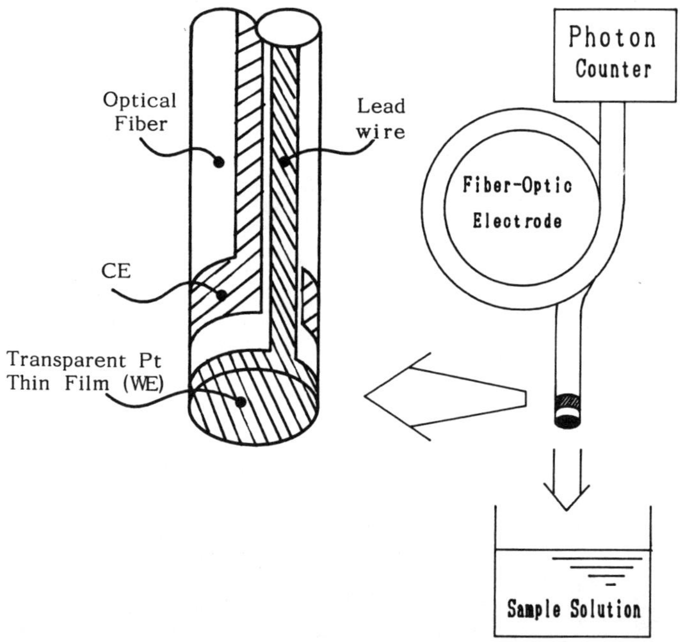

Figure 2. Schematic illustration of the configuration of a fiber-optic electrode for the generation and simultaneous transmission of electrochemiluminescence.

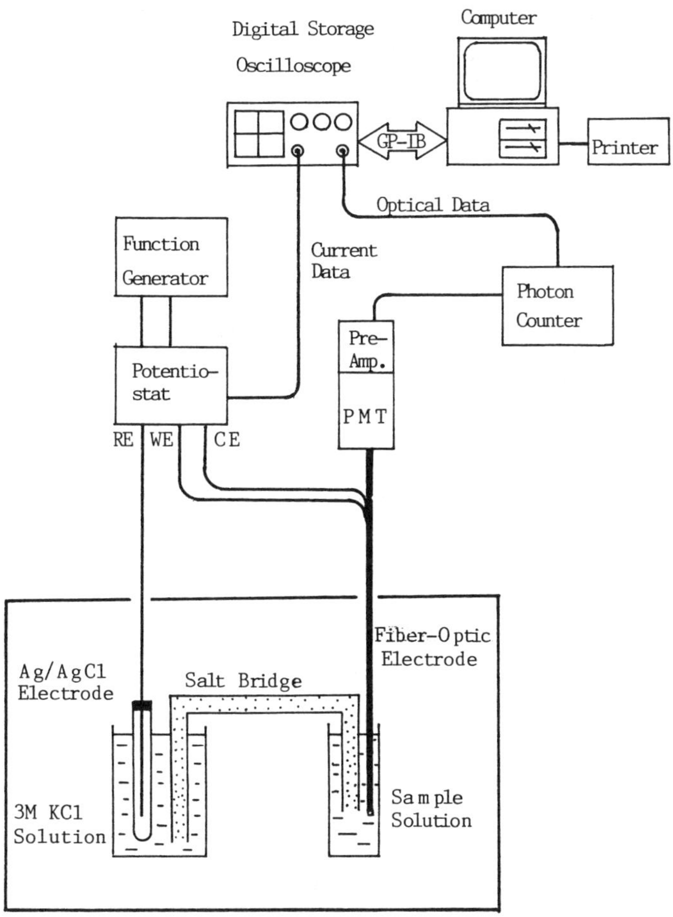

Figure 3. Experimental apparatus for the characterization of fiber-optic electrode. The cell volume was less than 2 mL, therefore the potential of the fiber-optic electrode was referred to the reference electrode (Ag/AgCl) through a salt bridge.

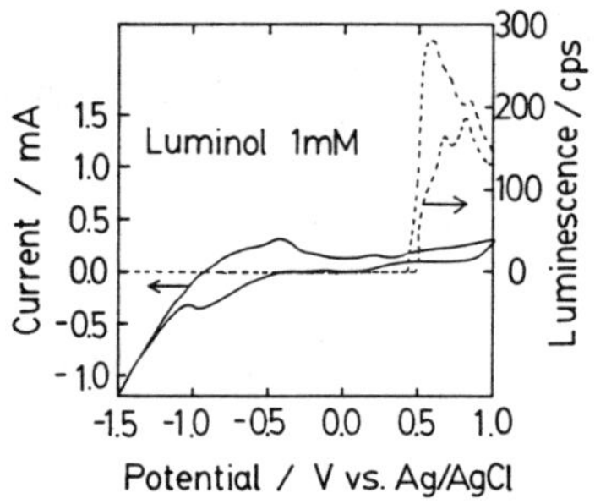

Figure 4. Cyclic voltammogram of a solution containing luminol.
Electrochemical luminescence of luminol was also shown with respect to the current-voltage relationship. Every solution was prepared in 0.1 M phosphate buffer (pH 7.0), and the scan rate was 100 mV/sec.

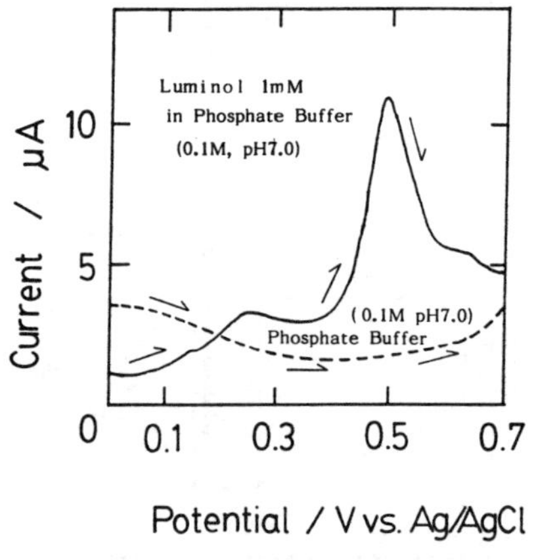

Figure 5. Differential pulse polarogram of luminol solution.
The fiber-optic electrode was employed for the oxidation of luminol. A phosphate-buffered solution (0.1 M, pH 7.0) was used as an electrolyte, and the scan rate was 5 mV/sec.

As negative potential application is a prerequisite electrolytic process for the luminol electrochemiluminescence, it seems to us that a greater luminescence should be generated by a potential step mode from a negative potential to a positive one. Electrolytic oxidation of luminol was studied by differential pulse voltammetry, and luminol was found to be oxidized at +0.65 V on the surface of the fiber-optic electrode at the neutrally buffered solution as shown in Figure 5. Furthermore, a greater luminescence was obverved at more negative potentials when the positive potential was fixed at +0.65 V. However, the platinum-sputtered fiber-optic electrode denatured rapidly at a potential less than -1.5 V. Fixing the negative potential at -1.5 V, and scanning positively gave greater luminescence at +0.65 V, at which potential luminol was oxidized. Figure 6 illustrates the relation between the luminescence intensity and a negative potential scan.

The interval of a step-wise potential application for the excitation of luminol was also investigated. Figure 7 shows that a duration of the negative potential application of 5 s and that of the positive potential application of 15 s was the best condition for the greatest luminescence generation. Under the optimum condition, electrochemiluminescence of luminol was detected with the fiber-optic electrode in the concentration range from 10^{-8} to 10^{-3} M as shown in Figure 8.

We have also studied the role of dissolved oxygen, because the dissolved oxygen may be easily reduced to hydrogen peroxide at a potential of -1.5 V. Monitoring of the dissolved oxygen was carried out by a Clark-type oxygen electrode. The relation between dissolved oxygen and luminescence is shown in Figure 9. After the removal of dissolved oxygen by bubbling nitrogen gas, little luminescence was observed. This finding led us to perform another potential-step excitation from 0 V to +0.65 V in the presence of hydrogen peroxide. The electrochemical luminescence generated by the application of a positive potential of +0.65 V was proportional to the concentration of luminol in the concentration range from 10^{-8} to x10^{-2} M in the presence of 2 mM hydrogen peroxide, as shown in Figure 10. Catalase (330 u/mL) addition annihilated the luminol electrochemiluminescence, while superoxide dismutase gave no effect on the electrochemical luminescent reaction. The wide dynamic range of the calibration curve was obtained due to the efficient generation and transmission of the electrochemiluminescence of luminol by a fiber-optic, electrode combination.

Kuwana et al. reported the electrochemical luminescence of luminol mainly in alkaline solutions by using a conventional three-electrode apparatus and a photon counter (17-19). The fiber-optic electrode described here, however, substantially improves the sensitivity in detecting the electrochemical luminescence of luminol in very low concentration at neutral pH. The micro-optical device coupled with the electrochemiluminescence seems promising as a signal-transducing and transmitting element for a homogeneous immunosensing system with high sensitivity (Fig. 1) because of the high quantum efficiency and the simple labeling of luminol.

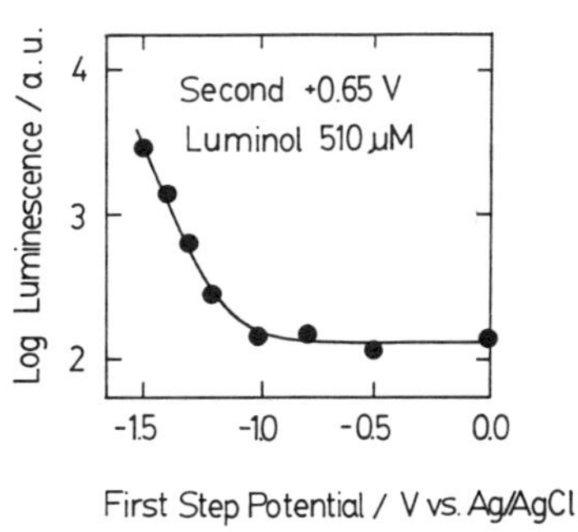

Figure 6. Relationship between the luminescent intensity and a potential scan in the negative direction.

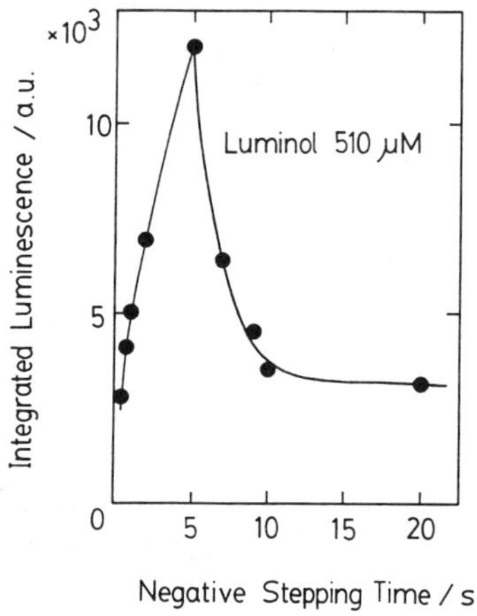

Figure 7. Optimization of pulse interval for the excitation of luminol.

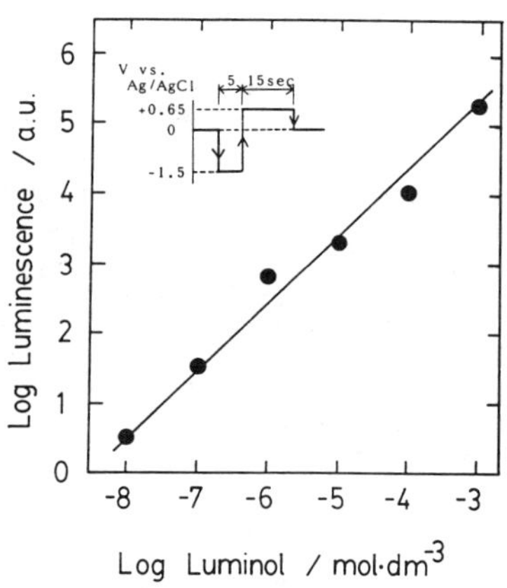

Figure 8. Relationship between luminescent intensity and luminol concentration.

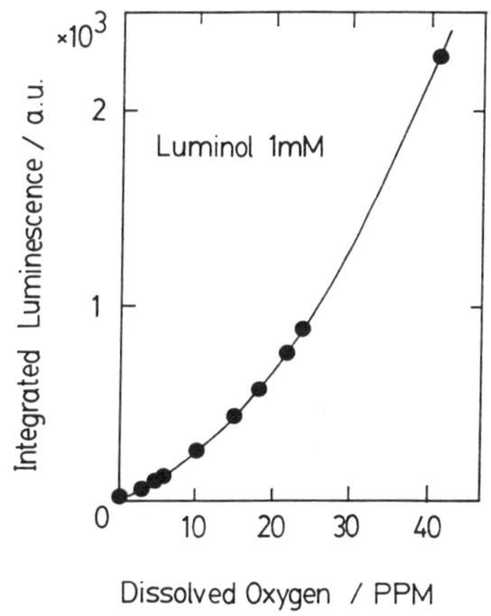

Figure 9. Effect of dissolved oxygen on luminol electrochemiluminescence.

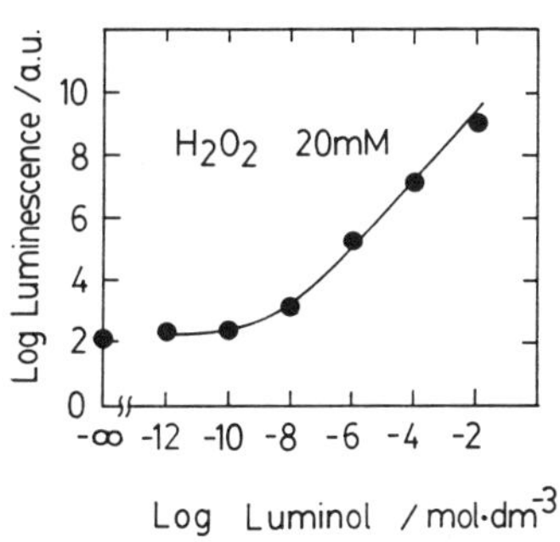

Figure 10. Calibration curve for luminol in the presence of hydrogen peroxide.

Conclusion

A variety of unique optical biosensing systems can be developed by paying attention to chemical and electrochemical luminescence. Microfabrication of an optical biosensing device can be carried out by taking optoelectric devices such as the photodiode and the fiber-optic. A fiber-optic electrode, the combination of fiber-optic and electrode, is especially suitable for the microinstrumentation of sensitive immunosensing device because of the sophisticated union of fiber optic and electrode for the efficient generation and transmission of electrochemiluminescence. Intensive research and development of the opto-electrochemical biosensors is being done to open up a new vista of biosensor technology.

Literature Cited

1. Seitz, W. R. *Anal. Chem.* 1984, **56**, 16A.
2. Petersen, J. I.; Vurek, G. G. *Science* 1984, **224**, 123.
3. Freeman, T. M.; Seitz, W. R. *Anal. Chem.* 1981, **53**, 98.
4. Lubbers, D. W.; Opitz, N. *Sensors & Actuators* 1983, **4**, 641.
5. Lippitoch, M.; Leiner, M. J. P.; Pusterhofer, J.; Wolfkeis, O. S. *Anal. Chim. Acta* 1988, **205**, 1.
6. Freeman, T. M.; Seitz, W. R. *Anal. Chem.* 1978, **50**, 1242.
7. Aizawa, M.; Ikariyama, Y.; Kuno, H. *Anal. Lett.* 1984, **17**, 555.
8. Ikariyama, Y.; Aizawa, M.; Suzuki, S. *Appl. Biochem. Biotech.* 1981, **6**, 223.
9. Aizawa, M.; Suzuki, S.; Kato, T.; Fujiwara, T.; Fujita, T. *J. Appl. Biochem.* 1980, **2**, 190.
10. Ikariyama, Y.; Suzuki, S.; Aizawa, M. *Anal. Chem.* 1982, **54**, 1126.
11. Ikariyama, Y.; Suzuki, S.; Aizawa, M. *Enzyme Microb. Technol.* 1983, **5**, 215.
12. Ikariyama, Y.; Suzuki, S.; Aizawa, M. *Anal. Chim. Acta* 1984, **156**, 245.
13. Ikariyama, Y.; Aizawa, M. *Proc. 2nd Sensor Symp.* (IEEE of Jpn.) 1982, 97.
14. Ikariyama, Y.; Kunoh, H.; Aizawa, M. *Biochem. Biophys. Res. Commun.* 1985, **128**, 987.
15. Heineman, W. R. *Anal. Chem.* 1978, **50**, 390A.
16. Kadish, K. M. Electrochemical and Spectrochemical Studies of Biological Redox Components : In *Advances in Chemistry Series 20*; American Chemical society : Washington, DC, 1982.
17. Kuwana, T. *J. Electroanal. Chem.* 1963, **6**, 164.
18. Kuwana, T.; Epstein, B.; Seo, E. T. *J. Phys. Chem.* 1963, **67**, 2243.
19. Epstein, B.; Kuwana, T. *Photochem. Photobiol.* 1965, **4**, 1157.

RECEIVED March 9, 1989

SENSORS BASED ON MASS OR THERMAL MEASUREMENTS

Introduction: Mass and Thermal Microsensors

The analyst is very familiar with techniques that depend on photon flux, wavelength measurements, or voltage and current analyses. There is less familiarity and focus on sensors that record thermal flux or mass changes. The papers to follow specifically address new advances in thermal and mass microsensors.

Each of the articles has sufficient tutorial material for the neophyte to understand how the systems are built and used. For the applied analytical chemist, the sytems offer a panorama of opportunities for application to process control streams, quality control labs, environmental testing, and clinical analysis. Most of the material is easy to absorb. You will certainly come away with a better idea of how the new electronic toolbox can be used to make your job as an analytical chemist easier. Finally, for the expert, there are some challenging theoretical issues that should stimulate future experimental work.

Classical thermal sensors involve thermometers, thermocouples, and thermistors. These devices measure temperature by its effect on the volume of a liquid, induced emf changes from a bimetallic contact, or changing resistance values.

Fiber optics, developed for the telecommunications revolution, have great applicability in chemical sensor development. Optical sensors based on fiber optics are discussed elsewhere in this volume. However, the first engineers to use fiber optics for telecommunications discovered that they were temperature sensitive. Their solution to the problem is to surround the fiber with two protective sheaths. The first involves a coating of silicone polymer that has a low thermal coefficient of expansion. The second coat is a tough protective polycarbonate jacket. After removing these coatings, it is possible for the scientist to construct a thermal sensor which depends on the effect of temperature changes upon the propagation of light down a fiber optic waveguide.

With care, physicists have demonstrated their ability to detect one microdegree changes. It is simple for the chemist to prepare a chemically specific biosensor by coating a fiber with an enzyme, and subsequently exposing it to a substrate solution. The changes that result from the heat generated can be related to substrate concentration.

The equipment required to support such a sensor is somewhat complex, but an inexpensive alternate is available based on piezoelectric/pyroelectric polyvinylidene fluoride films. Developed for use in sonar detectors, and adapted to civilian use in speakers and microphones, these films can be made into biosensors by coating one surface with an enzyme. The heat produced upon reaction with a substrate creates a charge flow that can be related to concentration. These two sensors are the subject of the first paper in the series, by Dessy, Burgess, Arney, and Petersen.

The other six papers are devoted to mass sensors that are based on surface acoustic wave (SAW) devices and their relatives. SAW interdigitized transducers were originally designed for use in tunable delay lines and filters for electron counter-measure warfare applications. Most TV sets have a SAW device in them to filter out unwanted signals. The first electrical engineer to work with them found that their electrical characteristics drifted as a function of ambient air conditions. We now know that this was due to changing water vapor and carbon dioxide concentrations. The engineers solution was to place the devices in a sealed metal can purged with an inert gas. A chemist, Henry Wohltjen, working in our lab, was the first to remove the top of the case and build gas and polymer film sensors.

SAW detectors work by interaction of a wave travelling down the surface of the device at acoustic velocities with chemicals on or near that surface. The wave is originally created by a 30 to 300 MHz AC signal applied to a set of interdigitized electrodes at one end of the device. The electrical field distorts the lattice of the piezoelectric material beneath the electrode, causing a SAW wave to propagate toward the other end through a space called the acoustic aperture. Upon arriving at the other end a duplicate set of interdigitized electrodes generate an AC signal as the wave passes underneath them. The signal can be monitered for amplitude, frequency, or phase shift. Changes in the material in the acoustic aperture involving weight increase or loss (mass loading) or elastic modulus are readily detected.

One popular application of SAW devices involves coating the acoustic aperture with a material that provides some selective physisorbtion or chemisorbtion for specific components in the ambient atmosphere. If an array of sensors, each with a different coating, is employed, an analyte mixture often produces a characteristic signature pattern, since each detector responds differently to each component in the mixture. The paper by Wohltjen, Ballantine and Jarvis describes such an approach.

A second approach uses a coating that reacts chemically with a component of the atmosphere. Zellers' paper reports the use of a square planar platinum complex of ethylene to measure the concentration of vinyl acetate in the surroundings. The latter replaces the ethylene in the original complex, resulting in an increase in mass loading.

Many different types of waves can be supported by piezoelectric devices. The SAW waves described above involve a retrograde elliptical motion of lattice particles in a plane perpendicular to the device (Rayleigh wave). Horizontal shear waves may also be generated, and Ricco, Martin, Niemczyk and Frye describe the use of such an acoustic wave device. Coating the acoustic aperture with ethylenediamine produces a sensor that has specificity for cupric ion.

Another variation in SAW sensor application relies on the ability of a porous film, with controlled pore size, to be selective for certain gas molecules. Such devices have been used by Frye, Martin, Ricco, and Brinker to measure diffusion coefficients, suggesting that SAW devices can be used as devices to measure fundamental molecular properties.

Ballantine and Wohltjen describe some interesting experiments that probe the glass transitions of very thin polymer films laid down on the SAW surface. Although this area is quite complex and not completely understood, the evidence suggests that very intimate information can be obtained in this way, facts that are most difficult or impossible to extract in any other way.

Finally, there are more classical devices that respond to mass loading of the surface. Conventional quartz crystal microbalance detectors may be made sensitive to biological materials by immobilization of an enzyme on the surface. In these devices mass loading causes a shift in the operating frequency of the crystal oscillator. Lasky and Buttry describe such a system involving the glucose/hexokinase system.

There is intense interest in the use of quartz crystals, piezolectric films, and Rayleigh and shear wave devices as bio-clinical detectors.

ENJOY!

Raymond E. Dessy
Department of Chemistry
Virginia Polytechnic Institute and State University
Blacksburg, VA 24061

RECEIVED March 24, 1989

Chapter 10

Fiber-Optic- and Polymer Film-Based Enthalpimeters for Biosensor Applications

Raymond E. Dessy, Lloyd Burgess, Larry Arney, and James Petersen

Department of Chemistry, Virginia Polytechnic Institute and State University, Blacksburg, VA 24061

Two novel enthalpimeters for biosensor applications are described. One of these employs a fiber optic interferometer, the other piezo/pyroelectric polyvinylidene fluoride films. Applications to urease and catalase systems are described.

Biosensors based on the heat produced by enzyme/substrate reactions have traditionally used microcalorimeters (1), thermistors (2), and Peltier or other macro devices (3,4) The area has been reviewed by Guilbault (5). The size, response time, and thermal mass of these detectors suggests that thermally responsive microsensors need to be explored. The ideal sensor would be inexpensive, and require simple, low cost support electronics. A fiber optic based sensor (Part A), and a pyroelectric polymer film based sensor (Part B) are described below.

A. A FIBER OPTIC BASED ENTHALPIMETER

Although there is currently great interest in the application of fiber optics as chemical sensors most of these applications have involved the extrinsic properties of the waveguide materials. In these applications the fibers are used as light pipes for the conduction of guided optical waves employed in absorbance and fluorescence measurements. However, changes in the intrinsic properties of such fibers offers a new realm for the development of microsensors. Such phenomena have been investigated by physicists in their development of acoustic, magnetic, thermal, and pressure sensors (6-9). Following is a report on the successful application of this approach to a biosensor based on a fiber optic enthalpimeter.

0097-6156/89/0403-0143$06.00/0

The instrument, Figure 1, involves a two-arm Mach-Zehnder interferometer constructed from mono-mode fiber optic waveguides. One arm of the interferometer is coated with an immobilized enzyme, while the other is used as a reference. Both arms of the interferometer are firmly held in the middle of a conduit that is incorporated into a flow injection analysis (FIA) system, serving as a substitute for the normal absorbance, fluorescence, pH, or electrochemical detectors. A common source of phase coherent light is launched down both arms of the interferometer. The exit beams from both fibers are superimposed in the far field on a linear array optical detector. This superposition produces the classical two-slit pattern of light and dark bars due to interference. As sample boluses containing substrates specific to the enzyme pass the sensing area, heat is produced around one fiber. This heat, partitioned between the flowing stream and the fiber, produces a thermal and stress/strain effect on the coated fiber which changes its light propagation characteristics. Although shifts in the resulting bar pattern can be detected by a single optical detector, a linear array has many advantages. If an integral number of periods of the bar pattern, n, illuminate the active area of the array a Fourier transform of the data from the scanned array will produce a new real and imaginary data set. The ratio of the real and imaginary values at the spacial frequency n gives directly the phase angle difference between the two beams and simultaneously eliminates noise contamination since this is found at other spacial frequencies. The phase angle difference can be related to sample concentration.

INSTRUMENT DESIGN

An inexpensive Spectra-Physics Model 120 HeNe laser was used as a source of coherent radiation. This radiation was equally divided between the two arms of an interferometer by use of a 3 dB coupler (10). This was constructed from two lengths of ITT T-1601 single mode optical communication fiber. This fiber exhibits a numerical aperture of 0.10 and will operate at single-mode propagation down to 580 nm. These fibers have an outer sheath of DuPont Hytrel 7246, and an inner sheath of General Electric RTV 615 (Figure 2). These may be stripped off by mechanical and chemical means in the coupler area. The residual glass fiber consists of a 4 micron core surrounded by an 80 micron glass cladding. If the two fibers are twisted together under tension the outer cladding can be largely stripped away in an HF bath. This stripping is continued until about 50% of the light launched down one fiber evanescently couples into its companion waveguide. Standard fiber positioners and microscope objectives may be used to deploy these fiber optic components.

Downstream, the unstripped sections of fiber were held in a Plexiglas channel 20mm x 1mm x 1mm in dimension.

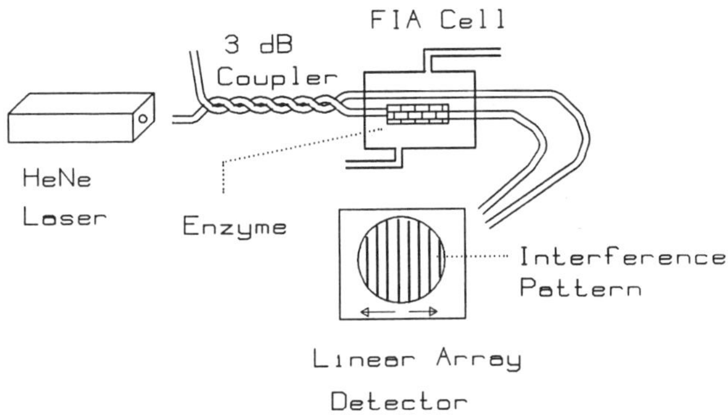

Figure 1. The Mach-Zehnder interferometer/enthalpimeter.

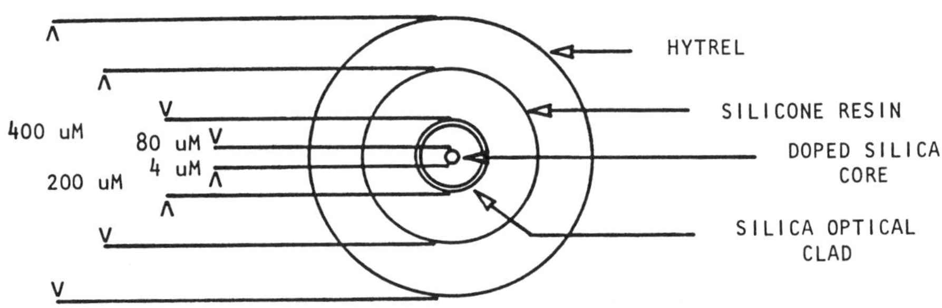

Figure 2. Cross section of ITT T-1601 optical fiber.

One arm of the interferometer was coated over a 10mm length with the desired enzyme immobilized in a gluteraldehyde copolymer (5, 11). The channel was covered with a Plexiglas plate and the fiber entrances sealed with RTV silicone. The channel was incorporated in a FIA system.

The exit beams of the two arms were optically combined onto the surface of a Reticon S-series 512 element linear array. A Reticon RC-1024SA interface board provided signal control, clocks, and integrate/sample-hold circuitry for the processing of the pulse packets emerging from the common video line of the detector chip. A 25 usec successive approximation analog-to-digital converter (ADC) was used to digitize this data stream. Each data set was subjected to Fourier transform using software extracted from Digital Equipment Corporation's Laboratory Applications package. All other control and data manipulation was written in PolyForth (Forth, Inc.). The computer was a DEC LSI-11 connected to a large laboratory network hosted by PDP-11/23's.

Purified enzymes were obtained from Sigma Chemicals; catalase (E.C.1.11.1.6) and urease (E.C.3.5.1.5). They were immobilized on the fiber using gluteraldehyde (Fisher). Phosphate buffers (0.1 M) were used for fiber storage and as carriers in the FIA system. Thymol (50 mg/L) was used to inhibit bacterial and fungal growth in the buffer.

RESULTS AND DISCUSSIONS

Changes in the light propagation characteristics of a fiber due to temperature arise from (a) thermally induced length changes, (b) the temperature dependence of the core refractive index, and (c) elasto-optic changes in the core due to thermally induced stress/strain changes.

Plots of phase angle difference in the interferometer arms vs. time are related to heat-production vs. time, and this in turn is related to the concentration of the species responsible for heat production. Typical instrument output for the urea/urease system is shown in Figure 3. Calibration curves can be constructed as shown in Figure 4. The system is quite stable, and reasonably sensitive. Minimum detectable levels of urea are 5 mM, compared to the 0.1-5 mM limits for traditional detectors. Over extended time periods (7 days) the relative standard deviation at 5 mM concentrations is better than 5%. The optimum FIA conditions were around 1.0 ml/min flow rate, with a sample loop of 0.1-0.25 ml.

Detailed modelling studies on heat transport in such systems, to be published elsewhere, are easy to develop since the thermal conductivity, thermal diffusivity, and the heat capacities of the various fiber components are known. The stress/strain effects are almost completely asserted in a thin surface layer of the Hytrel, because of the poor conductivity of this sheath, and the underlying viscoelastic RTV. However, Hytrel has an extremely high

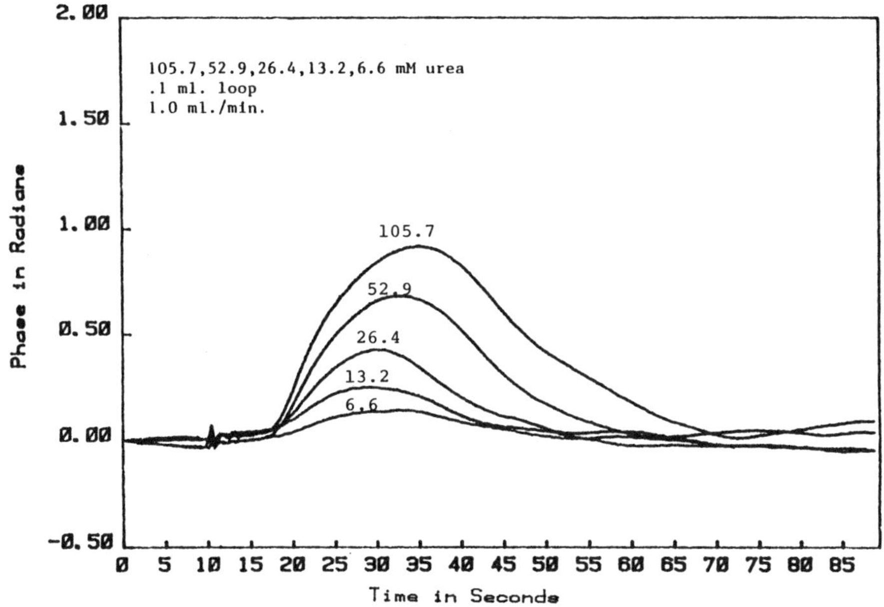

Figure 3. Interferometer output for various urea concentrations passing over a urease bed immobilized on one arm.

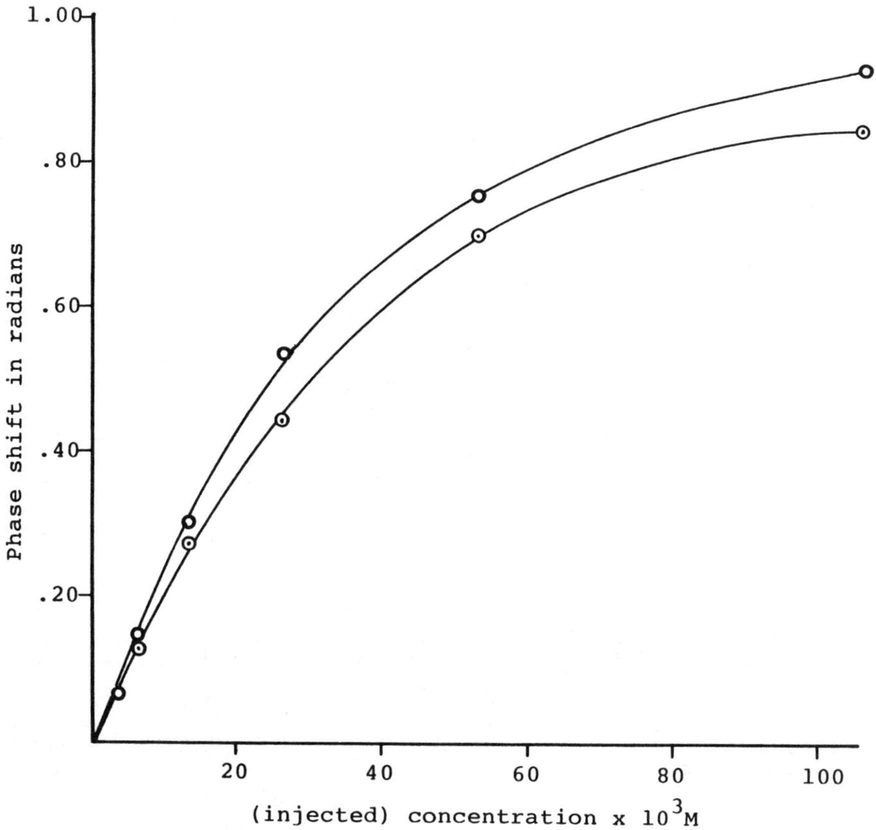

Figure 4. Calibration curve for urea/urease system at 0.8 and 1.0 ml/min.

thermal expansion modulus, which helps compensate for this deficiency (7). The jacketing does protect the detector system from effects due to interaction of the evanescent guided wave with its environment. The response time of the system is limited by the insulating characteristics of the jacket materials. Obvious extensions of the concept would involve different jacket configurations to enhance the stress/strain action, and/or coating with aluminum to enhance heat transport to the fiber surface. Simple calculations show that about half of the generated heat is lost to the surrounding water medium in the present configuration.

Studies on the peroxide/catalase system yielded response curves with very poorly defined maxima, and badly drifting base lines. The source is presumably the bubbles of gaseous oxygen produced by the reaction. These develop in the later stages of the reaction. Clinging to the fiber, these bubbles apparently result in mechanical and/or thermal effects that distort the expected profile. Although the addition of surface active agents (ethoxylated lauryl alcohol) moderates the distortion, and improves base line performance, the ill-defined peaks did not respond to peak-height or area measurements. The peaks were broad, did not return to any reliable base line position, and were typified by low signal/noise ratios. It proved impossible to derive consistent area or height information from repeat runs using either visual methods, or standard peak-picking algorithms. Reliable quantitative calibration curves could only be derived from evaluation of the slope of the phase angle difference curve as a function of time at a fixed arbitrary point near the beginning of the heat evolution.

These fiber optic sensor systems are extremely sensitive to environmental noise. Pneumatically supported optical tables were required. Pressure fronts developed from droplets falling from the exit of the FIA tubing are easily seen. Care is required in mounting the fibers in the FIA channel. Diminished response, or even inversion of the signal response, is easily induced by improper tensioning. Heat induced stresses compete with the radial and axial stresses resulting from the fixing of the two ends of the sensor area within the flow-cell. Heat may increase or decrease the total stress in the system. A crude analogy is the effect of heat on a piece of metal fixed at both ends. Although the systems have potential if the proper engineering developments are undertaken, alternate sensor systems are under study.

B. POLYMER FILM BASED ENTHALPIMETER

Modern flow systems place unusual constraints on sensor configuration. Rugged, inexpensive sensors that can easily adapt to the geometries imposed by FIA systems are essential. It is of interest to explore the possibility of fabricating a useful enthalpimetric sensor from films of piezo and pyroelectric (12) organic polymers. These materials are thin enough that simple cutting and molding would allow the sensor material to conform to the required geometry of the sample cell, rather than designing the cell geometry around a fixed sensor. The films are inexpensive, and available from several sources.

Polyvinylidene fluoride (PVDF) films, ranging in thickness from a few microns to a millimeter, can be made piezo and pyroelectric by special treatment after manufacture (13). The materials are simultaneously subjected to an increased temperature, elongation stress along one or two axes, and a high DC field potential. As the material is allowed to cool, with the stress and poling field applied, orientation effects occur that result in a residual polarization of the film, rendering it piezo/pyroelectric. Chemically, charge injection probably occurs during this process. Physically, alignment of the electronegative fluorine atoms toward the positive pole occurs, leading to a large amount of beta phase PVDF which forms crystallites. There is some controversy about the actual phenomena involved in the observed piezo/pyroelectricity, and several effects may be concurrently responsible (13-17). Such materials show electrostriction. The Poisson ratio of the films is high (differential axial thermal coefficients of expansion). Finally, the crystallite concentration is temperature dependent. Regardless of the source, the films find widespread application in sonar transducers, audio speakers, and microphones because of their piezoelectric behaviour. Their pyroelectric properties have led to imaging device applications.

In principle a biosensor could be fabricated from the material by immobilizing an enzyme on one surface, and flowing a substrate solution past the film sensor. A simple charge amplifier might be used to measure the difference between the two surfaces of the sensor. This, in turn, could be related to concentration. In practice a somewhat different approach needs to be taken.

INSTRUMENT DESIGN

Films of PVDF 40 microns thick from Solvay were used throughout most of the study. These films have a thin metal coating that is vapor deposited on each surface to facilitate the electrical polling process, and which may be used to make electrical contact in sensor applications. For the present studies these films were configured as a bimorph, or double laminate, by placing like surfaces of two pieces of film face-to-face. The bimorph was pressure mounted in an FIA flow cell made of Plexiglas (Figure 5). The intent of the bimorph architecture was to eliminate piezoelectric effects from the mounting process, reduce electromagnetic interference, and to help compensate for ambient temperature changes. The bimorph has three electrical points of contact; the mating interior surfaces of the bimorph, and the two exterior surfaces. The interior surfaces were commonly grounded and the other surfaces provided a reference and sample side. The sample side formed the floor of a flow channel that was machined into the Plexiglas mount. The channel was 25mm x 5mm x 0.5mm. The enzyme was immobilized onto the roof of the channel (Figure 6). Attempts at placing the enzyme directly onto the sample film surface led to poorer response, both in signal level and noise level. This is due to the poor thermal transport characteristics of the hydrated biopolymer.

The signals from the sample and reference surfaces were first buffered via the non-inverting inputs of two electrometers, AD515s, which had very high input impedance and low input bias currents. The output of the two electrometers were used as inputs to an instrument amplifier, AD625. Output from the instrument amplifier was digitized with a 25 usec successive approximation ADC, and noise reduced by use of a 21-point Savitzky-Golay (18) weighted digital filter. Several approaches to quantization were made. The best results were obtained by simple peak height measurements. All control and manipulation software was written in PolyForth. The computer was a DEC LSI-11 attached to the laboratory network.

RESULTS AND DISCUSSION

The peroxide/catalase system was selected for study because of the difficulties found with the fiber optic sensor described above. The FIA carrier contained Triton-X as a surface active agent to reduce bubble formation. Typical FIA flow characteristics were 4 ml/min, and a sample loop of 80 uL. Typical sensor output is shown in Figure 7. The calibration curve developed for the peroxide/catalase system is linear from 0.004M to 1.0M (Figure 8). The error bars shown in the calibration curve were calculated by multiplying the standard deviation by the 95% confidence limit for a set of data with four degrees of freedom. The correlation coefficient of the line is 0.995.

The thermal flux from the enzyme/substrate reaction crosses the aqueous layer and rapidly diffuses through the 200 Angstrom thick aluminum coating on the top layer of the bimorph. Modelling of the heat transport within the film, to be published elsewhere, shows that the thermal transport process may be treated as a semi-infinite solid; i.e., an infinitely thick solid with one surface. This is reasonable since the PVDF and Plexiglas have quite similar heat transfer properties, with the thermal diffusivity of Plexiglas being somewhat higher. The model produces thermal time gradients in agreement with those experimentally observed, and satisfactorily predict the timing and amplitude of the phase change observed as the thermal boundary passes from one layer of the bimorph to the other. The shape of each response curve is easily explained. There

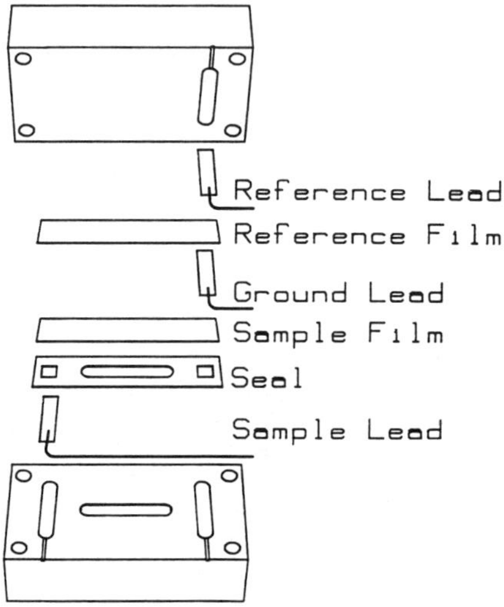

Figure 5. Exploded view of polyvinylidene fluoride film based enthalpimeter.

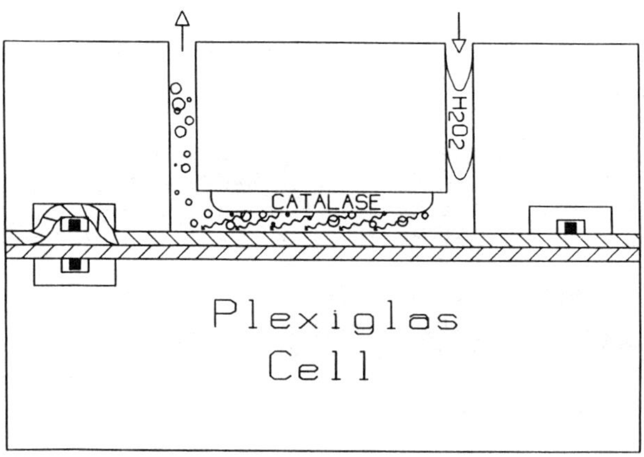

Figure 6. Cross section of polyvinylidene fluoride film based enthalpimeter.

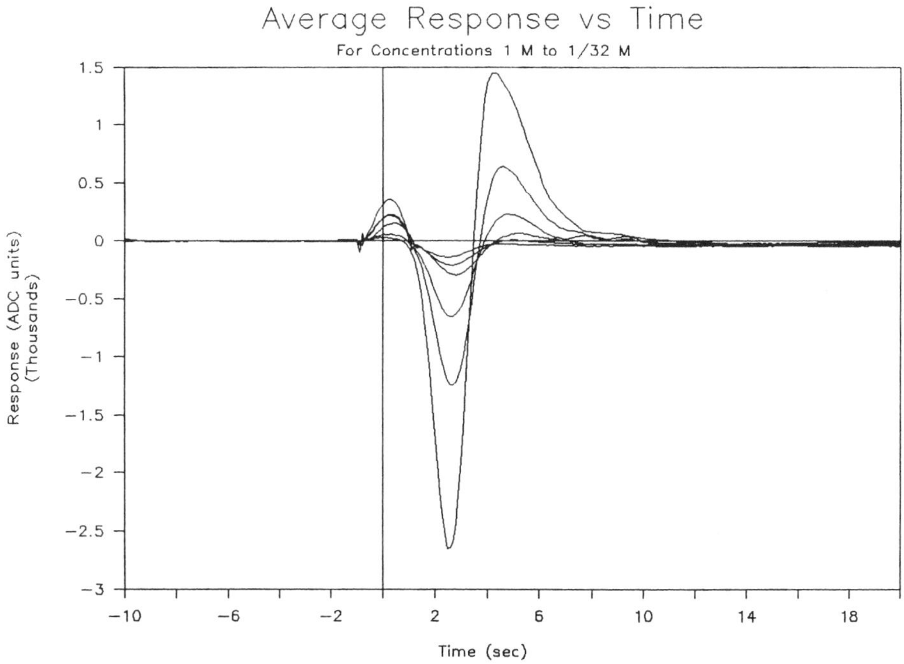

Figure 7. Polymer film enthalpimeter output for various concentrations of hydrogen peroxide passing over a catalase bed immobilized on the roof of the flow channel.

is an initial small "upward going" excursion that is caused by piezoelectric effects within the film resulting from the pressure surge in the stream injected by the FIA valve switching process. As this passes the sensor the amplifier output returns to baseline. This peak is rather variable, but can serve the same function as the air peak in a GC trace. Subsequently, as the thermal front moves into the first layer of the bimorph it becomes hotter than the reference side, and a "downward going" excursion is observed. As the thermal diffusion process continues a point is reached where both films are the same temperature, and a "zero-crossing" occurs. Then the reference film becomes warmer than the sample film, as the latter is cooled by the flowing FIA stream, causing an "upward going" excursion. Finally, both film layers return to ambient temperature, and the output of the amplifier returns toward zero baseline. The modelling suggests that improvement in the system might be made by inserting a thermal insulator between the two layers of the bimorph, and this is being studied.

CONCLUSIONS

Both detectors have use in the biosensor arena. The catalase and urease systems span a large enthalpy range bracketing systems such as lactase dehydrogenase, hexokinase, glucose oxidase, and cholesterol oxidase. Figure 9 compares the two sensor systems, based on the peroxide/catalase FIA system parameters. Throughput and sample volume advantages lie with the polymer film sensor. The ratio of signal full-width-at-half-height/bolus transit-time is 2-4 for the optic sensor, and about 1.3 for the film sensor. The source of this sluggishness in the optical system lies not in the thermal properties and response of the fiber core, but the fact that it is surrounded by a bulky external jacket with poor thermal diffusivity and an inner rubbery jacket. The fiber engineers deliberately choose the jacket construction to minimize thermal interferences in communication applications. Indeed, studies on the stripped fibers indicate that thermal or optical pulses, one second in width, lead to responses in a stripped fiber with a FWHH of less than 1.5 seconds. Hytrel coated fibers have FWHH responses about 4 times longer. Literature studies indicate that metal coated fibers will respond to DC heating pulses induced via the metal jacket up to 30 kHz (9). The obvious solution is to fabricate a special sensing fiber with a thinner jacket and more thermally sensitive core. This will reduce the surface area available for coating with the enzyme. In order to obtain a useful analytical response it may be necessary to deploy the biopolymer on the inner wall of a tube concentric with the fiber, similar to the PVDF film sensor described above. The sample/reference arm problem that this engenders might be resolved by use of a single arm interferometer (19).

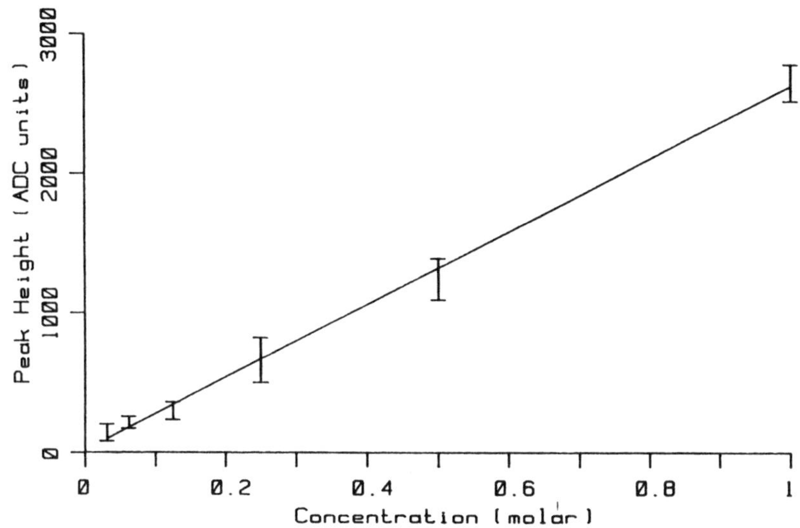

Figure 8. Typical calibration curves for a polymer film enthalpimeter using the peroxide/catalase system.

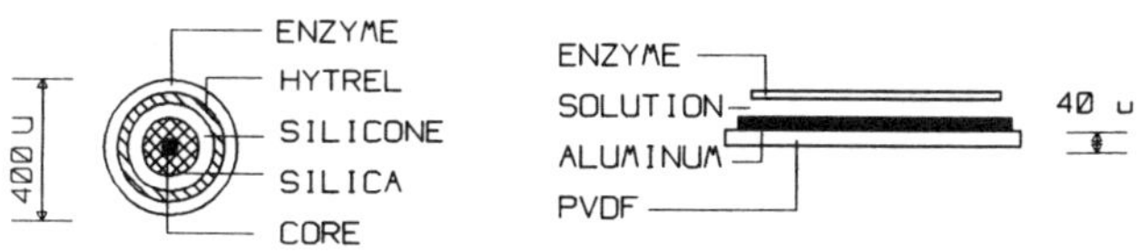

0.2 cm^2	AREA	1.0 cm^2
250 uL	SAMPLE	80 uL
1.5 ml/m	FLOW	4 ml/m
60 uL	DEAD VOL.	70 uL
10 s	BOLUS TRANSIT	1.5 s
40 s	FWHH RESPONSE	2 s

Figure 9. A comparison of the fiber optic and PVDF film enthalpimeter.

Interference at the output of this device is between the two orthogonal polarizations in a single waveguide with high internal lateral stress. However, mounting such fibers will require further engineering. Until an instrument vendor takes up these challenges, the polymer film sensor is an appropriate and sensible alternative. It is easily configured to unusual geometries, requires simple and inexpensive support electronics, and is reasonably sensitive.

LITERATURE CITED

1. Johansson, A., Lundberg, J., Mathiason, B. Mosbach, K., Biochim. Biophys. Acta, 304, 214 (1973).
2. Pennington, S., NSF/RA 760032 Enzyme Technology Grantees-Users Conference; PE 256 584 (1975).
3. Cooney, C., Weaver, J., Tannenbaum, S., Faller, D., Shields, A., Jahnke, M., "Enzyme Engineering", Vol. 2, Plenum Press, New York (1974).
4. Rich, S., Ianello, R.,Anal. Chem., 51(2), 204 (1979).
5. Guilbault, G., Danielsson, B., Mandenlus, C., Mosbach, K., Anal. Chem., 55(9), 1582 (1983).
6. Hocker, G., Applied Optics, 18(9), 1445 (1979).
7. Lagakos, N., Bucaro, J., Jarzynski, J., Applied Optics, 20(13), 2305 (1981).
8. Maurer, G., Cole, J., Optics Letters, 7(11), 561 (1982).
9. Schuetz, L., Cole, J., Jarzynski, J., Lagakos, N., Bucaro, J., Applied Optics, 22(3), 478 (1983).
10. Sheem, S., Giallorenzi, T., Optics Letters, 4(1), 29 (1979).
11. Attiyat, A., Christian, G., Biotechnology Laboratory, 2(2), 8 (1984).
12. Cady, W. G., "Piezoelectricity", McGraw Hill Book Co., New York(1946).
13. Kawai, H., Japanese Journal of Applied Physics, 8, 975 (1969).
14. Anderson, R., Kepler, R., Ferroelectrics, 33, 91 (1984).
15. Bergman, J., McFee, J., Crane, G., Applied Physics Letters, 18, 203 (1971).
16. Kepler, R., Anderson, R., Journal of Applied Physics, 49, 4918 (1978).
17. Kepler, R., Anderson, R., Lagasse, R., Ferroelectrics, 57, 151 (1984).
18. Savitzky, A., Golay, J., Anal. Chem., 36, 1627 (1964).
19. Eickhoff, W., Optics Letters, 6(4), 284 (1981).

RECEIVED March 17, 1989

Chapter 11

Vapor Detection with Surface Acoustic Wave Microsensors

Hank Wohltjen[1], David S. Ballantine, Jr.[2], and N. L. Jarvis[1]

[1]Microsensor Systems, Inc., P.O. Box 8, Springfield, VA 22150
[2]Department of Chemistry, Northern Illinois University, DeKalb, IL 60115

Progress continues to be made toward the development of practical vapor detection instruments using coated SAW sensors. This paper reviews some of the characteristics of these versatile devices and their current capabilities. Factors affecting sensor performance, such as temperature and coating selection, are considered in terms of the sensitivity and selectivity of the SAW device. Special attention is given to a 4-SAW array instrument that is able to "fingerprint" and identify a variety of chemical vapors.

The ability to generate elastic waves in piezoelectric materials using lithographically patterned microelectrodes was first reported by White and Voltmer in 1971 (1,2), and the potential application for devices using elastic waves as chemical and physical sensors was soon recognized. Some of the types of elastic waves that can be readily generated on piezoelectric materials include horizontally polarized shear waves, Lamb waves, and Rayleigh surface waves. While each type of wave has distinct advantages for specific applications, the Rayleigh surface acoustic wave is unique in that the crystal lattice displacement associated with the acoustic wave is confined mainly to the surface region, typically to a depth of only a few acoustic wavelengths. When first introduced, the surface acoustic wave, or SAW, device was used mostly for the generation, delay, and filtering of radio frequency signals. Since the first reported use of a SAW device for chemical vapor detection in 1979 (3) these devices have seen increased use as chemical sensors for a variety of applications. This growing interest is a result of the many attractive features offered by the SAW device including sensitivity,

0097–6156/89/0403–0157$06.00/0

ruggedness, and small size. SAW chemical sensors are currently being investigated for their applicability to problems in the fields of environmental monitoring, clinical analysis, and industrial process control.

The objectives of this paper are threefold. First, we will review the basic operating principles of SAW chemical vapor sensors including a theoretical model that predicts the sensitivity of SAW device response. Secondly, design considerations that affect the performance of SAW sensors will be presented. Finally, some recent results from various SAW sensors and an experimental four sensor SAW array will be discussed.

Theory of Operation

SAW chemical sensors often use a delay line configuration that consists of a set of lithographically patterned interdigital electrodes (IDTs) deposited onto the surface of an optically smooth piezoelectric substrate. When a time varying potential is applied to the "transmitting" set of IDTs, a Rayleigh wave is created that propagates along the device surface. This surface wave is converted back into an electrical signal by the second set of "receiving" IDTs. This process is illustrated in Figure 1. When a SAW delay line is used as the feedback element in an oscillator circuit, the frequency of the circuit is linearly related to the surface wave velocity. The SAW velocity is strongly perturbed by surface mass loading.

Perturbations of the medium adjacent to the device surface result in variations in the phase, amplitude, and velocity of the surface acoustic wave. Specifically, these properties will be affected by changes in the density, viscosity, or elastic properties of the medium in contact with the surface. Since the acoustic wave has an electric potential wave associated with it as well, the SAW can also be used to probe the dielectric and conductive properties of this surface medium. By far, the largest number of chemical sensor applications of SAW devices take advantage of the mass sensitivity of SAW oscillators.

The signal provided by a SAW oscillator vapor sensor can be described by the following theoretically derived relationship (4):

$$\Delta f=(k_1 + k_2) f_o^2 h\rho' - k_2 f_o^2 h(4\mu'/V_r^2)((\lambda'+\mu')/(\lambda'+2\mu')) \qquad (1)$$

where Δf is the observed SAW oscillator frequency change, k_1 and k_2 are material constants for the piezoelectric substrate, f_o is the unperturbed resonant frequency of the SAW oscillator, h is the coating thickness, p' is the coating density, μ' is the coating shear modulus, λ' is the coating Lamé constant, and V_r is the Rayleigh wave velocity on the piezoelectric substrate (3158 m/sec for ST-quartz). This relationship assumes that the SAW device coating is isotropic, non-conducting, and non-piezoelectric. Furthermore, the relationship is valid only for very thin films (e.g. < 0.2% of the acoustic wavelength). Depending on the elastic properties of the coating, thicker films can produce frequency shifts that are substantially different than their mass density and equation (1) would predict. When organic polymer coatings are applied to SAW devices it is sometimes possible to neglect the second half of equation (1) because the appropriate numerical value for the shear modulus of the coating is significantly smaller than the numerical value of the Rayleigh wave velocity raised to the second power. Under these conditions, equation (1) can be approximated by:

$$\Delta f = (k_1 + k_2) f_o^2 h p' \qquad (2)$$

The material constants, k_1 and k_2, for the commonly used ST-quartz piezoelectric SAW devices are -8.7×10^{-8} and -3.9×10^{-8} $m^2 s/Kg$, respectively (5). The product of coating thickness (h) and density (p') is equivalent to the mass per unit area present on the device surface. It should be clear that applying a coating to the SAW device surface will shift the resonant frequency of the SAW oscillator by an amount related to the mass per unit area applied to the device. When vapors sorb into the coating, there is a further incremental change in the mass loading experienced by the SAW device. It is this incremental change in mass of the coating caused by exposure to chemical vapors that makes chemical sensing with SAW devices possible.

Equation (2) indicates that the magnitude of frequency shift obtained from a given mass loading will increase with the square of the operating frequency of the device. This has important implications for the physical scaling laws that govern the performance limits of SAW sensors (6). Some of these are tabulated in Table I. The physical size of a given SAW device design is inversely proportional to its resonant frequency. As the resonant frequency of a SAW devices increases, the active surface area of the device decreases with a $1/f^2$ dependence while the mass sensitivity is increasing with f^2. SAW oscillator noise is more difficult to predict, but typically it increases with f. The net result is that the absolute mass that is detectable by a SAW device exhibits approximately a $1/f^3$ dependence. Thus, increasing the SAW oscillator frequency by a factor of 10 results in a device that is theoretically capable of detecting a mass change that is 1000 times smaller.

Table I. SAW Device Scaling Laws

PARAMETER	SCALING FACTOR
Linewidth (length, width)	$1/f$
Device Area (length x width)	$1/f^2$
Sensitivity	f^2
Noise	f
Mass Detection Limit	$1/f^3$

The theoretical relationship between operating frequency and mass sensitivity predicted by equation (2) has been tested experimentally (7). SAW devices with operating frequencies of 31, 52, 112, and 158 MHz were coated with layered films of different thickness and mass using the Langmuir-Blodgett technique. Each layer of film produced a highly reproducible mass loading of 570 ± 20 ng/cm^2, and the resulting frequency shifts for films up to 45 layers thick were recorded. These results are presented in Figure 2 (7). The experimentally measured mass sensitivities are compared in Table II with sensitivities predicted by equation (2). The strong dependence of mass sensitivity on operating frequency is apparent. The data show a systematic discrepancy between the theoretical and experimental values, with the theoretically predicted sensitivity being higher by a factor of 1.1 to 2.5.

Table II. Theoretical VS Experimental SAW Sensitivity

Device Freq. (MHz)	Theor. Sens. (KHz/layer)	Exp. Sens. (KHz/layer)	Ratio (Theor./Exp)
31	0.74	0.3	2.5
52	2.08	1.5	1.4
112	9.65	5.8	1.7
158	19.2	18.0	1.1

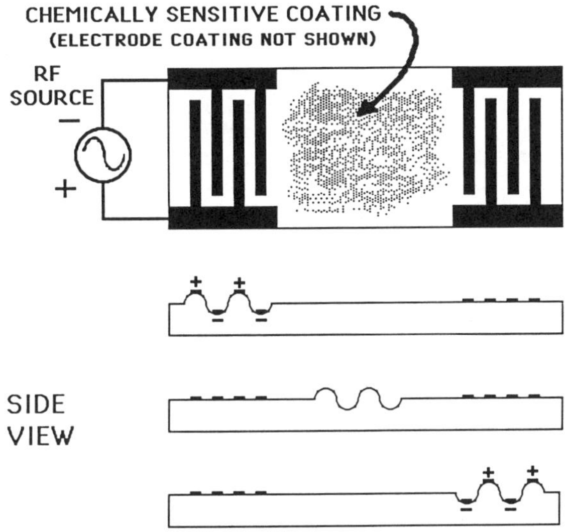

Figure 1. Simplified SAW delay line configuration.

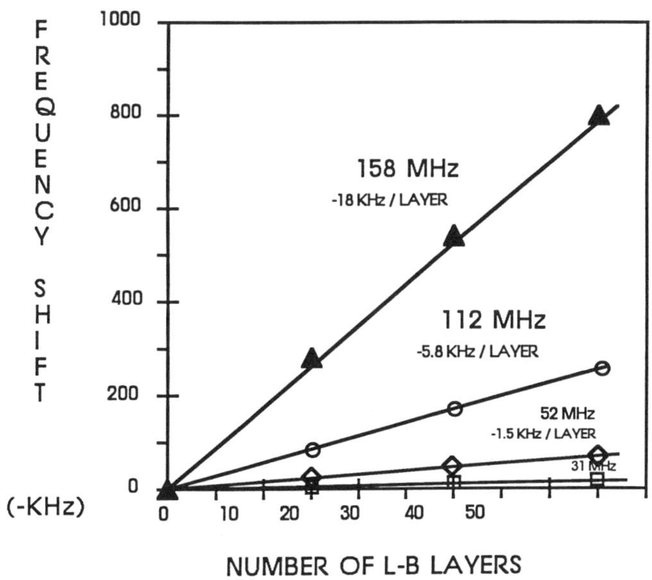

Figure 2. SAW oscillator frequency shift caused by L-B film mass loading for various SAW oscillator frequencies.

This systematic difference may be the result of elastic effects. As noted in equation (1), the frequency shift obtained for a given coating will be affected by the elastic properties of the film and the effect of non-negligible film stiffness will be to offset the effect of mass loading. If we assume that the discrepancy is due to elastic effects, then solving equation (1) using the experimental values results in an estimated shear modulus for the L-B film in the range of 8 - 16 x 10^9 N/m^2. While the shear modulus of the L-B films used in this study could not be determined by another method, it is useful to note that the estimated modulus value is higher than expected for typical organic polymers at low measurement frequencies. For example, Lucite (polymethylmethacrylate) has a shear modulus of about 7 x 10^8 N/m^2 and natural rubber (polyisoprene) has a shear modulus of about 5 x 10^6 N/m^2 (8). It is important to remember that viscoelastic materials such as organic polymers often exhibit rheological properties that are profoundly frequency dependent and, as such, the estimates of the L-B film elastic modulus obtained with the very high frequency SAW device may indeed be quite accurate. One can only conclude that care is required when interpreting data obtained from SAW sensors since the mass loading and elastic effects of the sensor coating combine to produce the observed sensor response.

Practical SAW Sensor Considerations

Device Design. The resonant frequency of the SAW device is determined by the velocity of the acoustic wave in the substrate and the spacing between the fingers of the IDTs (see Figure 1). The IDTs must be made of a conducting material (often aluminum or gold). In addition, the number of finger pairs in the IDT affects the quality factor (or Q) of the device. Up to a limit, having more finger pairs reduces the effective bandwidth of the device and results in lower insertion loss and a more stable SAW oscillator. The acoustic aperture (i.e. the finger overlap distance) of the IDT determines its electrical impedance. A variety of piezoelectric substrate materials have been investigated for use as chemical sensors. These include quartz, lithium niobate, and zinc oxide on silicon. ST-cut quartz has proved to be a popular substrate because it is chemically quite stable and affords a reduced temperature sensitivity combined with reasonably good piezoelectric coupling efficiency. While most of this discussion is focussed on SAW delay line devices, it is important to note that some investigators have found that the SAW resonator structure is also very useful for SAW sensor applications (9)

A dual SAW delay line oscillator is illustrated in Figure 3. It consists of two SAW delay lines fabricated on the same ST-quartz substrate. One delay line is coated with the chemically sensitive film and the other is left uncoated. The frequencies of the two delay line oscillators are mixed and filtered to provide a frequency equal to the difference of the two oscillator frequencies. This scheme has several advantages when used in chemical vapor sensor applications. First, it helps to compensate for SAW frequency drift caused by ambient temperature and pressure fluctuations. Additionally, the difference frequency is much lower than the frequency of the oscillators themselves. This permits SAW vapor response measurements using inexpensive digital circuitry.

Sensitive and Selective Coatings. The mass sensitive SAW device becomes chemically sensitive only after a suitable chemically selective coating has been deposited onto the device surface. The coating preferentially concentrates the vapor of interest onto the device surface through physical (e.g. adsorption or absorption) or chemical interactions. Coatings should exhibit a number of characteristics including oxidative and thermal stability, good adhesion to the device surface, and permeability to vapors. Sensitivity is a measure of a sensor's ability to respond to small changes in vapor concentration and, for the SAW device, it is defined with units of Hz/ppm or Hz/mg/m^3. Selectivity is defined as the ability of the sensor to preferentially detect the vapor of interest. It is usually reported as a selectivity ratio, or the ratio of the coatings response to the vapor of interest and the response to another, potentially interfering vapor. Both sensitivity and selectivity are heavily dependent upon the physical and chemical properties of both the coating and the vapor. Thus the selection of the SAW sensor coating material often determines the sensor performance for a given application.

Two strategies are currently in use to improve the sensitivity and selectivity of SAW sensor systems. One involves the selection of coatings that exhibit a unique chemical selectivity for the analyte vapor. It is often observed that for a particular vapor/coating interaction, the specificity and reversibility of the interaction are inversely related. That is, highly specific interactions often involve large bonding energies that are not spontaneously reversible. Such interactions are useful in dosimeter applications (10,11). This selectivity/reversibility paradox has led some investigators to explore weaker, solubility type interactions characteristic of GC stationary phases. Coatings that rely on solubility interactions exhibit reversible behavior at the expense of reduced selectivity and sensitivity. It is believed that selectivity can be enhanced by the use of SAW sensor arrays.

Coating Deposition Methods. The properties of coatings applied to SAW sensors are diverse and as a result, a variety of coating methods have been explored. Coating thickness is often reported in terms of the frequency shift (i.e. KHz) produced when the coating is applied to the SAW device. The Langmuir-Blodgett method mentioned previously is somewhat restrictive because it requires a coating material that is capable of forming a compressible film at the air-water interface. Successive dipping of the device results in the transfer of one layer at a time to the device surface with each pass through the air-water interface. Thus, films of precisely known mass and thickness can be reproducibly deposited. Descriptions of the L-B transfer method can be found elsewhere (7).

Other methods such as spraying, spin coating, and solvent casting are more flexible. These methods require the preparation of solutions of the coating material in appropriate, usually volatile, solvents. In the spray technique, the solution is aspirated to generate a finely dispersed aerosol that can be deposited onto the device at a controlled rate. The solvent evaporation method is similar except that the solution is deposited directly onto the device surface with a pipette, brush, or "Q-tip". In both cases, the coating is left behind on the surface when the solvent evaporates. Spin coating requires a more concentrated viscous solution that is used to flood the device surface. The device is then spun at high speed to remove the excess solution. The process can result in thin films (e.g. 2000 Å thick) of excellent uniformity. Other film deposition techniques include sputtering (metal films), sublimation, and chemical vapor deposition. It is important to note that the coatings should be applied over the IDT electrodes as well as the space in between if the theoretical SAW device sensitivity is to be obtained. In some instances this is not possible, particularly if the coating is electrically conductive and will short-out the IDT electrodes.

SAW Sensor Operational Performance. A typical SAW sensor response is illustrated in Figure 4. In this case, a poly(ethyleneimine) (PEI) coated 158 MHz dual

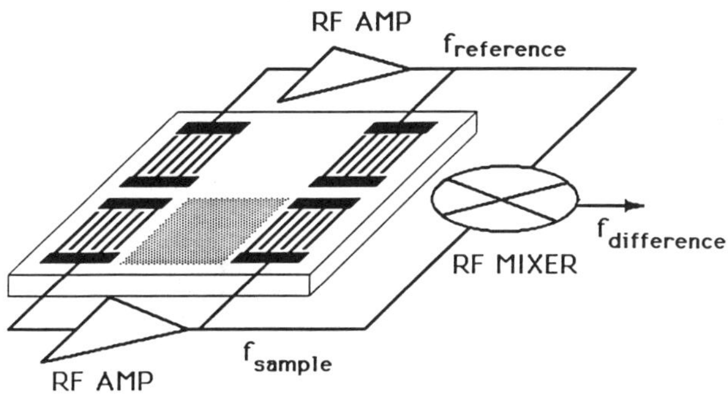

Figure 3. Dual SAW delay line sensor configuration.

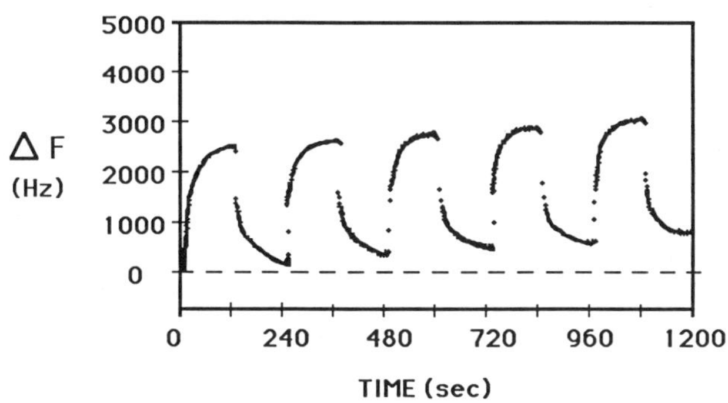

Figure 4. Typical response of a PEI coated 158 MHz SAW sensor to consecutive pulses of 1000 mg/m^3 water vapor at room temperature.

SAW device was exposed to repeated pulses of 1000 mg/m^3 water vapor at room temperature. As the sensor is exposed to vapor, the water is sorbed into the polymer coating thereby increasing the mass loading on the SAW oscillator and increasing the difference frequency between the coated and uncoated SAW oscillators. Upon removal of the vapor, the frequency difference decreases and returns to the original baseline value. Such reversibility of response is essential for continuous monitoring applications. The repeatability of response is very good, typically better than 2% at this concentration.

The response time of the sensor is fairly rapid, achieving about 80% of its equilibrium response in about 1 minute. The sensor response time is usually concentration dependent, with higher concentrations producing equilibrium responses in a few seconds or less and lower concentrations taking a few minutes or more to respond. The response time is affected by many factors including the dead volume of the detector cell, coating morphology, coating glass transition temperature, and thickness. Since vapor diffusion times into the coating are inversely related to the film thickness, thinner coatings can significantly improve response times. Higher frequency SAW devices are attractive because they can function well with thinner films and smaller cell volumes.

A portion of the sorption isotherm for the PEI/water sensor is shown in Figure 5. The sensor exhibits a reasonably linear response over a concentration range of more than three orders of magnitude, with a lower limit of detection less than 100 mg/m^3. Another 158 MHz dual SAW device was coated with "fluoropolyol" (FPOL) a fluorinated epoxy prepolymer described in (7). As with the PEI coated device, the amount of coating applied to the SAW sensor resulted in a frequency shift of approximately 250 KHz. The sorption isotherm for dimethylmethylphosphonate (DMMP) is presented in Figure 6. The sensor exhibits a great initial sensitivity, as indicated by the steep slope of the curve at lower concentrations, but the sensitivity decreases at higher concentrations, probably because the sorption sites in the coating are becoming filled. The detection limit for the FPOL coated sensor to DMMP is about 0.2 mg/m^3. The type of sorption isotherm and the absolute sensitivity of the device is clearly dependent on the particular coating selected and the vapor to be detected.

Temperature plays an important role in the behavior of SAW vapor sensors. The sensitivity of the SAW device itself to temperature can be minimized by selecting a

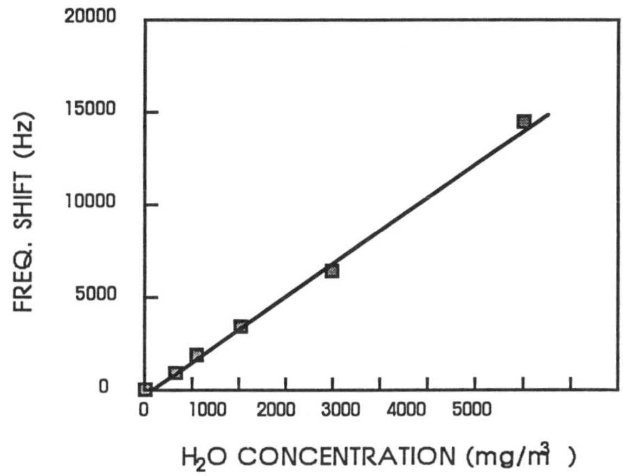

Figure 5. Sorption isotherm for water vapor on PEI coated 158 MHz SAW sensor at room temperature.

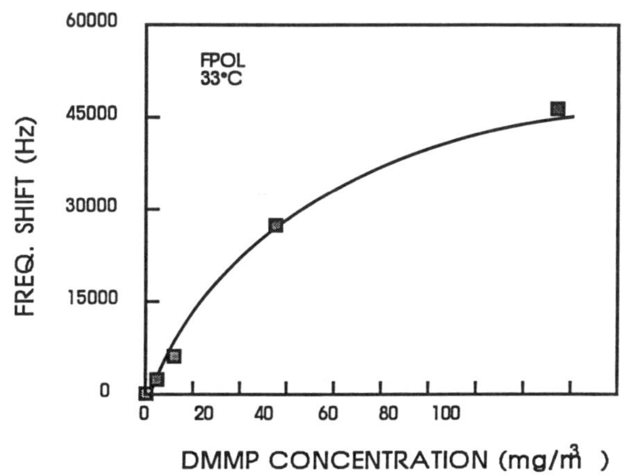

Figure 6. Sorption isotherm for DMMP vapor on FPOL sensor at 33 degrees C.

piezoelectric substrate that has a small temperature coefficient (such as ST-quartz) and by using a dual configuration as mentioned previously. Even after taking these steps, temperature drift can still remain, as the slowly shifting baseline in Figure 4 illustrates. Reduction of SAW oscillator temperature drift by improved sensor packaging and design is an active area of research. It is expected that dual SAW sensor temperature sensitivities of less than 1 Hz/deg.C are readily achievable. Current 158 MHz dual SAW sensors exhibit temperature drifts on the order of 100 Hz/°C. This compares with a typical noise level of about 16 Hz for the dual 158 MHz sensor. Thus, until improved temperature compensation techniques are developed, SAW sensor temperatures should be held constant to ± 0.2 deg.C for best results.

Temperature also affects SAW vapor sensor performance due to the nature of the vapor/coating interaction. The rate of diffusion of vapor into the coating increases with temperature as shown in Figure 7. This translates into more rapid response times at higher temperatures. The equilibrium distribution of vapor into the coating decreases with temperature, however, and the corresponding response is considerably reduced. These temperature effects can be dramatic, resulting in response variations of greater than 10% for a 5 deg.C temperature change. Because of this, temperature control of the SAW sensor may be required to optimize the response time, sensitivity, and precision of the vapor concentration measurement.

SAW Array Sensors

Prior work has shown that empirically determined solubility properties can be very useful in explaining and predicting the magnitude of mass change expected from a particular vapor/coating interaction (10,11). This knowledge can be used to select coatings having enhanced sensitivity to a given vapor. An example of this approach is presented in Figure 8 which shows the response of an FPOL coated sensor to a variety of vapors (7). The target vapor of interest in this case is dimethylacetamide (DMAC), and the sensor exhibits an order of magnitude greater sensitivity to DMAC compared to the other vapors tested. For example, the selectivity ratio for DMAC vs tributylphosphate (the second most sensitive response) is (358/11.9), or ·30, and the selectivity ratio for DMAC vs water vapor is 4475.

While such dramatic selectivity is achievable in some situations, it is unlikely that unique selectivity for a single vapor will be obtained. In addition,many sensor applications call for sensitivity and

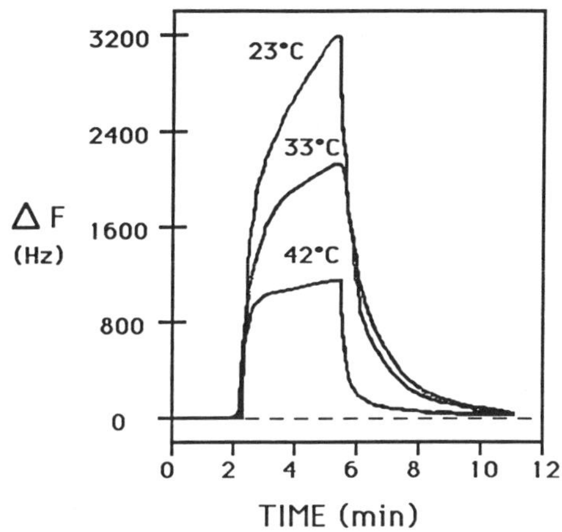

Figure 7. Effect of temperature on the response time and response magnitude of FPOL coated 158 MHz SAW sensor.

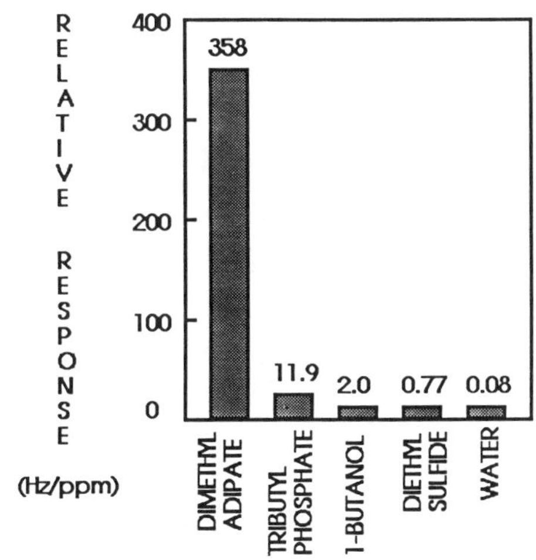

Figure 8. Sensitivity of FPOL to a variety of vapors. Numbers correspond to response sensitivity in Hz/ppm.

selectivity for more than one vapor. Such versatility of response is not possible with a single coated sensor. For these applications, a sensor array offers an attractive alternative.

A complete chemical vapor detection system using a 4-SAW sensor array is illustrated in Figure 9. The required RF amplifiers and support electronics for each SAW device are connected together with the SAW package on a small, credit card sized, printed circuit board to form a complete vapor sensor subsystem. Electrical power, output signal and vapor connections are made to each of the four boards in the array that is housed in a small (~1/3 cu.ft.) box containing a regulated AC power supply, 4-channel microcomputer controlled frequency counter, sampling pumps, solenoid valves, and associated plumbing. Every two seconds, the four channel frequency counter measures and reports the difference frequency from each of the four SAW sensors. Data are transmitted over an RS232C serial communications line at 9600 baud. The solenoid valves are used to control the introduction of ambient air to the sensors. Charcoal and molecular sieve traps are included to purge the air of organic vapors and water to provide a "zero" air source to the array. Periodically, the scrubbed air stream can be passed over the SAW array to establish a zero vapor concentration reference. In a typical sampling cycle, the sensors are exposed to the scrubbed air stream followed by a period of exposure to ambient air. The difference between the array response to ambient air and scrubbed air is then calculated and recorded. Sequencing of the valves and the data acquisition process is controlled by the system microcomputer.

The 158 MHz SAW devices used in this array are illustrated in Figure 10. These ST-quartz devices have an acoustic wavelength of 20 µm. Transmitting and receiving IDTs are made of aluminum having 75 and 100 finger pairs respectively, and an acoustic aperture of 70 wavelengths (i.e. 1400 µm). The devices are epoxied onto a 12 pin gold TO-8 header and have bonded gold-wire connections. Each delay line has an active surface area of about $8mm^2$, and the sensor package has a cell volume of about 0.12 cm^3. Experimental studies have demonstrated that these devices produce a frequency shift of about 365 Hz when perturbed by a surface mass loading of 1 nanogram, with a typical difference frequency noise level of less than 16 Hz RMS over a one second measurement interval. This results in a signal-to-noise ratio of about 23 to 1 for a 1 nanogram mass change.

Some typical array sensor responses are presented in Figure 11. The four 158 MHz devices in the array

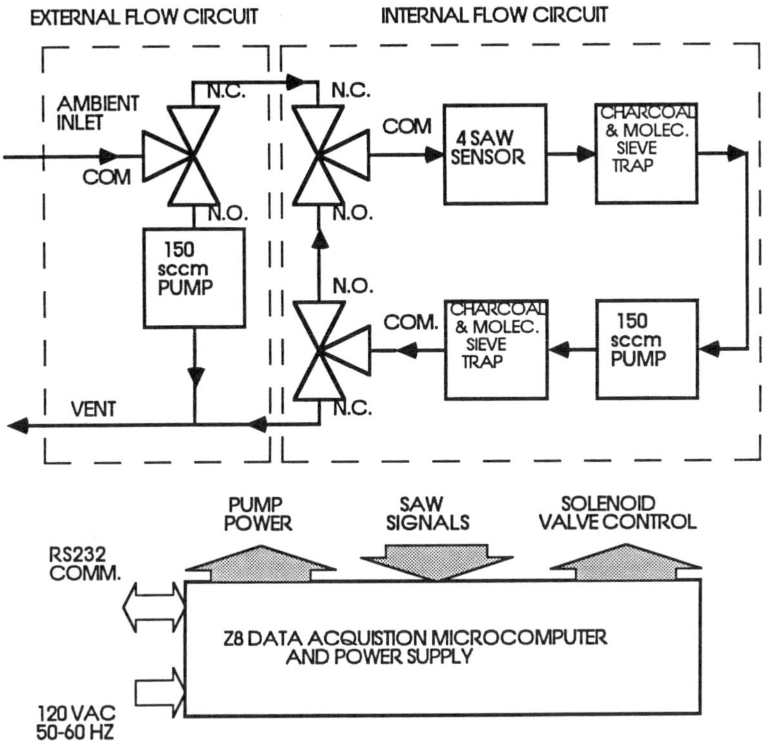

Figure 9. Four SAW array sensor system configuration.

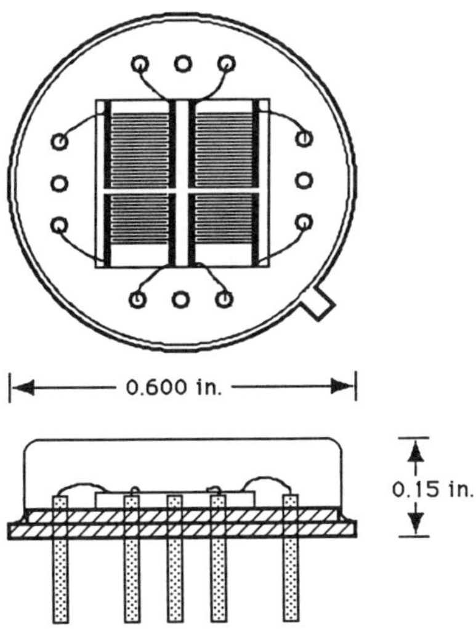

Figure 10. Layout of the 158 MHz dual SAW sensor "chip".

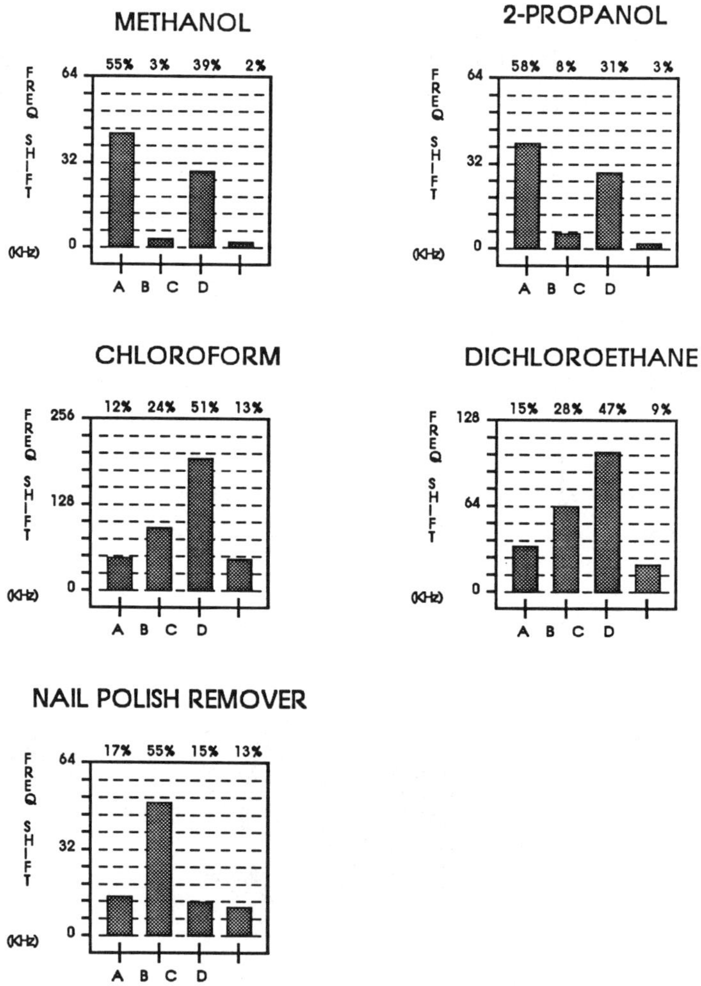

Figure 11. Typical response patterns obtained using the 4-SAW array to different vapors.

were coated with materials of varying selectivity using a spray deposition method. Each coating was applied until it produced a shift of approximately 250 KHz in the resonant frequency of each SAW device. Channel A was coated with poly(ethyleneimine) (PEI); channel B was coated with fluoropolyol (FPOL); channel C was coated with ethyl cellulose (ECEL); and channel D was coated with a chloroform soluble extract of TENAX GC.

When exposed to various organic vapors in air, the 4-SAW array produces a response pattern, or fingerprint, that is characteristic of that vapor. Analysis of the response pattern by pattern recognition techniques can produce both qualitative and quantitative information about the vapor being sampled (12). Inspection of the response patterns shown in Figure 11 indicates that similar vapor produce similar patterns. When exposed to two alcohols (methanol and 2-propanol), the array produced patterns that are very similar in appearance, suggesting that these two vapors belong to the same class of compounds. A close inspection of the patterns will reveal that they are not identical and indeed the differences are large enough to permit reliable discrimination of one alcohol from the other. Likewise, the halogenated hydrocarbons (chloroform and dichloroethane) produce patterns that are very similar to one another, yet are quite different from the response patterns for the alcohols.

The SAW sensor array can be used to detect and identify a particular mixture of vapors, such as nail polish remover, but determination of the constituents of the mixture (e.g. acetone, ethyl acetate, water, and "fragrance") is probably beyond the capability of current chemometric algorithms. Nevertheless, when used in conjunction with a library of array responses and a relatively simple pattern recognition algorithm such as k-nearest neighbors analysis, it is possible to identify vapors or combinations of vapors that the instrument has been previously trained to recognize. The array sensor approach combined with pattern recognition offers very attractive capabilities for vapor detection and alarm applications where the significant background interferents are known, or in process control applications where it is necessary to know if conditions in the process stream have changed. Both scenarios do not require detailed vapor composition analysis to provide the desired chemical information.

Pattern recognition techniques can be used not only to aid in the identification of patterns for vapor detection, but also to identify relationships between sensor response and various coating properties. This approach has been used to correlate coating responses with solubility properties of the vapors (13) and with chemical functionalities of the coatings (14,15). This information can be used in the selection and design of coatings with increased sensitivity and selectivity for a given vapor sensor application.

Future Developments

Progress in the development of SAW vapor sensor devices will proceed in several directions. Attempts to reduce SAW sensor noise will increase the emphasis on the use of SAW resonators and more sophisticated RF amplifier designs. Improved packaging schemes will be explored to reduce SAW temperature drift. Much of the temperature drift exhibited by dual SAW sensors is caused by unsymmetrical stresses imposed on the SAW devices as a result of thermal expansion of the package. Miniature temperature controllers for the SAW device may offer a brute-force fix for this problem.

Greater sensitivity will be obtained by using higher frequency SAW devices. Unpublished research conducted in our laboratory on 700 MHz SAW resonator vapor sensors supports this conclusion. Operation of very high sensitivity SAW vapor sensor devices at frequencies near 1.1 GHz should be possible in the near future. Greater sensitivity may also be obtained from clever new device structures that use different types of acoustic waves. White et.al. have reported on the use of a thin membrane Lamb wave device that exhibits a theoretical sensitivity that is an order of magnitude greater than that of a Rayleigh SAW of the same frequency (16).

The development of new selective coatings will continue to yield the greatest overall improvements in SAW vapor sensor performance. The best organic polymer stationary phase type coatings used on SAW sensors are only able to trap about 1% of the vapor molecules presented to them. Clearly, there is much room for improvement. In addition, the list of vapor analytes for which effective SAW coatings have been found will continue to grow.

This paper has dealt exclusively with SAW sensors that exploit the mass sensitivity of the device to achieve chemical vapor detection. Schemes to exploit the SAW sensitivity to coating conductance changes (17) or elastic modulus changes should afford new opportunities for imaginative chemical vapor sensor designs. Finally, the field of liquid phase chemical analysis may also yield to surface acoustic wave devices that utilize plate waves and horizontally polarized shear waves to minimize acoustic losses in the liquid (18).

Literature Cited

1. White, R.M.; Voltmer, F.W., Appl. Phys. Lett. 1965 7,314.
2. White, R.M., Proc. IEEE 1971 58(8), 1238.
3 Wohltjen, H.; Dessy, R.E., Anal. Chem. 1979 51(9), 1458.
4. Wohltjen, H.,Sens. & Actuators 1984 5, 307.
5. Slobodnik, A.J.; Delmonico, R.J.; Conway, E.D., eds., Microwave Acoustic Handbook, Vol. 2 1974 Air Force Cambridge Research Laboratory Report #AFCRL TR-74-0536, p 197.
6. Wohltjen, H.; Snow, A.W.; Ballantine, D.S.,Proc. of Int'l. Conf. on Sold State Sens. & Actuators 1985, p 66.
7. Wohltjen, H.; Snow, A.W.; Barger, W.R.; Ballantine, D.S., IEEE Trans. Ultrason., Ferroelectrics & Freq. Contr. 1987 UFFC-34, 172.
8. Ristic, V.M. Principles of Acoustic Devices; Wiley-Interscience: New York, 1983; p 40.
9. Martin, S.J.; Schweizer, K.S.; Schwartz, S.S.; Gunshor, R.L., Proc. of IEEE Ultrason. Symp. 1984, p 207.
10. Zellers, E.T.; White, R.M.; Rappaport, S.M.; Wenzel, S.W., Proc. of Int'l. Conf. on Solid State Sens. & Actuators 1987,
11. Zellers, E.T.; (this proceedings).
12. Zaromb. S.; Stetter, J.,Sens. & Actuators 1984 6, 225.
13. Grate, J.W.; Snow. A.; Ballantine, D.S., Jr.; Wohltjen, H.; Abraham, M.H.; Mcgill, A.; Sasson,P. Anal. Chem. 1988, 60, 869.
14. Ballantine, D.S. Jr.; Rose, S.L.; Grate, J.W.; Wohltjen, H. Anal. Chem. 1986 58, 3058.
15. Carey, W.P.; Beebe, K.R.; Kowalski, B.R.; Illman, D.L.; Hirschfeld, T., Anal. Chem. 1986 58, 149.
16. White, R.M.; Wicher,P.J.; Wenzel, S.W.; Zellers, E.T., IEEE Trans. Ultrason., Ferroelectrics & Freq. Contr. 1987 UFFC-34, 163.
17. Ricco, A.J.; Martin, S.J.;Zipperian, T.E.,Sens. & Actuators 1985 8, 333.
18. Ricco, A.J.; Martin, S.J. Appl. Phys. Lett. 1987 50(21), 1474.(and this proceedings)

RECEIVED March 24, 1989

Chapter 12

Selective Measurement of Vinyl Acetate Vapor Using a Coated Surface Acoustic Wave Oscillator

Applications to Industrial Hygiene

Edward T. Zellers

Department of Environmental and Industrial Health, School of Public Health, University of Michigan, Ann Arbor, MI 48109–2029

A coated surface-acoustic-wave (SAW) sensor capable of real-time, selective measurement of vinyl acetate vapor in the presence of several olefin and non-olefin co-contaminants is described. The coating film employed consists of the solid platinum-ethylene π-complex, trans-$PtCl_2$(ethylene)(pyridine), occluded in a polyisobutylene matrix. Exposure to vinyl acetate results in displacement of ethylene from the complex and formation of the vinyl acetate-substituted complex. Subsequent regeneration of the original reagent is possible by treatment with ethylene gas, in situ. A lower detection limit of 5 ppm of vinyl acetate is achieved for operation at 46°C. The industrial-hygiene applications of the sensor are discussed.

A major problem faced by occupational health professionals is the lack of sufficient data to adequately characterize worker exposures to toxic chemicals. As a rule, decisions made regarding the potential hazard posed by a given air contaminant or compliance with established health standards are based on a minimum number of air-monitoring measurements. While it is well-known that personal monitoring (i.e., collecting air samples in the breathing-zone of the worker) provides the most accurate measure of exposure, the influence of spatial and temporal variability can lead to contaminant concentrations ranging over several orders of magnitude even for well-controlled industrial processes (1). The ability to accurately describe the resultant exposure distributions for an individual or group of workers on the basis of a few measurements is questionable. Without accurate exposure estimates occupational physicians and epidemiologists cannot effectively correlate exposure levels with adverse health effects or establish meaningful risk estimates. This, in turn, undermines efforts by regulators to establish appropriate exposure limits for the protection of worker populations.

Several factors contribute to this situation not the least of which is the dependence on conventional sampling methods for contaminant collection. These methods employ a portable sampling

0097–6156/89/0403–0176$06.00/0

pump to draw air through a small, lapel-mounted collector such as a tube filled with a granular adsorbent. Subsequent analysis of trapped contaminants yields the time-averaged air concentration to which the individual was exposed. Virtually all of the sampling methods developed by the National Institute for Occupational Safety and Health (NIOSH) are of this type (2). Unfortunately, the costs, both in labor and capital, associated with operating and maintaining sampling equipment, chemically analyzing collected samples, and cataloging exposure data tend to discourage the collection of large numbers of samples. As a result, both the size and quality of the exposure data base are reduced. The recent proliferation of passive monitoring techniques has been motivated, in part, by the prospect of reduced operating costs. Yet, even with these devices, costs of analysis and data management remain high.

In recent years, there has been a marked increase in the number and sophistication of direct-reading instruments for monitoring toxic chemicals in the workplace (3). Compact instruments are now available that combine gas detection capabilities with on-board microprocessors to provide several desirable operating features: they are small enough to collect breathing-zone measurements, they operate continuously, they can be equipped with alarms to provide immediate warning of excessive exposure levels, and they provide modest data reduction functions. In addition, some of the instruments incorporate datalogging electronics that permit collection of individual exposure histories over the entire sampling interval. Direct computer interfacing also eases the data management burden.

While these features greatly facilitate monitoring worker exposures, most instruments suffer from limitations associated with the types of sensing components employed. For example, the majority of the truly portable instruments used for monitoring organic gases and vapors detect contaminants by catalytic combustion, semiconductor charge-transfer, or photo-ionization. The sensitivity of these detectors is generally satisfactory, but measurement of specific organic compounds or classes of compounds is not possible: the presence of even simple mixtures of organic contaminants can preclude an accurate determination of the exposure hazard due to any of the mixture components. Portable instruments employing electrochemical detection or impregnated-paper-tape photometry provide improved selectivity, but their use is limited chiefly to reactive inorganic gases. Thus, while the "intelligence" of available instruments is quite advanced, their overall utility is compromised by their limited sensing capabilities.

Microfabricated chemical sensors offer the potential to overcome these limitations in industrial-hygiene monitoring instruments. The small size, low power requirements, and low production costs of these devices will facilitate their incorporation into miniaturized instrumentation suitable for real-time personal-exposure monitoring. The key challenge to their successful implementation, however, is designing sensors that respond selectively to a single target chemical or class of chemicals.

Selectivity and reversibility are often competing characteristics of a chemical sensor. Whereas low-energy, spontaneously reversible interactions (e.g., physical adsorption or absorption) are desirable for prolonged sensor life, they generally lack a high degree of selectivity. Higher-energy interactions (e.g., bond formation) are more selective but also tend to be less reversible: a sensor employing this type of interaction would eventually become saturated. The use of sensor arrays in conjunction with pattern recognition methods has been suggested as a way to retain a high degree of selectivity using reversible sensor-analyte interactions (4). In this case, the individual sensors need only be partially selective as long as the collective pattern of responses permits differentiation of the target analyte(s) from potential interferents. Using multiple sensors also adds versatility since the same array could be used for detecting several different species. However, the ultimate discriminating power of the array, particularly for mixtures of structurally similar chemicals, will still depend on the selectivity of the individual sensors. The approach taken in the design of chemically sensitive sensor layers will thus be dictated by the constraints of the particular application. For industrial-hygiene monitoring over a single workshift (i.e., 8-10 hr) it is feasible to consider non-reversible interactions since the sensor could be replaced or regenerated between sampling periods. For longer-term monitoring reversibility may be the more important consideration.

The research described here concerns one type of sensor, the coated surface-acoustic-wave (SAW) oscillator. In this device, a high-frequency mechanical wave propagates along the surface of a piezoelectric substrate which has been coated with a chemically-sensitive film. Using a simple feedback amplifier to connect the input and output transducers of the device results in oscillation at a fixed resonant frequency. Small changes in the mass or elastic properties of the coating film caused by interactions with gases or vapors result in a change of the velocity of the propagating wave, and hence of the oscillator frequency.

Equation 1, derived from the perturbation theory of Auld (5,6), gives an approximate expression for the change of the oscillator frequency upon deposition of a thin, non-conducting, low-modulus, isotropic coating film,

$$\Delta f = (k_1+k_2) f_o^2 h\rho \tag{1}$$

where Δf is the change of frequency (Hz), k_1 and k_2 are negative substrate-dependent constants of the order 10^{-8} (m^2/kg-sec), f_o is the resonant frequency (Hz), ρ is the density of the coating film (kg/m^3), and h is the coating-film thickness (m). Since the quantity $h\rho$ is equivalent to the mass per unit area, to a first approximation the change of frequency upon deposition of a thin coating film is a linear function of the mass per unit area of the coating material. This relationship can be used to estimate the mass or thickness of the deposited film as well as to correlate changes in the coating mass with the air concentrations of interacting gases and vapors. Note that the sensitivity of the device to a change of coating mass depends on the square of the initial operating frequency, f_o. For a 30-MHz oscillator on ST-quartz with an active area of 1.5 cm^2, such as that used here, Equation 1 predicts a change of 840 Hz for every μg change of surface mass.

For a given operating frequency, the coated-sensor response is a function of the strength of the vapor-coating interaction. A key advantage of the SAW sensor compared to several other microsensors is the versatility imparted by the ability to use a wide range of coating materials. Coated-SAW sensors have been used successfully for the detection of a range of organic and inorganic gases and vapors, including, NO_2 (7,8), H_2 (9), H_2O (10), H_2S (11), SO_2 (12), dimethyl acetamide (13), and cyclopentadiene (14) by using coating materials with strong affinities for the particular gas-phase analytes. SAW-sensor arrays employing partially selective polymer coatings have also been investigated for detecting organic vapors (4,15). Detection limits in the part-per-billion concentration range have been reported by several authors (11-13). However, these studies have generally focused on the detection of a single target compound; very limited testing of selectivity among relevant potential interferences has been performed.

In this chapter, we describe a coated-SAW sensor capable of real-time measurement of vinyl acetate (VA). VA is an industrially important chemical used as a monomer in the production of thermoplastic polymers, adhesives and coatings (16,17). The principal health effects associated with exposure to airborne VA are irritation of mucous membranes, lacrimation, and central nervous system impairment (18). The American Conference of Governmental Industrial Hygienists (ACGIH) recommends Threshold Limit Values (TLVs) for VA of 10 parts-per-million (ppm) based on an eight-hour time-weighted-average (8-hr TWA), and 20 ppm for short-term (15-min TWA) exposures (19).

Characterizing occupational exposures to VA is complicated by the presence of comonomers and organic-solvent co-contaminants used with VA in industrial processes. For example, VA is often combined with vinyl chloride or one of several alkyl acrylates to modify the physical properties of the final polymer product. Common organic solvents used in VA polymer production or coating formulations include ethyl acetate and methanol (16,17). Given the limited selectivity of currently available portable industrial-hygiene instruments, we were interested in developing a SAW chemical sensor capable of continuous detection of VA in the presence of its likely industrial co-contaminants.

The reagent used in the sensor coating, *trans*-$PtCl_2$(ethylene)(pyridine), is an example of a class of square-planar metal-olefin π-complexes wherein the central metal atom and the four symmetrically bound ligands occupy the same spatial plane. The unique behavior of these compounds in ligand-substitution reactions has been the subject of extensive investigation (20,21). Of particular interest here is the observation that nucleophilic attack by olefins results only in displacement of the initially bound olefin (ethylene in this case) with retention of the original configuration of the ligands in the olefin-substituted product.

Pyridine, as well as other amines, are thus referred to as _trans_-directing ligands.

Bonding between the metal and the olefin in these complexes involves donation of electron density from the olefin to the metal in a σ-type bond as well as "back-bonding" from the metal to the olefin through a π-type bond. Introduction of electron-releasing groups onto the olefin strengthens the σ-bond and weakens the π-bond with respect to ethylene, and electron withdrawing substituents strengthen the π-bond and weaken the σ-bond. The contribution of each bond type to the net strength of the metal-olefin interaction depends on the energies, symmetries and overlaps of the orbitals on both the metal and the olefin (_22_). These factors affect the ability of a given olefin to displace ethylene from the complex, as well.

The _trans_-amine ligand also influences the rate and mechanism of the reaction. The relative σ-bonding strength of the amine is determined chiefly by its Lewis basicity. For two ligands that are in a _trans_ configuration about the metal there is, roughly speaking, a competitive donation of electron density to the metal, with the stronger σ-donor reducing the weaker σ-donor's bond to the metal. When the amine substituent is pyridine or a substituted pyridine, reaction proceeds through an associative (i.e., bimolecular) transition state (_23_). In contrast, when the amine substituent is aniline, a combination of associative and dissociative (i.e. unimolecular, or solvent-assisted) pathways are involved (_24_). The presence of traces of coordinating solvents (e.g., ethanol) enhances the dissociative pathway. The difference in reaction mechanisms is attributable to the ability of vacant π-orbitals on the pyridine to accept excess electron density that accumulates on the metal during the course of the reaction. Such an interaction is not possible with aniline due to the unavailability of π-accepting orbitals. The predominance of the bimolecular mechanism operating in the pyridine-containing complexes strongly favors their choice for sensor coatings since the existence of a solvent-assisted reaction pathway would lead to changes of the reaction rate in the presence of other solvents and a reduction in selectivity.

As shown in Figure 1, the entering and leaving olefins, and the _trans_-pyridine ligand comprise the base of the trigonal bipyramidal transition state. The coplanarity of the metal and these three ligands allows for translation of the electronic effects from one ligand to the other via the platinum 5d-orbitals during the reaction. Steric factors can also have a strong effect on the rates of these reactions: crowding in the transition state, for example, by bulky substituents on the amine or either of the olefins, can raise the transition state energy and reduce the reaction rate.

Experimental verification of these effects can be found in one study of the substitution of optically active 2-methyl-2-butene (mbn*) in the compound $PtCl_2$(mbn*)(4-X-pyridine) by non-optically active 2-methyl-2-butene (mbn), _cis_-1,2-dichloroethylene (dce), and 2,3-dimethyl-2-butene (dmb) (_23_). For a given amine ligand, the rate of reaction was in the order mbn > dce > dmb, with the reaction rates decreasing by a factor of nearly 200 in going from mbn to dmb. The slower rate for the dce relative to mbn was attributed to the electron-withdrawing chlorine atoms on the dce which reduce its nucleophilic strength. The slower rate observed for dmb relative to both mbn and dce was attributed to the steric hindrance of the additional methyl group in the dmb (note: the _cis_ configuration of the dce apparently reduces the steric effects of the chlorine atoms). For a given olefin, the rate of substitution increased with the electron-withdrawing strength of the 4-substituent on pyridine (i.e., as the basicity of the pyridine decreased).

We wanted to determine whether the difference in olefin-substitution reaction rates observed in solution would translate into selective detection of certain olefins using the solid organoplatinum complex on the surface of a SAW sensor. The results described here are for the pyridine-containing complex only. The effect of using various substituted-pyridines on the reactivity of these complexes is currently being explored.

Experimental

Two matched 30-MHz SAW delay-line oscillators were fabricated on a 2-inch ST-quartz wafer. Each IDT consisted of 50 pairs of evaporated Au/Cr electrodes (2600 Å thick) having equal widths and spaces of 25 μm and an acoustic aperture of 6000 μm. The distance between the IDT centers was approximately 2 cm, resulting in an active area of 1.5 cm^2 for each device. A variable series inductor was used to cancel the reactive component of the impedance in each IDT. Two cascaded wide-band amplifiers provided the gain necessary to maintain oscillations in each device. A mixer and low-pass filter were used to obtain the difference frequency of the oscillators, which was monitored with a digital frequency counter (Hewlett-Packard 5384A) and logged on a personal computer (Hewlett-Packard Vectra).

For exposure experiments the sensor was placed in a 0.24-L stainless-steel exposure chamber and electrical connections to the external circuitry were made through coaxial connectors in the floor of the chamber. Heating tape was used to control the chamber temperature, and a thermocouple was used to monitor the temperature at the surface of the sensor. Test atmospheres of the solvent vapors were generated by passing a low flow of nitrogen gas through a bubbler containing the liquid solvent and then into a dilution-air stream. For vinyl chloride (Matheson), a syringe pump was used to introduce the pure gas into the system. A flow of 8 L/min was maintained through the exposure chamber at all times with an in-line rotameter. This flow rate corresponds to a theoretical mixing time of 8 s (_25_). The remainder of the airflow passed through an infrared gas analyzer (Foxboro, MIRAN 1A) used for verification of solvent-vapor concentrations. The relative humidity was kept between 17 and 25% for all experiments. Dilution-air flow, temperature, and humidity were controlled with a Miller-Nelson Research HCS 301 control unit.

The trapping reagent, _trans_-$PtCl_2$(ethylene)(pyridine), was synthesized by adding one molar equivalent of pyridine (Aldrich) to an aqueous solution of Zeise's salt, $KPtCl_3$(ethylene) (Alfa), resulting in immediate formation of a solid precipitate (21). Recrystallization from methylene chloride/petroleum ether yielded yellow prisms, m.p.=113°C (dec); elemental analysis, percent found (theor.): C, 22.5(22.5); H, 2.5(2.4); N, 3.7(3.7). The VA-substituted compound was synthesized by adding an excess of VA (Aldrich) to a methylene chloride solution of $PtCl_2$(ethylene)(pyridine) and refluxing for 1.5 hr (22). Addition of excess petroleum ether yielded the pure orange-yellow solid, m.p.=88°C (dec). Both of the complexes are soluble in aromatic and polar organic solvents.

Earlier experiments revealed that combining the solid organoplatinum complex with the amorphous, rubbery polymer, polyisobutylene (PIB) (Scientific Polymer Products), resulted in more uniform coating films and more consistent results than those obtained using the solid complex alone. Microscopic examination of deposited films showed a reasonably uniform distribution of the organoplatinum crystals in the polymer film. PIB was found to absorb VA and other organic vapors rapidly and reversibly, with the amount absorbed being directly proportional to the vapor concentration (i.e., a linear sorption isotherm). Solutions containing 6 mg/mL of each component in 3:1 toluene:hexane were cast over the entire oscillator using a pipette. The mass of coating material deposited was then estimated using the net frequency shift of the oscillator upon evaporation of the casting solvent (via Equation 1). The reference device was coated only with PIB so that measuring the difference frequency would cancel the response due to sorption/desorption of the vapors by the polymer on the sensing oscillator (see Figure 2).

The conversion of $PtCl_2$(ethylene)(pyridine) to $PtCl_2$(VA)(pyridine) was confirmed by IR spectrophotometry. A thin-film of the coating material was deposited on a KBr disc and exposed to a saturated atmosphere of VA. IR spectra taken periodically over the course of the exposure indicated complete conversion to the VA-substituted product after about two hours.

Results and Discussion

The interaction of VA with the coating film can be considered a two-step process as shown in Equations 2 and 3. Upon exposure, the VA partitions into the coating film (on each oscillator) to an extent determined by the equilibrium air/film partition coefficient, K (the designations (g) and (p) denote gas-phase and polymer-phase VA, respectively). For the sensing oscillator, the sorbed VA then reacts with the organoplatinum compound at a rate determined by the effective VA concentration at the surface of the solid and the rate constant for the reaction, k. Since ethylene gas is released from the surface, the reaction proceeds to the right as long as VA is present (note, however, that this is a reversible reaction which allows eventual regeneration of the ethylene complex upon exposure to pure ethylene). As the VA reacts, its concentration in the polymer decreases proportionally, causing more VA to be sorbed by the polymer. While for every molecule of VA added to the surface there is a loss of one molecule of ethylene, the higher molecular weight of the VA results in a

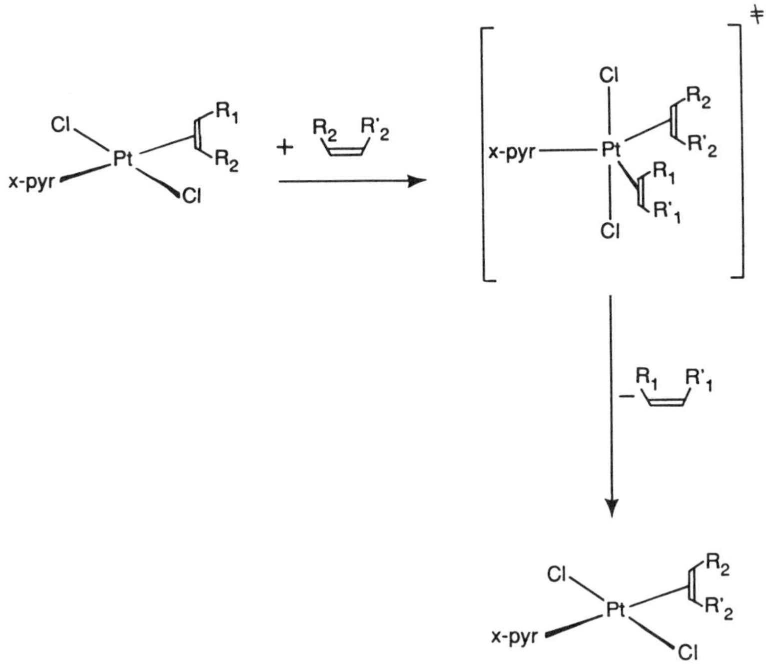

Figure 1. General olefin-substitution reaction pathway for $PtCl_2$(olefin)(x-pyridine) showing 5-coordinate associative transition state.

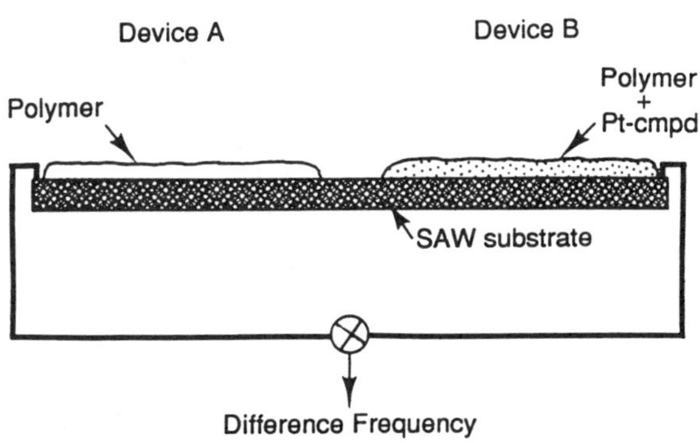

Figure 2. Dual oscillator configuration employed for measurement of vinyl acetate. The reference device (Device A) is coated only with PIB and the sensing device (Device B) is coated with a 1:1 mixture of PIB and $PtCl_2$(ethylene)(pyridine).

continual increase in surface mass and, consequently, a continual reduction in the frequency of the SAW oscillator (see Equation 1). By measuring the difference frequency of the two oscillators the frequency shifts due to sorption and desorption of VA are masked and the measured frequency shift is due only to the chemical reaction. Since the rate of frequency change is related to the instantaneous VA air concentration, real-time concentration measurement is possible.

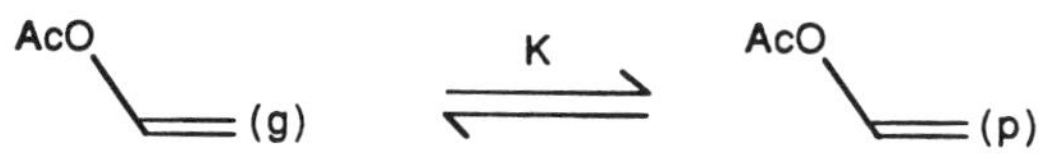

(2)

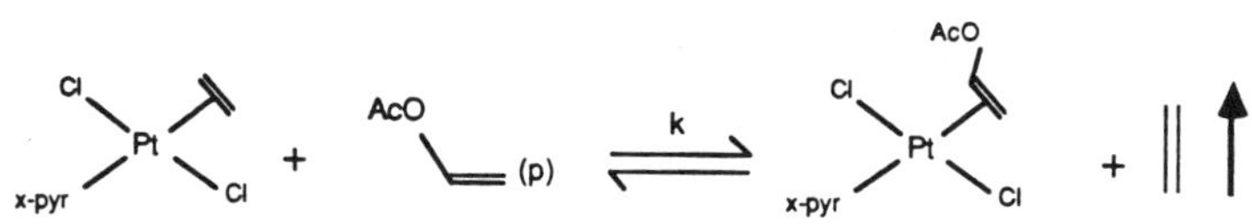

(3)

In Figure 3 the rate of frequency change is plotted versus the VA concentration at several temperatures. Each point represents the mean response for exposures of 10-20 min duration at each concentration (standard deviations are indicated by the error bars in the figure). The time required to establish a steady-state rate of frequency change was less than 60 s for all VA concentrations. The response to VA increases monotonically over the entire concentration range, but there is a continual decrease in sensitivity. This trend is consistent with the heterogeneous nature of the reaction: since there is a limited amount of available surface area on the solid, the surface becomes more saturated as the VA concentration increases. Eventually, a full monolayer would be formed at the surface and further increases in concentration would have no effect on the reaction rate provided multilayer adsorption was not important. It is clear from Figure 3 that this point has not been reached under the experimental conditions, although the slopes at the lower temperatures appear to be approaching a plateau at the higher concentrations.

Since the rate of frequency change is proportional to the rate of reaction with the organoplatinum complex, it is possible to compare the sensor response to theoretical models of gas-solid reactions. We find that the data in Figure 3 are described very well by a power-law kinetic model: a plot of the logarithm of the sensitivity versus the logarithm of the VA vapor concentration gives a straight line at all temperatures (r^2>0.98). The power-law model, which can be viewed as the reaction-rate analogue of the Freundlich adsorption model, is strictly applicable only to

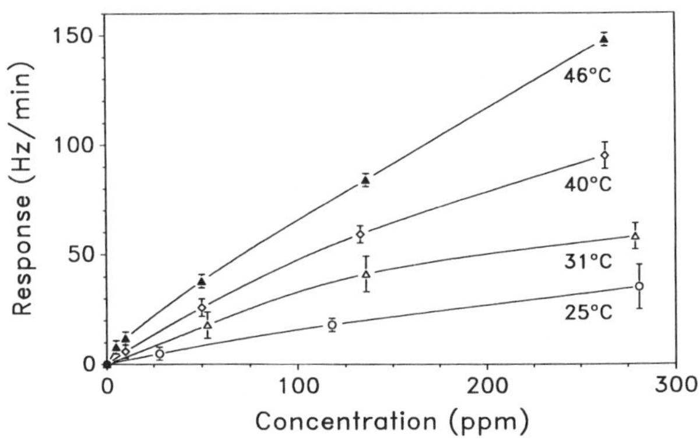

Figure 3. Rate of frequency change (response) vs. vinyl acetate concentration for operation at each of several temperatures.

catalytic reactions (26). While the trapping reaction here is not catalytic, for the short-term exposures conditions employed here the assumption of approximate catalytic behavior appears reasonable.

The Arrhenius temperature dependence of the reaction is reflected in the sensor response as shown in Figure 4 for exposure to 260 ppm of VA. The experimentally determined activation enthalpy was 12 kcal/mole which is similar to values reported for the reactions involving substituted butenes described above (23). This suggests that the surface reaction is the rate-limiting step in the overall interaction of VA with the coating, although, diffusion of the vapor through the solid may also be important (i.e., the diffusion coefficient of VA would be expected to show a similar temperature dependence). Apparently, transport of the VA through the polymer is fast relative to the chemical reaction since the rate increases even though the equilibrium mass of VA sorbed by the polymer decreases exponentially with increasing temperature.

At 25°C, a concentration of 30 ppm gives a signal-to-noise ratio of only 2. As shown in Figures 3 and 4, however, raising the temperature significantly improves the sensitivity. For operation at 46°C, a lower detection limit of 5 ppm is achieved (note: we define the limit of detection as 3σ/sensitivity, where σ (=3Hz/min) is the standard deviation of the noise). This is clearly adequate for detection of VA below both the 8-hr and 15-min ACGIH-TLV values (10 and 20 ppm, respectively). Of course, further improvements in sensitivity would be expected by employing an oscillator that operates at a higher frequency (see Equation 1).

For the relatively short-term exposures examined here, the response to VA remains constant even at high concentrations. But continued exposure eventually leads to a reduction in sensitivity due to the consumption of the trapping agent. The production of an increasingly thick product layer through which the VA (and the ethylene) must diffuse may also be a contributing factor (27). However, following exposure to VA, the sensor could be treated with ethylene gas to regenerate the original complex (i.e., the reverse of Equation 3). The upper solid curve in Figure 5 shows the sensor response upon initial exposure to a fresh coating at 40°C. The lower solid curve shows the response upon re-exposure to the same series of concentrations, illustrating the decline in response. Passing pure ethylene over the sensor caused a rapid and large decrease in the difference frequency (increase in the single-oscillator frequency) corresponding to replacement of the lighter ethylene for the VA. No further change in frequency was observed after about 30 min. Exposing the sensor again to VA gave the response represented by the dashed curve. The recovery in the response is evident. In fact, the regenerated coating gives a slightly higher response at all concentrations, although the difference is well within the noise level of the response. One explanation for this observation is that substitution by VA causes an expansion of the crystal lattice. The resultant increase in available reagent surface area and/or diffusion of the VA through the regenerated solid would explain the slight increase in reaction rate.

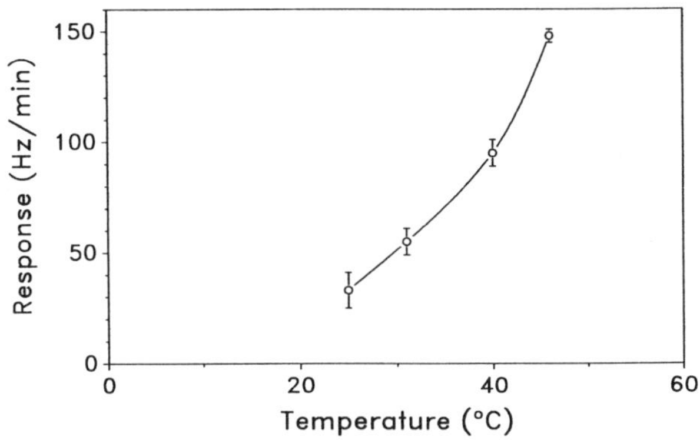

Figure 4. Rate of frequency change (response) vs. temperature for exposure to 260 ppm of vinyl acetate.

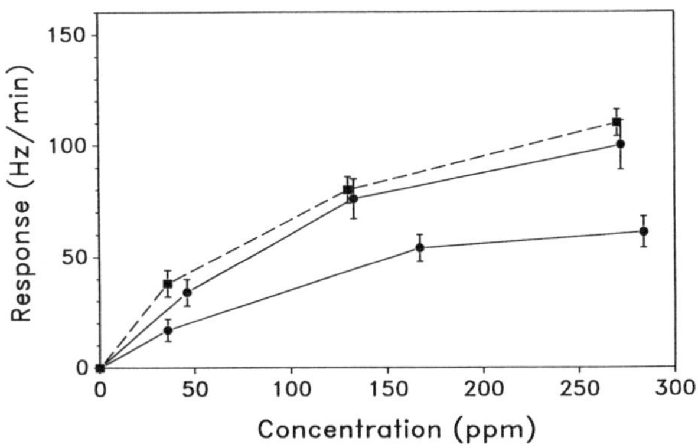

Figure 5. Response to vinyl acetate at 40°C for initial exposure to fresh coating film (upper solid curve), re-exposure to the same series of concentrations (lower solid curve), and exposure of the regenerated coating film (dashed curve).

The selectivity of the sensor was tested by simultaneous exposure to VA and each of several other gases and vapors encountered with VA in industrial processes. The chemicals examined included ethyl acetate, methanol, methyl acrylate, methylmethacrylate, mesityl oxide and vinyl chloride. Table I summarizes the results for exposures at 40°C to 50 ppm of VA and each of the other chemicals. As shown, there was no measurable effect on the response to VA. With the exception of methyl acrylate, none of these compounds reacted with the complex when exposed individually. The response to 60 ppm of methyl acrylate alone was 10 Hz/min which is about one-third that of VA at the same concentration. That methyl acrylate does not affect the response to VA during simultaneous exposure is remarkable since it indicates that the VA is exclusively favored in the competition for reaction with the solid. This observation, coupled with the fact that the strongly-coordinating methanol does not increase the response to VA, is consistent with previous data (23) indicating an associative pathway for the reaction.

Table I. Response (Hz/min) to 50 ppm of vinyl acetate at 40°C with and without co-contaminants present at the indicated concentrations

Co-contaminant	Response to 50 ppm VA mean (s.d.)
methanol (350 ppm)	28(3)
	27(3)
ethyl acetate (50 ppm)	23(3)
	25(4)
mesityl oxide (65 ppm)	30(5)
	28(4)
vinyl chloride (80 ppm)	28(3)
	28(4)
methylacrylate (60 ppm)	28(2)
	29(3)
methylmethacrylate (320 ppm)	25(4)
	27(3)

The selectivity for VA may be explained, in part, by the relative energies of the orbitals on the olefins and platinum. Meester et al. have reported orbital energies for platinum, VA and methyl acrylate in *trans*-$PtCl_2$(olefin)(pyridine) complexes based on spectroscopic data (22). Their results indicate a closer match for methyl acrylate in the energies of the unfilled olefin and platinum orbitals, but a closer match for VA in the energies of the filled orbitals. Apparently, the σ-bond interaction predominates in determining the relative reaction rates (and product stabilities) for these olefin substitutions. The fact that methylmethacrylate does not react at all is probably due to steric hindrance of the additional methyl group on the double bond. A similar argument can be made for the case of mesityl oxide. The failure of vinyl chloride to react is most likely due to the strong electron-withdrawing strength of the chlorine atom.

Conclusions

We have shown that real-time measurement of vinyl acetate vapor at low-ppm concentrations is possible using a SAW sensor coated with *trans*-$PtCl_2$(ethylene)(pyridine). Ideal selectivity for vinyl acetate is observed in the presence of several olefin and non-olefin industrial co-contaminants. The strong temperature dependence of the trapping reaction demands careful thermostatting of the sensor but may also provide a way of controlling the sensitivity and coating life: lower temperatures can be used for higher concentrations and vice versa. While the reagent is gradually consumed as exposure proceeds, regeneration is possible by simple chemical treatment, permitting repeated use of the coating. In addition to its practical applications in monitoring olefin vapors, the SAW sensor appears to be a useful probe for studying the rates and mechanisms of gas-solid reactions.

We have found that styrene vapor can also be measured with a SAW sensor using this coating reagent, without interference from acrylonitrile or several other solvent vapors used together with styrene in industrial processes. These results, as well as investigations of the effects of altering the electronic and steric properties of the *trans*-pyridine ligand in the complex, will be reported in a subsequent publication.

Acknowledgments

Funding for this work was provided by the University of Michigan, School of Public Health. The author wishes to acknowledge the technical assistance of Ms. Noralynn Hassold.

Literature Cited

1. Leidel, N; Busch, K. A.; Lynch, J. R. *Occupational Exposure Sampling Strategy Manual*; U.S. Department of Health, Education, and Welfare (CDC-NIOSH): Cincinnati, 1977; p 73.
2. *NIOSH Manual of Analytical Methods*, 3rd ed.; Eller, P. M., Ed.; U.S. Department of Health and Human Services (CDC-NIOSH): Cincinnati, 1984; Vol. 1, 2.
3. Nader, J. S.; Lauderdale, J. F.; McCammon, S. C. *In Air Sampling Instruments for Evaluation of Atmospheric Contaminants*, 6th ed.; Lioy, P. J.; Lioy, M. J. Y. Eds.; American Conference of Governmental Industrial Hygienists: Cincinnati, 1983; pp V1-V118.
4. Ballantine, D. S.; Rose, S. L.; Grate, J. W.; Wohltjen, H. *Anal. Chem.* 1986, *58*, 3058.
5. Wohltjen, H. *Sensors and Actuators* 1984, *5*, 307.
6. Auld, B. A.; *Acoustic Fields and Waves in Solids*; Wiley: New York, 1973; Vol. 2, pp 275-280.

7. Venema, A.; Nieukoop, E.; Vellekoop, M. J.; Ghijsen, W. J.; Barandz, A. W.; Nieuwenhuizen, M. S. IEEE Trans. Ultrason. Ferroelec. and Freq. Control 1987, 34, 149.
8. Nieuwenhuizen, M. S.; Nederlof, A. J.; Barandz, A. W. Anal. Chem. 1988, 60, 230.
9. D'Amico, A.; Palma, A.; Verona, E. Sensors and Actuators 1982/83, 3, 31.
10. Brace, J. G.; SanFelippo, T. S.; Joshi, S. G. IEEE Solid-State Sensors Workshop Tech. Digest, 1986, pp 44-48.
11. Vetelino, J. F.; Lade, R.; Falconer, R. S. IEEE Trans. Ultrason. Ferroelec. and Freq. Control 1987, 34, 157.
12. Bryant, A.; Lee, D. L.; Vetelino, J. F. Proc. IEEE Ultrasonics Symposium, 1981, p 171.
13. Wohltjen, H.; Snow, A. W.; Barger, W. R.; Ballantine, Jr., D. S. IEEE Trans. Ultrason. Ferroelec. and Freq. Control 1987, 34, 172.
14. Snow, A.; Wohltjen, H.; Anal. Chem. 1984, 56, 1411.
15. Rose-Pehrsson, S. L.; Grate, J. W.; Ballantine, D. S., Jr.; Jurs, P. C. Anal. Chem. 1988, 60, 2801.
16. Radian Corporation Polymer Manufacturing - Technology and Health Effects; Noyes: Park Ridge, NJ, 1986; Chapter 24, 26.
17. The Encyclopedia of Basic Materials for Plastics; Simonds, H. R.; Church, J. M., Eds.; Reinhold: New York, 1967; pp 487-491.
18. Documentation of the Threshold Limit Values and Biological Exposure Indices; American Conference of Governmental Industrial Hygienists: Cincinnati, 1985; pp 426-427.
19. TLVs - Threshold Limit Values and Biological Exposure Indices for 1988-89; American Conference of Governmental Industrial Hygienists: Cincinnati, 1988; p 37.
20. Herberhold, M. Metal π-Complexes; Elsevier: Amsterdam, 1974; Vol. 1, Pts. 1, 2.
21. Orchin, M.; Schmidt, P. J. Inorg. Chim. Acta. Rev. 1968, 123.
22. Meester, M. A.; van Dam, H.; Stufkins, D. J.; Oskam, A. Inorg. Chim. Acta. 1976, 20, 155.
23. Miya, S.; Kashiwabara, K.; Saito, K. Inorg. Chem. 1980, 19, 98.
24. Hupp, S. S.; Dahlgren, G. Inorg. Chem. 1967, 15, 2349.
25. Nelson, G. O. Controlled Test Atmospheres; Ann Arbor: Ann Arbor, MI, 1976; pp 99-101.
26. Doraiswamy, L. K.; Sharma, M. M. Heterogeneous Reactions: Analysis, Examples, and Reactor Design; Wiley: New York, 1984; Vol. 1, Chapter 2.
27. Levenspiel, O. Chemical Reaction Engineering; Wiley: New York, 1972; Chapter 12.

RECEIVED March 9, 1989

Chapter 13

Liquid-Phase Sensors Based on Acoustic Plate Mode Devices

A. J. Ricco, S. J. Martin, T. M. Niemczyk[1], and G. C. Frye

Sandia National Laboratories, Albuquerque, NM 87185

The response of piezoelectric devices propagating shear horizontal acoustic plate modes (SH-APMs) has been modeled and experimentally characterized for variations in surface mass, liquid rheological properties, and solution dielectric coefficient and electrical conductivity. The nature of the SH-APM and its propagation characteristics are outlined and used to describe a range of interactions at the solid/liquid interface. Sensitivity to sub-monolayer mass changes is demonstrated and a Cu^{2+} sensor is described. The APM device is compared to the surface acoustic wave device and the quartz crystal microbalance for liquid sensing applications.

Chemical sensors based on acoustic wave (AW) devices have been studied for a number of sensing applications, the majority of which fall in the category of gas and vapor detection (1-8). Recently, the use of these sensors in liquid environments has been explored (9-13). AW sensors utilize various types of acoustic waves, including the surface acoustic wave (SAW), the shear-horizontal acoustic plate mode (SH-APM) (10-13), and the Lamb wave (also a plate mode) (3,14). Even though most studies of these piezoelectric sensors have centered on SAW devices (1,2,4-8), differences in the propagation characteristics of the various acoustic modes make some better suited than others for a given sensing application.

An interdigital transducer on the surface of a piezoelectric material can excite and detect waves which propagate along the surface (e.g. the SAW) or through the bulk (e.g. the Lamb wave and the SH-APM) of the substrate. AW sensors typically include an input transducer to generate the wave, an interaction region in which the propagating wave is affected by its environment, and an output transducer to detect the wave. Thus, unlike the quartz crystal microbal-

[3]Current address: Department of Chemistry, University of New Mexico, Albuquerque, NM 87131

0097–6156/89/0403–0191$06.00/0

ance (QCM) (15-20), AW sensors are two-port devices in which the mode propagates many wavelengths between input and output transducers. This distinction, combined with higher operating frequencies, make AW devices significantly more sensitive than the QCM to changes in surface mass. In addition, those AW devices which utilize bulk modes have the considerable advantage that all electrical connections can be made to the face of the crystal *not* immersed in solution.

In this chapter, we describe the SH-APM and its propagation characteristics, pointing out the advantages of this device for liquid sensing. Experimental characterizations of the APM's response to surface mass changes and to the viscous and acoustoelectric coupling which occurs between the device and the contacting solution are presented as well. The detailed mathematical derivations necessary to fully describe viscous coupling and mass sensitivity, which are beyond the scope of this chapter, will be published elsewhere (Martin, S. J.; Ricco, A. J.; Niemczyk, T. M.; Frye, G. C., *Sensors & Actuators*, submitted for publication).

An important application of the APM device in the area of chemical sensing is the indication of the presence, and in optimal cases the concentration, of specific dissolved species. We have reported the use of the APM device to monitor electrodeposition (21), electroless deposition (13), and the corrosion of metal films (22). In these applications, the device responds non-specifically to the presence of mass accumulated on (or removed from) the surface. Although such applications clearly demonstrate the use of the APM device to monitor surface mass changes while in direct contact with solution, they are not examples of specific chemical sensors.

Through suitable chemical modification of the surface, the APM device can be sensitized to the presence of species in solution. The general approach we are taking is to derivatize the quartz surface of the APM device with molecules well known as ligands for metal ions in solution. The ligands are covalently attached to the surface *via* siloxane bonds using a reagent which has the ligand of interest attached to the alkyl group of a trialkoxysilylalkyl moiety (23-25). Ethylenediamine (en), a bidentate ligand which forms complexes with many of the transition metals (26) including copper (27), has formation constants such that two ens will bind each Cu^{2+} ion (28). In addition, the nitrogens which bind the metal are sufficiently basic that addition of acid protonates them, freeing the Cu^{2+} and allowing a check on the reversibility of the reaction. Thus, the application of the APM to chemical sensing is demonstrated in the form of a sensor for aqueous Cu^{2+}.

Propagation and Interactions of Acoustic Plate Modes

Fundamentals. In an AW device, an alternating voltage applied to an interdigital transducer on a piezoelectric substrate generates an alternating strain field, which in turn launches an acoustic wave (29). Depending on the frequency of the alternating voltage, the physical properties of the substrate, and the orientation of the transducers on the substrate, the AW energy may be confined largely to within one wavelength of the surface, as in the case of the SAW, or it may propagate through the bulk of the substrate, as in the case of the APM. Regardless of the path it follows, an acoustic mode having significant amplitude at the substrate surface can have its propagation characteristics altered by changes in the nature and/or quantity of a material in intimate contact with the device surface. The cumulative effects of such interactions over the wave propagation path are changes in AW amplitude and phase delay at the output transducer.

The SH-APM sensors discussed in this chapter utilize thin, crystalline, ST-cut quartz plates which act as acoustic waveguides, confining wave energy between the upper and lower surfaces of the plate as the wave propagates between input and output transducers. Though not strictly valid for a single crystal, treatment of the substrate as an isotropic medium greatly simplifies calculations and leads to reasonable predictions of wave propagation characteristics and perturbations thereon. A schematic representation of an acoustic wave device with a propagating SH-APM is shown in Figure 1a. For simplicity, only one pair of interdigitated fingers is shown for each transducer; actual transducers typically consist of 25-75 such pairs. Note that the wavelength (λ) of the APM as it propagates between input and output transducers is twice the center-to-center spacing (labeled d/2) of the transducer fingers. Together with the propagation velocity (v_n) of the APM, the overall transducer periodicity (d) determines the frequency (f_n) at which a given mode propagates: $f_n = v_n/d$.

An SH plate mode may be thought of as a superposition of SH plane waves multiply reflected at some angle between the upper and lower faces of the quartz plate. The substrate faces impose a transverse resonance condition which results in each APM having displacement maxima at the surfaces. Figure 1b shows a cross sectional view (normal to the propagation direction of Figure 1a) of the particle displacement for the four lowest-order SH modes, with mode indices n = 0 through 3 representing the number of displacement nodes between substrate faces. As illustrated by Figure 1b, each mode has equal displacement on both surfaces of the APM device, allowing the use of either surface for liquid sensing. Each of these acoustic modes has a slightly different unperturbed velocity; the frequency of most efficient coupling between transducer and substrate for the nth mode may be written:

$$f_n = \frac{v_o}{d}\left[1 + \left(\frac{nd}{2b}\right)^2\right]^{\frac{1}{2}} \tag{1}$$

where b is the substrate thickness. Thus, thinning the substrate increases the frequency spacing between modes.

Because the effect of external perturbations (i.e. causes of sensor response) on the oscillation frequency of a given mode may depend on the displacement distribution for that mode, interpretation of APM sensor results is simplified by exciting only a single mode. This is accomplished by using a transducer with a bandwidth which is less than the frequency separation between modes. The transducer bandwidth decreases as the number of finger pairs, N, increases, while the frequency separation between modes increases in proportion to $(d/b)^2$. Thus, increasing the value of the factor

$$R = N(d/b)^2 \tag{2}$$

enhances separation between adjacent modes.

Mass Sensitivity. Extreme sensitivity to changes in surface mass set AW devices apart from all other types of sensors. When mass is bound

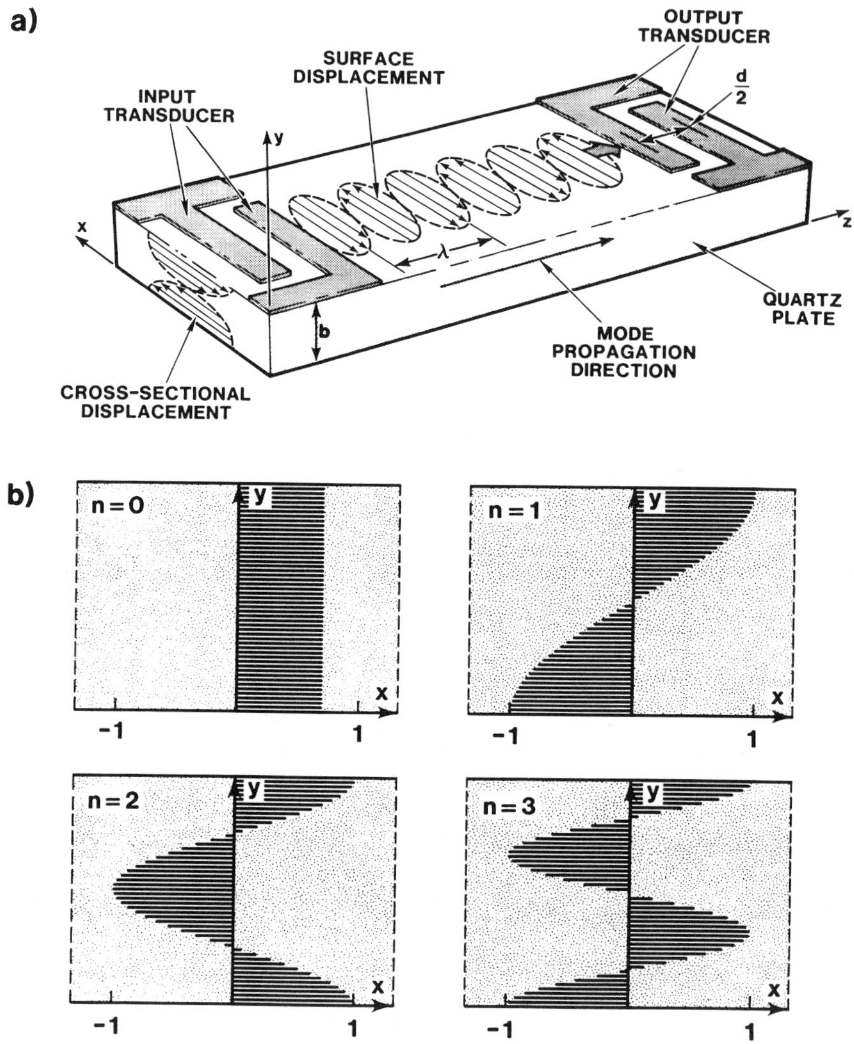

Figure 1. **a)** Schematic of an APM device showing the shear horizontal displacement of the mode as it propagates between input and output transducers. Each transducer consists of 50 - 75 interdigitated electrode pairs, one of which is shown. **b)** Cross-sectional displacement profiles for the four lowest order shear horizontal (SH) plate modes, n = 0 through 3. The n = 1 mode is depicted in the device of part a).

strongly to the surface of an APM device, it oscillates synchronously with the quartz surface under the influence of the passing plate mode. The extent to which mechanical surface perturbations influence APM propagation is proportional to the ratio of surface particle velocity squared to acoustic power flow through the substrate. This makes the n = 0 mode, which has smaller surface displacement (and hence particle velocity) relative to the higher order modes, less sensitive to mass changes. A simple and highly accurate way to measure wave velocity is to utilize the APM device as the feedback element of an oscillator loop (*vide infra*). For the nth mode, the frequency shift Δf_n caused by a change in surface mass/area of magnitude ρ_s is approximated by:

$$\frac{\Delta f_n}{f_n} \equiv -c_f \rho_s \simeq -\frac{\nu_n}{\rho b} \rho_s \tag{3}$$

in which c_f is the frequency sensitivity to surface mass, ρ is the density of the quartz substrate, $\nu_o \equiv ½$, and $\nu_n \equiv 1$ for $n \geq 1$. Equation 3 predicts that frequency will decrease linearly with accumulated mass density. Also, sensitivity is predicted to depend inversely on plate thickness. As with Lamb waves (3,14), changes in elastic properties of a surface layer can influence plate mode oscillator frequency as well, but this effect is often negligible. It should be pointed out that Equation 3 will accurately relate frequency shift to mass changes only when no sizable changes in various physical properties of the solution (described below) accompany the mass change.

Liquid Entrainment. Because of its shear-horizontal surface particle displacement (in the plane of the device surface), the SH-APM propagates in contact with liquids without excessive attenuation. The in-plane oscillation of the quartz surface contacting the liquid does lead, however, to entrainment of a thin layer of liquid near the interface (11). This viscous coupling of liquid to the APM has two effects: (1) to alter the propagation characteristics (velocity and attenuation) of the APM, and (2) to alter the transduction efficiency for excitation and detection of APMs. By confining the liquid to the region between transducers, propagation effects are measured without the influence of transduction effects; this is necessary even if the side of the device opposite the transducers is used for measurement.

Continuity of particle displacement across the solid/liquid interface requires that the surface displacement of the APM generate motion in the liquid. Solution of the Navier-Stokes equation in the liquid, subject to this non-slip boundary condition at the solid/liquid interface, indicates that the liquid undergoes a shear motion which decays rapidly with distance from the surface (11). For an angular frequency of oscillation ω_n, the velocity field decay length δ of the liquid entrained by the plate mode is approximated by:

$$\delta \simeq \sqrt{2\eta/\rho_1\omega_n} \tag{4}$$

in which ρ_1 and η are liquid density and shear viscosity, respectively. Because the APM device operates at high frequency, the coupled liquid layer is very thin: about 50 nm in water at 158 MHz.

The high frequency of the APM necessitates consideration of viscoelastic response by the liquid. Modeling simple liquids as Maxwellian fluids with a single relaxation time, τ, gives good agreement with experimental data. When the Maxwellian fluid is driven in oscillatory flow with $\omega_n\tau \ll 1$, it responds as a Newtonian (ideal) fluid characterized by the shear viscosity η. For $\omega_n\tau \gg 1$, the oscillation rate exceeds the rate of molecular motion in the liquid and energy ceases to be dissipated in viscous flow, being stored elastically instead (30). Consequently, when driven at high frequencies, a Maxwellian fluid behaves as an amorphous solid with shear properties characterized by a shear modulus μ. The relaxation time associated with the transition from viscous to elastic behavior in a Maxwellian liquid is related to these parameters by $\tau = \eta/\mu$ (30).

Viscous coupling of an APM to an adjacent liquid results in both attenuation (α_n) of the modes and a change in propagation velocity (Δv_n). These can be estimated from a perturbation analysis and are given by:

$$\alpha_n \simeq \frac{\eta c_{v,n}}{2v_n} Re\left[\frac{\gamma_n}{1 + j\omega_n\tau}\right] \tag{5a}$$

and

$$\frac{\Delta v_n}{v_n} \simeq -\frac{\eta c_{v,n}}{2\omega_n} Im\left[\frac{\gamma_n}{1 + j\omega_n\tau}\right] \tag{5b}$$

where

$$\gamma_n^2 = \left(\beta_n^2 - \frac{\omega_n^2\rho_1}{\mu}\right) + j\frac{\omega_n\rho_1}{\eta} \tag{5c}$$

Here, *Re* and *Im* denote the real and imaginary parts of the quantity in brackets, respectively, $c_{v,n}$ is the mass sensitivity of plate mode velocity, and β_n is the APM wavenumber: $\beta_n = [(\omega_n/v_n)^2 - (n\pi/b)^2]^{1/2}$. The velocity shift arises from mass loading by the entrained liquid layer, explaining the dependence of this effect on $c_{v,n}$; attenuation arises from power dissipation in the liquid.

Acoustoelectric Coupling. Propagation of an APM through a piezoelectric waveguide generates a layer of bound charge at the device surface, and the evanescent electric field associated with this charge extends into the adjacent liquid, coupling to ions and dipoles in solution (31). This interaction is analogous to that observed between piezoelectric waves and charge carriers in semiconductors (32), which has been exploited to construct a SAW-based NO_2 sensor (5). Ion coupling decays exponentially with distance from the solid/liquid interface, extending several microns into the liquid (decay length $\simeq \lambda/2\pi$; the double layer capacitance behaves as a short circuit at the typical APM frequencies of > 100 MHz). Ion, dipole, and induced dipole motion resulting from this acoustoelectric coupling lead to perturbations in plate mode velocity and attenuation, which are related to the solution conductivity σ by (31):

$$\frac{\Delta v_n}{v_n} \simeq -\frac{K^2}{2}\left(\frac{\epsilon_s + \epsilon_o}{\epsilon_s + \epsilon_1}\right)\frac{\sigma^2}{\sigma^2 + \omega_n^2(\epsilon_s + \epsilon_1)^2} \tag{6a}$$

$$\frac{\alpha_n}{k} \simeq \frac{K^2}{2}\left(\frac{\epsilon_s + \epsilon_o}{\epsilon_s + \epsilon_1}\right)\frac{\omega_n\sigma(\epsilon_s + \epsilon_1)}{\sigma^2 + \omega_n^2(\epsilon_s + \epsilon_1)^2} \tag{6b}$$

in which K^2 is the electromechanical coupling coefficient, $k = 2\pi/\lambda$ is the wavenumber for the APM, and ϵ_s, ϵ_1, and ϵ_o are the dielectric coefficients of the substrate, the solution, and free space, respectively.

Comparison of the APM Device to Other Piezoelectric Devices. ***SAW Device.*** While SAW devices have proven to be very effective detectors for gas phase sensing (1-8), their performance in liquids is less than optimal. The substantial surface-normal displacement component characteristic of Rayleigh waves generates compressional waves in any solution contacting the device surface, carrying energy away from the wave and causing prohibitive attenuation (11,33). Because the propagation loss of a SAW due to a contacting liquid is proportional to frequency and to the length of the path over which liquid contacts the device surface (29), a low frequency (< 20 MHz) SAW device with a relatively short path length may function as a sensor in liquid environments (9). Since SAW sensitivity is proportional to frequency, however, such a device has sensitivity similar to that of bulk crystal oscillators.

Quartz crystal microbalance. Although bulk crystal oscillators of the type commonly used to monitor metal deposition processes *in vacuo* can also function in liquids (18-20), they have lower sensitivity, in terms of the size of the frequency shift recorded for a given mass density change, than APM and SAW devices, due to low operating frequency (5-15 MHz). The QCM utilizes a thin quartz disk with electrodes on each side, forming a single electrical port. As alluded to earlier, electrical contact must be made to both faces of the crystal, one of which contacts the solution under examination, adding complexity to the experimental apparatus. Resonance occurs when a wave propagating perpendicular to the crystal faces satisfies $b = n\lambda/2$, where b is the plate thickness and λ is the wavelength (15); typically, $n = 1$. This resonance condition restricts the operation of the QCM to lower frequencies, because there are limitations on how thin the substrate can be made. The variation in electrical input impedance which occurs at this resonant frequency can be used to control an oscillator circuit; circuitry is commercially available to instrument the single-port QCM. In general, however, two-port devices such as the SAW and APM are more easily instrumented as oscillators, particularly at high frequencies (34).

Instrumentation of APM Sensors. Two instrumentation arrangements were utilized in this study. In *oscillator measurements*, the device is used as the frequency control element of an oscillator circuit; perturbations in oscillation frequency $\Delta f_n/f_n$ are monitored in response to changes in liquid properties or surface perturbations. The APM device is utilized as the frequency control element of an oscillator circuit by serving as an amplifier feedback path. The elements of a simple oscillator loop are shown in Figure 2a. To form a stable oscillator, a signal traversing the closed loop must return to its starting point (1) having equal amplitude and (2) being phase shifted by a multiple of 2π radians; the latter requirement is facilitated by a tunable phase shifter. The bandpass filter prevents oscillation at other frequencies, such as that characteristic of the SAW. For oscillator measurements, the transducer region should be perturbed along with the intervening propagation path. When liquid contacts the surface of the substrate without transducers, cell design should be such that liquid is present on the region of the crystal opposite the transducers. As the oscillation frequency changes, the transducers will then "self-tune" to the changing frequency, preserving the linearity of response to various perturbations.

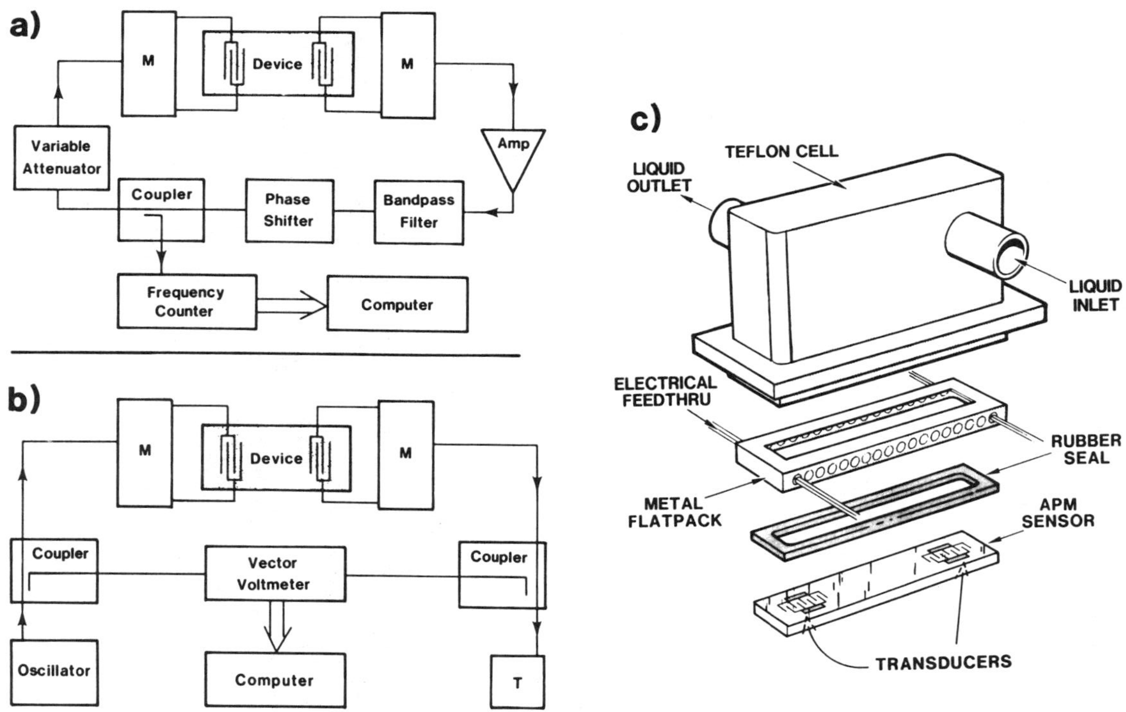

Figure 2. **a)** Instrumentation for the APM oscillator configuration (M = impedance matching network). **b)** Instrumentation used to measure changes in amplitude and phase between transducers for an input signal of fixed frequency and amplitude (T = 50 Ω termination). **c)** Exploded view of APM device mounted in a metal flatpack and fitted with a teflon flow-through cell to contain liquid in contact with the side of the device opposite the transducers.

In *propagation measurements*, a signal of fixed frequency and amplitude is input to the device using an external oscillator, while a vector voltmeter monitors changes in APM amplitude and phase between transducers, as illustrated in Figure 2b. These measurements allow determination of the perturbation in APM phase velocity and attenuation. Although less sensitive than the oscillator measurement, this technique is not subject to "mode hopping" (sudden jumps from one mode to another) which can afflict an APM oscillator in the event of rapid environmental perturbations, e.g. the transition from a dry surface to one covered with liquid. During propagation measurements, it is important to isolate the transducers, including the region opposite them, from the perturbation (Martin, S. J.; Ricco, A. J.; Niemczyk, T. M.; Frye, G. C., *Sensors & Actuators*, submitted for publication).

For a non-dispersive AW (e.g. a SAW), frequency and velocity perturbations are equal in magnitude. As noted by White (14), however, the sensitivity of $\Delta f_n/f_n$ and $\Delta v_n/v_n$ to a perturbation are not necessarily equal in a system involving dispersive modes, such as the APM or Lamb wave, because the phase velocity of each mode depends on frequency. Since SH modes are dispersive, the following relationship holds:

$$\frac{\Delta f_n}{f_n} = \frac{v_{gn}}{v_n}\frac{\Delta v}{v_n} \qquad (7)$$

where v_n and v_{gn} are the phase and group velocities, respectively, for the nth SH-APM. For propagating SH modes, $v_{gn} \le v_n$, so that fractional frequency perturbations are less than fractional velocity perturbations (For Lamb waves, the converse holds: $v_n \le v_{gn}$). Typically, $v_{gn}/v_n = 0.6 - 0.7$ for our devices.

Experimental Methods and Materials

Device Fabrication and Instrumentation. APM devices were designed at Sandia National Laboratories and fabricated by Crystal Technologies, Inc., Palo Alto, CA and Sawtek, Inc., Orlando, FL. The ST-cut of quartz was used, with propagation along the x-direction of the crystal; substrates measured 22.9 x 7.6 x 0.5 mm thick before lapping. Transducer finger dimensions scale with periodicity: finger width and separation are both d/4; finger length is 50d.

Two different transducer periodicities were examined. The first device, designed with d = 32 μm to propagate the SH-APM at 158 MHz, has a center-to-center separation between transducers of 7.36 mm. Each transducer is composed of 50 finger-pairs, photolithographically defined from 200 nm-thick Au-on-Cr metallization. The second device, designed with d = 50 μm to propagate the SH-APM at 104 MHz, has transducers comprised of 75 finger-pairs defined from 100 nm thick Al metallization. Center-to-center separation between transducers is 7.28 mm. The unmetallized side of all devices was lapped to obtain the desired plate thickness, then polished so that the upper and lower faces were parallel and optically smooth.

Each device was mounted in a 25.5 mm x 12.7 mm gold-plated steel flatpack (Isotronics) with a 20.5 mm x 3.7 mm opening, allowing liquid to contact the unmetallized side of the device as shown in exploded view by Figure 2c. The unmetallized face of the device was bonded (in the region surrounding the acoustic wave path) to the opening in the flatpack using a bead of RTV elastomer; the device thus contacts liquid only in the APM propagation path. Electrical contact was made between transducer bonding pads and flatpack feed-throughs by 76-μm diameter Au or Al leads attached with an ultrasonic bonder (Westbond). The flatpack was mounted in a brass test fixture containing impedance matching networks (Integrated Chemical Sensors, Inc., Newton, MA). Wires were soldered to the flatpack feed-throughs to make contact to the matching networks in the body of the fixture below.

Liquid was held in contact with the sensing surface by a teflon cell sealed by compression to the metal flatpack. An open cell, having an approximate internal volume of 1 ml, was used for some experiments, with liquids being added and withdrawn by means of a pipet. The closed cell shown in Figure 2c was used in other experiments with the liquid being drawn through by a peristaltic pump.

Instrumentation for the oscillator measurements, arranged as shown in Figure 2a, included two cascaded wide-band amplifiers (Hewlett-Packard 8447D), a band-pass filter (K&L Microwave 5BT-95/190-5N), two variable attenuators in series (Hewlett-Packard 8494A and 8495A), a tunable phase shifter (Merrimac PSL-4-160B or PSL-4-100B), a 10 dB directional coupler (Anzac DCG-10-4), a frequency counter (Hewlett-Packard 5384A), and a computer for data acquisition (Hewlett-Packard 9816). For the propagation measurements, a synthesized oscillator (Hewlett-Packard 8656A) was used along with a vector voltmeter (Hewlett-Packard 8405A) and the HP 9816 computer in the arrangement of Figure 2b.

Chemicals, Surface Derivatization, and Miscellaneous Apparatus. Water was doubly distilled. Electrolytes and solvents were commercially available reagent grade, used as received. Solution conductivities were measured at 1 kHz using a 1 cm path length conductivity cell (YSI Models 32 and 3445). Viscosity standards were prepared from water and reagent-grade glycerol. Derivatization of APM device surfaces was accomplished by immersing the device (cleaned by rinsing in trichloroethylene, acetone, *2*-propanol, and water, then soaking in conc. HNO_3) in a 2% (v/v) solution of *N-2*-aminoethyl-*3*-aminopropyltrimethoxysilane (Petrarch Systems, Inc.; distilled under reduced pressure prior to use) in refluxing dry toluene for 1 h. The device was then rinsed with toluene and annealed at 150 C for 1 h. Silver shot (Alfa, 99.999%) was thermally evaporated from a resistively heated tungsten basket at a rate of 0.6 - 0.8 Å/s; base pressure in the cryo-pumped vacuum system was 2×10^{-8} Torr. Evaporation was monitored with an Inficon XTC quartz crystal microbalance operating at 6 MHz.

Results and Discussion

Mode Resolution. By using three devices with different substrate thicknesses, transducer periodicities, and numbers of transducer finger pairs, an estimate of the minimum requirements for resolution of adjacent acoustic plate modes was obtained. Modes were essentially unresolvable using Device 1 (158 MHz), which had a 191-μm thick substrate, 32-μm transducer periodicity, and 50 finger pairs (R = 1.4 in Equation 2). Device 2, also operating at 158 MHz, had b = 152 μm, d = 32 μm, and N = 50 (R = 2.2), and resolved modes well.

Mode resolution was best with Device 3 (104 MHz), having b = 203 μm, d = 50 μm, and N = 75 (R = 4.5).

Mass Sensitivity. ***Calibration.*** The mass sensitivity of an APM device was calibrated by vacuum deposition of silver onto the (unmetallized) quartz surface. Device 3 was incorporated in an oscillator loop (Figure 2a) and the frequency monitored during thermal evaporation of a Ag film onto the substrate. The frequency shift was compared to the film thickness measured by the commercial QCM. Heating of the device was negligible: after removing power from the evaporation filament, no significant shift in frequency was observed as the APM device cooled. The APM frequency shift is plotted *vs*. the mass density of deposited silver in Figure 3a for modes 0 through 3. As expected (Equation 3), the device is approximately twice as sensitive when higher order modes are excited as for the n = 0 mode. The mass sensitivity measured for Device 3 is 9.5 cm^2/g (0.99 Hz-cm^2/ng) for the n = 0 mode and 19.4 cm^2/g (2.0 Hz-cm^2/ng) for the average of the next three higher-order modes. The corresponding sensitivities estimated from Equation 3, 9.3 cm^2/g and 18.6 cm^2/g, are in excellent agreement.

The accuracy with which the mass sensitivity *in vacuo* reflects that when the device is in contact with liquid was investigated with an etching experiment. After evaporating an 80.5 nm-thick silver film onto Device 3, it was incorporated in an oscillator circuit and an open cell on the device surface was filled with 1.0 ml of water. After stable oscillation of the n = 1 mode was achieved, 0.1 ml of a 2:2:1 H_2O:H_2SO_4:HNO_3 etchant was added to the cell. The 1470 ppm frequency shift measured as the film dissolved over a 15 min period yields a mass sensitivity of 17.4 cm^2/g, approximately 6% less than the value measured in vacuum.

A similar vacuum deposition experiment was carried out using Device 2. In this case, the measured mass sensitivities were 31 cm^2/g (4.9 Hz-cm^2/ng) for n = 0 and 65 ± 4 cm^2/g (10 Hz-cm^2/ng) for the average of n = 1, 2, and 3. The sensitivity of this device, which is within a factor of 2 of that for a 97 MHz SAW device (130 cm^2/g or 13 Hz-cm^2/ng), is significantly larger than the sensitivity calculated from Equation 3. This is probably because plate modes begin to couple with surface modes, enhancing surface particle velocity and thus mass sensitivity, when b/λ exceeds approximately five (12).

Comparison of APM to Related Devices. The APM device has significantly higher mass resolution than the QCM, in terms of frequency shift for a given change in mass/unit area, as a result of the APM's higher operating frequency. The commercial (6 MHz) QCM used to monitor Ag deposition has a mass sensitivity of 14 cm^2/g (0.084 Hz-cm^2/ng) (15), while the highest APM device sensitivity is 65 cm^2/g (10 Hz-cm^2/ng) for Device 2. With a measured short-term frequency stability of 5 Hz, APM Device 2 has a minimum detectable mass limit of approximately 0.5 ng/cm^2. We estimate the commercial QCM to have short-term frequency stability of approximately 2 Hz, corresponding to a 24 ng/cm^2 mass change.

As mentioned above, the mass sensitivity of a 97 MHz SAW device exceeds that of the best APM sensitivity by a factor of two. However, because a 97 MHz Rayleigh wave propagating on ST-quartz suffers over 45 dB of increased attenuation when water covers just 3

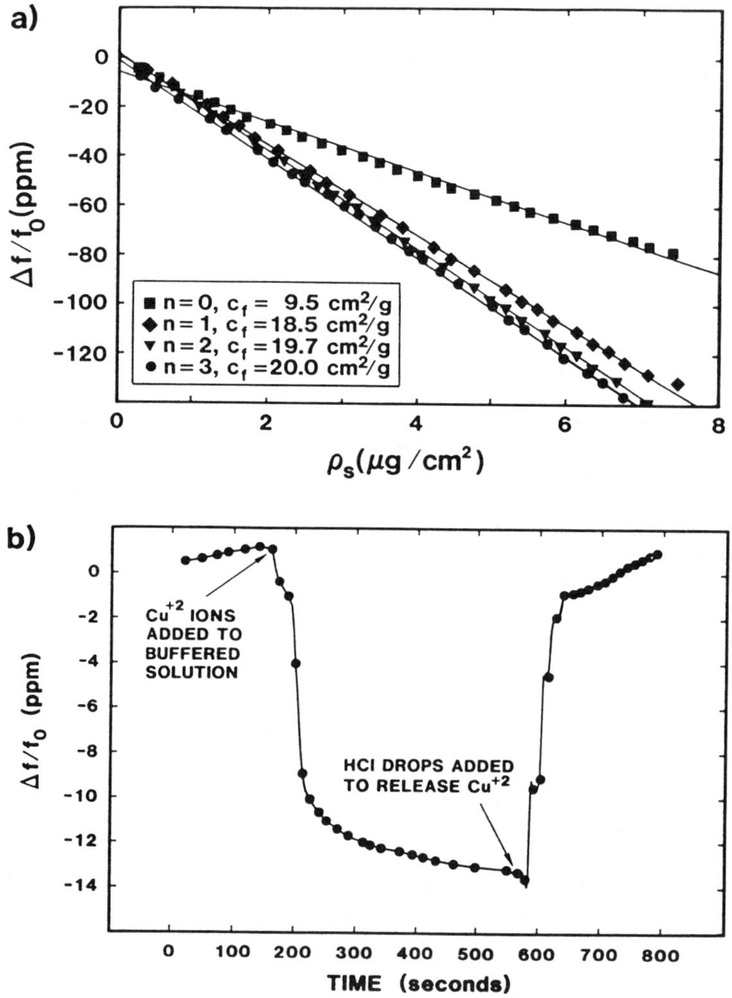

Figure 3. **a)** Shift in oscillation frequency of APM Device 3 as silver is deposited onto it *in vacuo*. Symbols are data and lines are linear least-squares fits. **b)** APM sensor response during the binding and release of aqueous Cu^{2+} ions ($[Cu^{2+}]$ = 0.25 mM) by the ethylenediamine-derivatized surface. HCl was added to give a pH near 3, removing the bound Cu^{2+}.

mm of the acoustic path (13), such a device is difficult to use for liquid sensing. In contrast, less than 8 dB of attenuation typically results for the SH-APM when liquid is added to a dry device surface (13), leaving ample signal with which to make measurements.

Detection of Aqueous Cu^{2+}. The high mass sensitivity of the APM device suggests that chemical sensors can be developed utilizing interactions which lead to mass changes as small as a fraction of a monolayer. Derivatization of the quartz APM device surface with the en-based ligand, under the reaction conditions given in the Experimental Methods and Materials section, should result in surface coverages of between one and several molecular layers (23); the surface-immobilized ligand will be denoted en_{im}. An open liquid cell affixed to the device was filled with 0.1 M KNO_3; this solution maintains ionic conductivity at a constant level, eliminating acoustoelectric effects. The device was included in the oscillator loop of Figure 2a. After achieving stable oscillation, 0.25 ml of 1 mM $CuSO_4$ in 0.1 M KNO_3 were added to the cell, resulting in a final Cu^{2+} concentration of approximately 0.25 mM. The data of Figure 3b show that aqueous Cu^{2+} binds readily and reversibly to the en_{im}-derivatized APM device surface. A frequency shift of -13 ppm occurred over a period of roughly 400 sec following addition of Cu^{2+} to the cell; this shift corresponds to a bound ion density of about $3 \times 10^{15}/cm^2$. Based on additional experiments, essentially all the available binding sites are occupied at this concentration, so the coverage of en_{im} is thus about $6 \times 10^{15}/cm^2$. To learn how much of the Cu^{2+} binding might be attributable to the quartz surface itself, a control experiment was performed using an underivatized device. In this case, Cu^{2+} was added to give a final concentration of approximately 100 μM; the result was a frequency shift of roughly -2 ppm, about six times smaller than the shift recorded for the en_{im}-treated surface.

To examine the reversibility of the Cu^{2+} binding, 0.25 ml of 10 mM HCl in 0.1 M KNO_3 were added to the cell, giving a final pH near 3 and protonating the N atoms of the en_{im}. The data of Figure 3b show the return of the oscillation frequency to near its initial value, indicating nearly complete release of the bound Cu^{2+} ions as a result of the acid addition. The binding and release of Cu^{2+} were found to be repeatable with this device, indicating that the derivatized APM device functions as a reversible detector of low concentrations of Cu^{2+} in solution.

Viscous Coupling. APM velocity shifts and attenuation arising from liquid entrainment were measured for Device 1. The changes in propagation characteristics were measured while glycerol/water mixtures having viscosities between 1 and 62 centipoise (cP) were placed in an open cell which confined the liquid to the region between transducers. The device and cell were maintained at 20 °C while α and $\Delta v_n/v_n$ were measured using the instrumentation of Figure 2b; the measured variations are plotted vs. $\sqrt{\eta}$ in Figure 4a and b. It should be noted that although the density of the solution increases by approximately 20% over the range of viscosities shown in Figure 4, the effect of this change is negligible compared to the effect of changing viscosity. At low values of viscosity, the liquid behaves as a Newtonian fluid with $\Delta v_n/v_n$ and α proportional to $\sqrt{\eta}$. For viscosities exceeding about 10 cP, relaxation times become comparable to the wave period (6 ns) and viscoelastic behavior results (11).

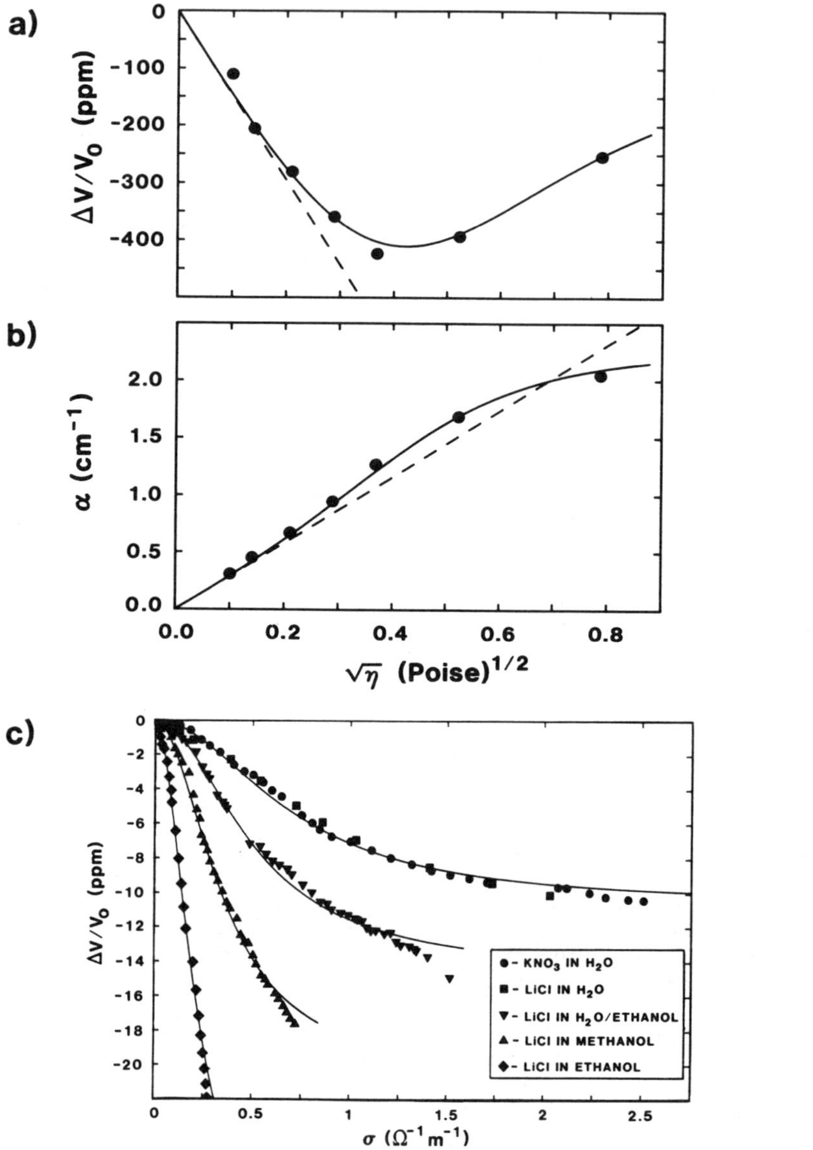

Figure 4. Perturbations in **a)** plate mode velocity and **b)** attenuation due to liquid entrainment by the APM device surface. Data are for glycerol/water mixtures which only contact the surface between transducers. The solid lines are calculated from a viscoelastic model for the liquid; the dashed lines are calculated using a Newtonian model. **c)** Perturbation in oscillator frequency due to acoustoelectric interactions between APM and ions and dipoles in solution. Solid lines are calculated using best-fit values for the dielectric coefficient of each solvent.

The perturbation theory estimates are in good agreement with the data over the viscosity range tested, as shown by the solid lines calculated from Equations 5 using the experimentally determined value of $c_{v,n}$ for this device. A shear modulus $\mu = 3.1 \times 10^8$ dyne/cm^2 was found to give the simultaneous best fit of Equations 5 to the experimental data shown in Figure 4a and b. Interestingly, this modulus implies a liquid relaxation time which agrees well with measured dielectric relaxation times (11). The dashed lines are the predictions when viscoelastic effects are neglected and the liquid is treated as a Newtonian fluid, i.e., $\mu = \infty$ in Equations 5.

Acoustoelectric Coupling. The acoustoelectric interaction between plate modes and ions was investigated by measuring the wave velocity shift and attenuation as a function of the ionic conductivity of a solution contacting Device 1. The salt (KNO_3 or LiCl) concentration of a solution pumped across the device was varied using the closed cell configuration of Figure 2c while monitoring solution conductivity. To determine the effect of the solution dielectric constant, this experiment was repeated using solvents with widely varying dielectric constants.

The variation in APM velocity as a function of solution conductivity σ is shown in Figure 4c for four solvents having ϵ_1/ϵ_o ratios of 79.3 (H_2O), 52.5 (H_2O/ethanol), 33.0 (methanol), and 24.6 (ethanol). The behavior of KNO_3 and LiCl in water are indistinguishable, arguing against specific ion effects. The highest conductivity shown for KNO_3 in H_2O corresponds to a concentration of about 0.25 M. Changes in density and viscosity over the range of conductivities examined are insignificant on the scale of Figure 4c. As Figure 4c shows, the dielectric constant of the solvent significantly affects both the rate at which velocity changes with conductivity and also the overall magnitude of the change. Attenuation of the APM due to ionic conductivity is too small to measure with precision using quartz devices. The solid lines in Figure 4c are velocity shifts calculated from Equation 6a using a single value of K^2 and the best-fit value of the dielectric constant for each solvent; ϵ_1/ϵ_o ratios of 74, 53, 36, and 21 give the curves shown for H_2O, ethanol/H_2O, methanol, and ethanol, respectively. The K^2 of 3.2×10^{-4} was obtained from a least-squares fit of Equation 6a simultaneously to all four sets of data. Independent measurements of K^2 during the vacuum evaporation of a metal film give values in the range 2.2-3.1×10^{-4}, suggesting the best-fit value is reasonable (31).

In many sensor applications, response resulting from conductivity variations are undesirable; they can be eliminated using several methods: (1) buffering solution ionic concentration; (2) depositing a conductive metal layer on the quartz surface to decouple ions in solution from the APM; or (3) adjusting solution conductivity to a value outside the range over which its effects are important.

Conclusions

In comparison to similar devices, the SH-APM offers advantages for sensing species in solution. These include considerably less attenuation of the propagating wave in comparison to the SAW, and enhanced sensitivity and the complete isolation of transducers from the solution in comparison to the QCM. The APM interacts with species in solution and at the solid/liquid interface *via* several mechanisms, including changes in surface mass and mechanical properties, viscous coupling to the adjacent layer of solution, and acoustoelectric coupling between the APM's evanescent electric field and ions and dipoles in solution. Knowledge of these interactions allows the determination of sub-monolayer mass changes (0.5 ng/cm^2) at the solid /liquid interface; combination of this sensitivity with surface derivatization techniques leads to chemical sensors for dissolved species.

Acknowledgments

The authors are grateful to P. A. Taylor, V. M. Hietala, and L. Romero of Sandia National Laboratories for helpful discussions and to B. J. Lammie of Sandia National Laboratories and I. Adhihetty of the University of New Mexico Department of Chemistry for valuable technical assistance. This work was supported by the U.S. Department of Energy under contract no. DE-AC04-76DP00789.

Literature Cited

1. Wohltjen, H; Dessy, R. *Anal. Chem.* **1979**, *51*, 1458.
2. Wohltjen, H. *Sensors and Actuators* **1984**, *5*, 307.
3. Zellers, E. T.; White, R. M.; Wenzel, S. W. *Sensors and Actuators* **1988**, *14*, 35.
4. Martin, S. J.; Schweizer, K. S.; Schwartz, S. S.; Gunshor, R. L. *Proc. 1984 IEEE Ultrasonics Symp.*, **1984**, p 207.
5. Ricco, A. J.; Martin, S. J.; Zipperian, T. E. *Sensors and Actuators* **1985**, *8*, 319.
6. D'Amico, A.; Palma, A.; Verona, E. *Sensors and Actuators* **1982**, *3*, 31.
7. Venema, A.; *et al. IEEE Trans. Ultrasonics, Ferroelectrics, and Freq. Contr.* **1987**, *UFFC-34*, 148.
8. Bryant, A.; Lee, D. L.; Vetelino, J. F. *Proc. 1981 IEEE Ultrasonics Symp.*, **1981**, p 171.
9. Roederer, J. E.; Bastiaans, G. J. *Anal. Chem.* **1983**, *55*, 2333.
10. Martin, S. J.; Ricco, A. J. *Proc. Int. Electron Devices Mtg.*, **1987**, p 290.
11. Ricco, A. J.; Martin, S. J. *Appl. Phys. Lett.* **1987**, *50*, 1474.
12. Hou, J.; van de Vaart, H. *Proc. 1987 IEEE Ultrasonics Symp.*, **1987**, p 573.
13. Ricco, A. J.; Martin, S. J. *Proc. Symp. on Electroless Dep. of Metals and Alloys*, **1988**, Vol. 88-12, p 142.
14. White, R. M.; Wicher, P. J.; Wenzel, S. W.; Zellers, E. T. *IEEE Trans. Ultrasonics, Ferroelectrics, Freq. Contr.* **1987**, *UFFC-34*, 162.
15. Sauerbrey, G. *Z. Phys.* **1959**, *155*, 206.
16. Thompson, M.; Dhaliwahl, G. K.; Arthur, C. L.; Calabrese, G. C. *IEEE Trans. Ultrasonics, Ferroelectrics, Freq. Contr.* **1987**, *UFFC-34*, 128.
17. Carey, W. P.; Beebe, K. R.; Kowalski, B. R. *Anal. Chem.* **1987**, *59*, 1529.
18. Nomura, T.; Okuhara, M. Anal. Chim. Acta **1982**, *142*, 281.
19. Bruckenstein, S.; Shay, M. *J. Electroanal. Chem.* **1985**, *188*, 131.

20. Kanazawa; K. K.; Gordon II, J. G. *Anal. Chem.* **1985**, *57*, 1770.
21. Ricco, A. J.; Martin, S. J. *Ext. Abstr. 171st Electrochem. Soc. Mtg.*, **1987**, Vol. 87-1, p 501.
22. Martin, S. J.; Ricco, A. J.; Sorensen, N. R. *Ext. Abstr. 171st Electrochem. Soc. Mtg.*, **1987**, Vol. 87-1, p 50.
23. Arkles, B. *Chemtech* **1977**, *7*, 766.
24. Leyden, D. E.; Collins, W. *Silylated Surfaces*; Gordon and Breach Science: New York, 1980.
25. Bookbinder, D. C.; Wrighton, M. S. *J. Electrochem. Soc.* **1983**, *130*, 1080.
26. Cotton, F. A.; Wilkinson, G. *Advanced Inorganic Chemistry*, 4th Edn.; John Wiley & Sons: New York, 1980; Chapters 3, 21, 22.
27. Hathaway, B. J.; Billing, D. E. *Coord. Chem. Rev.* **1970**, *5*, 143.
28. McIntyre Jr., G. H.; Block, B. P.; Fernelius, W. C. *J. Am. Chem. Soc.* **1959**, *81*, 529.
29. Auld, B. A. *Acoustic Waves and Fields in Solids*; John Wiley & Sons: New York, 1973; Vol. 2.
30. Matheson, A. J. *Molecular Acoustics*; John Wiley & Sons: New York, 1971; pp. 82-83.
31. Niemczyk, T. M.; Martin, S. J.; Frye, G. C.; Ricco, A. J. *J. Appl. Phys.* **1988**, *64*, 5002.
32. Gunshor, R. L. *Sol.-State Electron.* **1975**, *18*, 1089.
33. Calabrese, G. S.; Wohltjen, H.; Roy, M. K. *Anal. Chem.* **1987**, *59*, 833.
34. Parker, T. E.; Montress, G. K. *IEEE Trans. Ultrasonics, Ferroelectrics, Freq. Contr.* **1988**, *UFFC-35*, 342.

RECEIVED March 9, 1989

Chapter 14

Monitoring Thin-Film Properties with Surface Acoustic Wave Devices

Diffusion, Surface Area, and Pore Size Distribution

G. C. Frye, S. J. Martin, A. J. Ricco, and C. J. Brinker

Sandia National Laboratories, Albuquerque, NM 87185

The ability of surface acoustic wave (SAW) devices to monitor adsorption of N_2 onto the surfaces of porous films and diffusion of species into polymer films has been demonstrated. Calculations based on the N_2 adsorption isotherms illustrate how sol-gel solution chemistry can be used to tailor the surface area and pore size distribution of thin films. BET surface areas from unity to over 30 cm^2/cm^2 of film have been obtained on various samples with median pore diameters from less than 0.4 nm to greater than 6 nm. SAW frequency transients occurring during the diffusion of small molecular species into polymer films have been used to determine diffusion coefficients from 10^{-9} to 10^{-15} cm^2/sec.

When a surface acoustic wave, also called a Rayleigh wave, travels along a substrate surface, the wave velocity and amplitude are affected by changes occurring at the surface. In general, since the acoustic energy is concentrated within one wavelength of the surface, SAW devices are inherently more sensitive to these surface changes than bulk crystal oscillators which have mechanical energy distributed throughout the substrate. As an example of the sensitivity of SAW devices, surface mass changes as small as 100 pg/cm^2 (1) can be detected. This extreme sensitivity has been applied to the construction of several SAW-based chemical sensors (2-7).

In addition, since the SAW is sensitive to minute perturbations occurring in thin films which are in intimate contact with the surface, SAW devices can be used to monitor physical and chemical processes occurring in these overlayers. Based on this effect, SAW devices have recently found applications in the characterization of the properties of thin films (8-10). In this paper, we report on the utility of SAW devices to characterize: (1) the surface area and pore size distribution of porous thin films based on N_2 adsorption isotherms and (2) diffusion coefficients (D) for thin polymer films based on absorption transients (i.e., mass absorbed as a function of time) as indicated by SAW velocity transients (i.e., SAW velocity

0097-6156/89/0403-0208$06.00/0

shift as a function of time). For the first application, the SAW device enhances the detection limit for the amount of adsorbed N_2 by more than four orders of magnitude over conventional techniques currently used for this type of characterization. For the second application, the use of thin films decreases the diffusional length scale from that obtained with bulk samples used in some conventional gravimetric techniques. This results in a dramatic decrease in the time required to obtain D values.

Background

Surface Acoustic Wave Devices. A SAW device typically consists of input and output interdigital transducers on a piezoelectric substrate such as quartz or lithium niobate (see Figure 1). When an alternating voltage is applied to the input transducer, the alternating strain generated in the piezoelectric substrate launches the SAW. The SAW probes the mechanical properties of an overlaying thin film as it travels across the substrate before being converted back into an electrical signal at the output transducer. Changes in mechanical properties (e.g., density, stiffness) of the film result in changes in wave propagation velocity and amplitude.

As illustrated in Figure 1, a simple and highly sensitive method for monitoring these changes is to use the SAW device as the feedback element of an oscillator circuit. In this way, changes in the frequency of oscillation can be directly related to changes in the velocity of propagation of the wave. For situations where only mass changes in an overlaying film are important, frequency shift can be related to the mass loading/area of film (m) using (1):

$$\frac{\Delta f}{f_o} = \kappa \frac{\Delta v}{v_o} = -\kappa c_m f_o m \qquad (1)$$

in which κ is the fraction of the acoustic wave path length between transducers covered by the film, c_m is the mass sensitivity of the device (1.3×10^{-6} cm^2-s/g for quartz (11)) and v_o and f_o are unperturbed wave velocity and oscillator frequency, respectively. To optimize the sensitivity of the device, all the films used in this study covered the entire active area of the SAW device ($\kappa = 1$). Relationships to account for frequency shifts due to changes in other properties, such as film conductivity (12) or stiffness (13), are given elsewhere.

Since frequency can easily be measured to within 1 Hz, 10 ppb changes in frequency can be monitored for the 97 MHz devices used in this study. This gives a limit of mass resolution of approximately 80 pg/cm^2. Even though the frequency stability is less than 10 Hz (11) under ideal situations (i.e., minimal environmental perturbations), the noise level observed during an experiment is sometimes on the order of 1 ppm (100 Hz). This noise level probably results from fluctuations in the total pressure or concentration of the sorbing species in the gas phase over the device. For applications where the total frequency response is small, improvements in the experimental gas flow system may lead to a significant reduction in this noise level.

Characterization of Porous Samples. Adsorption isotherms are a useful analytical method for characterizing porous samples. This measurement consists of monitoring the amount of a molecular species taken up by the sample as a function of the relative saturation of the species in the gas phase over the sample. This relative saturation is defined as p/p_o, where p is the partial pressure of the test species and p_o is its saturation vapor pressure at the temperature of the run. In a typical measurement, p/p_o is increased from zero to a value near 1 (onset of bulk condensation) and then decreased back down to zero. For many porous samples, hysteresis is observed between the adsorption and desorption branches of the isotherm obtained in this manner (14).

The surface area of a sample can be determined from an adsorption isotherm using a well known model developed by Brunauer, Emmett and Teller (15). This BET analysis is based on using one binding energy for the adsorption of the first monolayer on the surfaces and a second binding energy for adsorption of subsequent layers. A linear form of the resulting relationship (15,16) is given by:

$$\beta \equiv \frac{p/p_o}{n(1 - p/p_o)} = \frac{1}{n_m c} + \frac{(c - 1)}{n_m c} \frac{p}{p_o} \qquad (2)$$

in which n is the number of adsorbed molecules, n_m is the number of adsorbed molecules corresponding to one monolayer on the available surfaces, and c is a constant which depends on the two binding energies. Linear behavior in the plot of β vs. p/p_o is typically observed for p/p_o values between 0.05 and 0.3 (14,16). The slope (S) and intercept (I) values are used to calculate $n_m = 1/(S + I)$ and $c = 1 + S/I$. The surface area (A) of the sample is then calculated using $A = n_m a_m$, where a_m is the adsorbed area per molecule. The standard adsorbate used for this analysis is N_2 at its boiling point (77 K) due in part to the relative constancy of a_m at 0.162 nm^2/molecule on a wide variety of sample materials (14).

A pore size distribution (PSD) of a sample is a measure of the cumulative or differential pore volume as a function of pore diameter. PSDs can be calculated from adsorption isotherms based on an analysis which accounts for capillary condensation into pores. This analysis (14,16) uses a model of the pore structure combined with the Kelvin equation (17) to relate the pore size to the value of p/p_o at which pore "filling" occurs. Due to limitations in this technique, only pores with diameters from about 3 to 50 nm, called mesopores (14), can be characterized. This pore size range, however, is typical of many porous samples of interest. For samples with pores smaller or larger than this range, alternative techniques, such as mercury intrusion for large pores (14,16), are typically more suitable.

Determination of Diffusion Coefficients. Even though diffusional characteristics of polymers can be complex, there are many situations where simple concentration-independent Fickian diffusion occurs for small molecular species (18,19). For a polymer film of constant thickness L, the concentration C(x,t) of species at position x (measured perpendicular to the surface) and time t is governed by the one-dimensional Fick's law (20):

$$\frac{\partial C}{\partial t} = D \frac{\partial^2 C}{\partial x^2} \tag{3}$$

where D is the diffusivity. With the film on an impermeable substrate, the boundary conditions are: (1) $\partial C/\partial x = 0$ at $x = 0$ (i.e., the substrate acts as a diffusion barrier) and (2) $C(L,t) = C_o$ for $t \geq 0$ (i.e., the concentration at the gas/film interface is held constant). A uniform concentration in the film prior to the start of diffusion is typically assumed: $C(x,t) = C_o'$ for $t < 0$. If the diffusion coefficient is constant, Equation 3 can be solved using these boundary conditions to yield the following analytical expression (20) for the amount of absorbed species as a function of time (M(t)):

$$M(t) = M_{max}\left[1 - \frac{2}{\pi^2}\sum_{n=1}^{\infty}\frac{e^{-\psi^2 Dt}}{(n-\frac{1}{2})^2}\right] \tag{4}$$

where $\psi = \pi(n-\frac{1}{2})/L$ and M_{max} is the incremental amount absorbed at film saturation ($t \rightarrow \infty$). This expression predicts that M(t) is proportional to $\sqrt{t}$ until $M(t)/M_{max} > 0.6$. Thereafter, "reflection" of species at the impermeable SAW substrate decreases the net flux into the film. If the SAW frequency shift is linear with amount absorbed ($\Delta f(t) = kM(t)$, where k is a constant), Equation 4 can be used to obtain an expression for the transient frequency response $\Delta f(t)$ of the SAW oscillator (hereafter referred to as a frequency transient).

Experimental

SAW Devices. SAW devices were designed at Sandia National Labs and manufactured on crystalline ST-cut quartz substrates by Crystal Technologies (Palo Alto, CA). Transducers, consisting of 50 finger-pairs with a periodicity (Λ) of 32 μm, were formed photolithographically from 200 nm-thick Au-on-Cr metallization. Fingers are 8 μm wide and 1.7 mm long. Center-to-center separation of transducers is 7.36 mm. Each SAW device was mounted in a standard 25 x 13 mm flatpack using two beads of RTV silicone rubber to damp reflections from the ends of the crystal. An ultrasonic bonder was used to attach 25 μm gold wires between the device and the flatpack leads. The flatpack was mounted in a brass test case.

Thin Film Formation. The porous films used in this study were prepared from sol-gel solutions by dip-coating SAW devices at 20 cm/min, followed by drying and heating in air at 400 °C for 5 min. The sol-gel system used, denoted Four-Component, contains SiO_2, B_2O_3, Al_2O_3 and BaO in ratios of 71:18:7:4 (by weight). The reaction conditions were similar to those described elsewhere (21) except that this solution did not contain Na_2O. Thin films were formed using the as-prepared solution as well as a solution which had been "aged" at pH 3 and 50 °C until gelation occurred (about three weeks). The gel was ultrasonically disrupted to form a fluid sol suitable for dip-coating. This aging process is known to result in larger polymeric species in the sol-gel solution (22) which should result in a larger pore structure in the deposited film. Film thicknesses and refractive indices were obtained using Gaertner Model L119 or Rudolph Auto ELIV ellipsometers. Percent porosity was evaluated from the refractive indices using the Lorentz-Lorenz (23) relationship assuming a skeletal refractive index of 1.45.

The polymer film used to study diffusional properties was formed by spin coating at 4000 rpm using Dupont PI-2545. This formulation is one of a class of polyamic acid solutions which converts to a fully aromatic polyimide coating when heated. After being partially imidized at 115 °C for 30 min, a final film thickness of 1.8 μm (measured using a surface profilimeter) was obtained.

Oscillator Circuitry. As shown in Figure 1, the oscillator circuitry consists of a wide band amplifier (HP 8447D) and variable attenuators (HP 335C and 355D) to adjust the net gain, a tunable bandpass filter (K&L 5BT-95/190-5N) to prevent spurious oscillation at other frequencies, and a phase shifter (Merrimac PSL-4-100B) to tune the oscillation frequency. A 20 dB directional coupler splits off a fraction of the power to a frequency counter (HP 5385A).

The synchronous frequency, at which the transducers most efficiently excite a SAW, is given by $f = v/\Lambda$. For our devices, v = 3100 m/s and Λ = 32 μm, resulting in f = 97 MHz. The optimum operating frequency is determined by network analysis using an HP 8656A signal generator and an HP 8405A vector voltmeter.

Gas Flow System. The experimental gas flow system used is shown in Figure 2. A stainless steel lid containing gas inlet and outlet was sealed to the flatpack using a teflon gasket. For N_2 adsorption runs, the device test case and a stainless steel coil on the gas inlet side were placed in a Dewar flask and submerged in liquid N_2. For diffusion runs, the case and coil were placed in a well-insulated chamber equipped with a heat exchanger fed by a temperature bath (Haake A81). SAW device response was monitored using the computer (HP 9816) controlled test apparatus shown in Figure 2. A Data Acquisition/Control Unit (HP 3497A) was used to control flow rates of various N_2 and He streams using Tylan Model FC-260 mass flow controllers and Fluorocarbon Corp. series DV-224 solenoid-operated teflon valves. The N_2 was dried by passage through a bed containing Drierite and molecular sieves. To prevent contamination, all fittings and tubing were teflon or stainless steel. Before a run was started, the test films were purged by flowing N_2 over the device at 165 °C (aged Four-Component film) or 25 °C (all other runs).

To measure an adsorption isotherm, the oscillation frequency was monitored as the computer adjusted the partial pressure p of N_2 in the gas phase over the device at 77 K. This was done by controlling the relative flow rates of a He mix-down stream and a N_2 carrier stream. Since He is nonadsorbing at liquid N_2 temperature (16), $p = 0$ for the mix-down stream, while for the carrier stream, $p = p_o$ since the device temperature is maintained at the boiling point of N_2. The value of p/p_o was incremented every 3 sec from a value of 0 at the start of the run to a value around 0.95 and then back to 0. It was found that two hours for a full adsorption isotherm was sufficiently slow to maintain adsorption equilibrium (10).

For diffusion runs, the mix-down stream was N_2 at 25 °C and the carrier stream was N_2 saturated with an organic species by passage

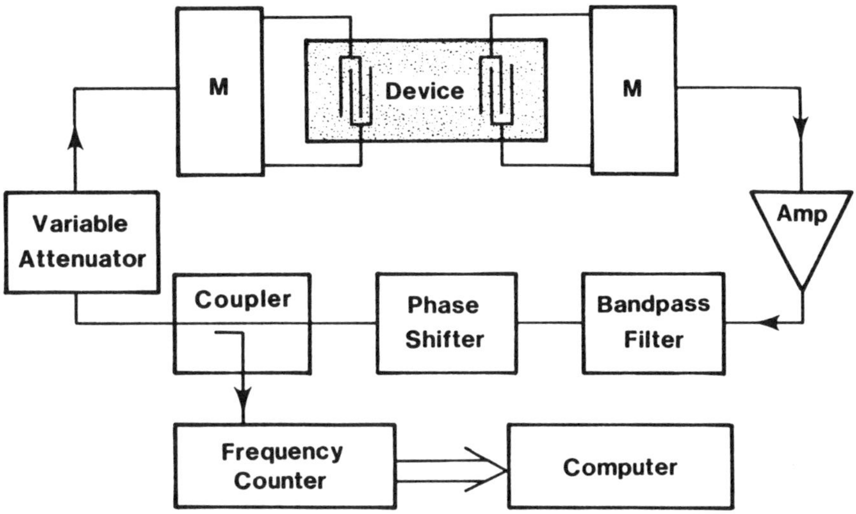

Figure 1. Schematic of a SAW device as the feedback element of an oscillator circuit. Input and output interdigital transducers are used to excite and detect a SAW in the piezoelectric ST-cut quartz substrate.

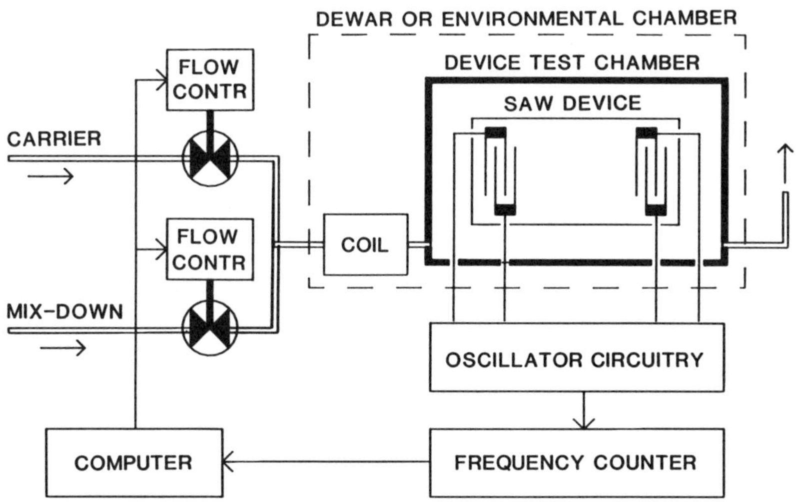

Figure 2. Schematic of the experimental set-up for control of vapor flow and measurement of SAW oscillator frequency.

through a bubbler containing the test liquid. To maintain the carrier stream at $p = p_o$, the same temperature for the bubbler and the device is required. Therefore, the bubbler was immersed in the temperature bath. After a stable oscillation frequency was obtained, the p/p_o value was changed stepwise and the frequency transient that occurred was monitored until a new stable frequency was obtained.

Results and Discussion

Characterization of Sol-Gel Films. Adsorption isotherms for the two Four-Component films are shown in Figure 3. Since the only effect giving SAW device response should be the mass loading that occurs due to adsorption onto the surfaces of the film, the experimental frequency shifts (right axis) have been converted into an adsorbed mass/area of film (left axis) using Equation 1. For comparison, the adsorption of one monolayer of N_2 onto the flat SAW substrate gives a mass loading of 28.7 ng/cm^2 ($1/a_m = 6.2 \times 10^{14}$ molecules/cm^2) which should result in a 3.6 ppm frequency shift.

With this in mind, it is clear that the film from the unaged solution (Figure 3a) exhibits a small amount of adsorption. The adsorption isotherm shape resembles what has been called a Type II isotherm system (14), typical of nonporous samples. The apparent hysteresis in this isotherm is probably the result of long-term drift over the course of the two hour experiment. This relatively small shift (1-2 ppm) may be the result of temperature changes due to increases in the concentration of O_2 (condensed from the atmosphere) in the liquid N_2 surrounding the test case. A linear BET plot is obtained from these data ($0.05 < p/p_o < 0.3$) and the resulting surface area is calculated to be 1.3 cm^2/cm^2 of film. This rather unconventional unit of film surface area/nominal film area is appropriate here since it is the sample area, and not its mass, which is the well-known parameter. The experimental surface area value is very close to the value of 1.0 cm^2/cm^2 expected for a flat, nonporous film on the SAW substrate. The slightly higher value may be due to surface roughness rather than film porosity. Furthermore, the lack of measurable porosity with N_2 (no pores > 0.4 nm) is consistent with the refractive index value (1.454) being indistinguishable from the value expected for a dense glass of this composition (1.45).

The film formed from the aged solution exhibits over 30 times as much adsorption as the unaged film (see Figure 3b). The shape of the isotherm, especially the hysteresis loop at higher p/p_o values, resembles a Type IV isotherm system (14), typical of samples containing mesopores. The BET surface area of 33 cm^2/cm^2 of film indicates a large amount of porosity for a film which has a thickness of only 148 nm.

As discussed previously, pore size distributions can be obtained from adsorption isotherms due to the dependence of capillary condensation on pore size. A PSD obtained from the desorption branch of the data in Figure 3b using standard calculation procedures (16) is shown in Figure 4. A relatively unimodal distribution is obtained with a median diameter of around 6 nm. As shown in Table I, results with films having shorter aging times indicate that the median pore diameter increases with aging time. This is expected based on the effects of aging on the size of the polymeric species in the precursor sol-gel solution (22).

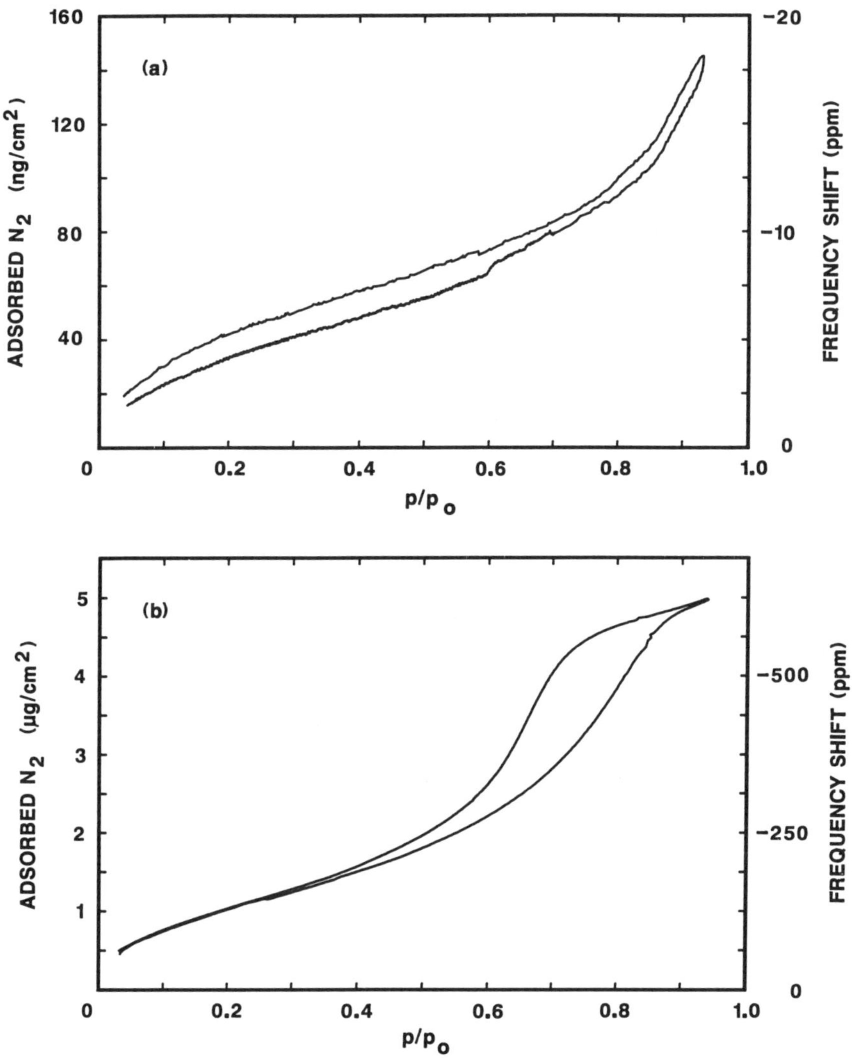

Figure 3. Adsorption isotherms obtained using coated SAW devices for (a) unaged and (b) aged Four-Component sol-gel films.

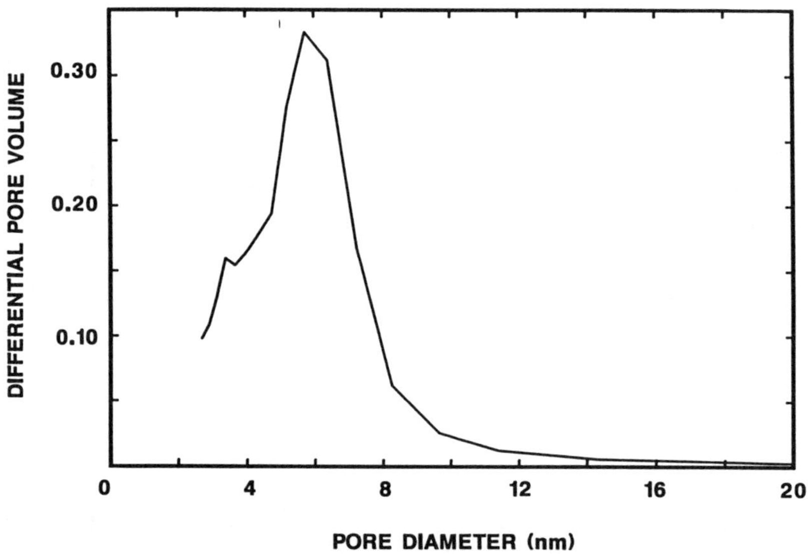

Figure 4. A pore size distribution obtained from the desorption branch for the aged Four-Component film (Figure 3b).

Table I: Effect of Aging Time on Film Porosity

Aging Time (Days)	Adsorption Isotherm	Median Pore Diameter	Volume Percent Porosity: Refractive Index Method	Volume Percent Porosity: Adsorption Method
0	Figure 3a	<0.4	0	0
7	Not shown	3.4	42	24
14	Not shown	4.0	50	33
21	Figure 3b	6.2	56	52

For the sample in Figure 3b, a percent porosity of 52% (by volume) is calculated from the film thickness and the amount of N_2 adsorbed/ condensed at high p/p_o values. This is very close to the value of 56% calculated (23) from the refractive index (1.18). Since N_2 adsorption only probes pores which are both large enough to accept an N_2 molecule and are accessible from the gas phase, while the refractive index is sensitive to all the pores in the film, it is reasonable that the N_2 adsorption result is somewhat lower than the refractive index result. Films with smaller pores would be expected to have a larger fraction of their porosity inaccessible to N_2. As shown in Table I, this is verified by the larger discrepancy between these two porosity measures for films formed from solutions at shorter aging times.

One area of continuing research with regards to the SAW-obtained adsorption isotherms is the shape of the isotherm at low p/p_o values ($p/p_o < 0.3$). A good indicator of this shape is the value of the BET parameter c obtained by fitting the data to Equation 2. For N_2 on bulk silica samples (e.g., powders), c values are typically on the order of 80 to 150, which is reflected in a more dramatic rise in adsorption during the initial stages of a run ($p/p_o < 0.2$) than in Figure 3. The c values obtained from the data presented here are 16 for the dense film and 21 for the porous film. Some factors being considered to account for these differences are: (1) differences in the chemical nature of the surfaces, (2) incomplete purging of the sample prior to analysis or (3) inadequate sealing of the gas system resulting in a small amount of N_2 "leaking" into the gas stream (i.e., during the He purge at the start of the run, some N_2 may already be adsorbed on the film surfaces). To evaluate the last two possibilities, a new test case formed on a vacuum flange is being fabricated. This case should provide an excellent seal as well as being able to withstand higher purge temperatures.

The results obtained with these films indicate that this SAW-based technique is a powerful tool for characterizing porous thin films. It is based on a standard industry practice involving the use of N_2 adsorption isotherms. The main advantage of the technique is that the SAW device decreases the detection limit for the amount of adsorbed N_2 by several orders of magnitude over conventional techniques used for this type of characterization (16). This increased sensitivity is critical for thin films, due to the small total surface area present. For example, conventional instruments typically require a total sample surface area on the order of 10,000 cm^2

in order to obtain a N_2 adsorption isotherm. For thin films, where the surface area may only be an order of magnitude greater than the nominal film area, this would require the impractical addition of 1000 cm^2 of test film. The SAW technique, however, only requires the active SAW area (0.15 cm^2) for a measurement, even for nonporous films.

Characterization of Diffusion in Polymer Films.

Frequency transients obtained with the polyimide-coated SAW device upon exposure to methanol at various p/p_o values are shown in Figure 5a. At each of the points marked, the relative flow rates of the dry N_2 mix-down stream and the saturated carrier stream were changed to obtain a higher p/p_o value. After each change in ambient vapor concentration, a sorption transient is observed as species enter the film to reestablish equilibrium with the gas phase concentration. The positive frequency shift indicates that mass loading effects do not dominate in this case as they did with the silicate films: increased mass loading gives negative frequency shifts. This positive shift, which was observed for a variety of test species (Frye, G. C.; Martin, S. J.; Ricco, A. J. Sensors & Materials, submitted), is probably due to an increase in the stiffness of the polymer at these low p/p_o values. This stiffening has been observed by other researchers with other polymer systems (24). Swelling of the polymer, which would result in an increase in the film thickness, may also play a role in giving this positive shift.

From careful inspection of the data in Figure 5a, it can be seen that the frequency transients are faster at higher p/p_o values. These differences can be quantified by obtaining D values for each step. Equation 4 can be used to extract D values from these transients if it is assumed that: (1) Fickian diffusion occurs, (2) the diffusion coefficient over each small change in p/p_o can be taken as constant and (3) the SAW response is linear with concentration in the polyimide film ($\Delta f(t) \propto M(t)$). Even though these assumptions do not hold in all cases, they appear to be valid for small p/p_o steps, such as were used in obtaining the data in Figure 5a. A nonlinear least squares fit of Equation 4 to the data was used to obtain values for: (1) D, (2) t_o, the time for the start of the diffusion (adjusted to account for the time required for flow through the tubing leading to the test case) and (3) Δf_{max}, the total frequency shift in the step.

An example of a fit of Equation 4 to an experimental frequency transient is shown in Figure 5b. The experimental data for the final p/p_o step from 0.051 to 0.061 are shown by the points; the line is the best fit of Equation 4 to the data. The data are plotted versus $\sqrt{t}$ to show the linear response for the initial stage of diffusion. It is clear that this analysis fits the data well. The rms error for this fit is less than 0.7% of Δf_{max}. The D value obtained at 25 °C is 8.4×10^{-11} cm^2/sec.

Similar fits to the data in the other p/p_o steps were obtained. Based on the calculated D values, a concentration dependence on D was observed. Diffusivity values vs. the average p/p_o value for each step are shown in Figure 5c. A factor of four increase in D is seen for this relatively small range of p/p_o values. The data appear to be linear over the range studied. It has been observed with small

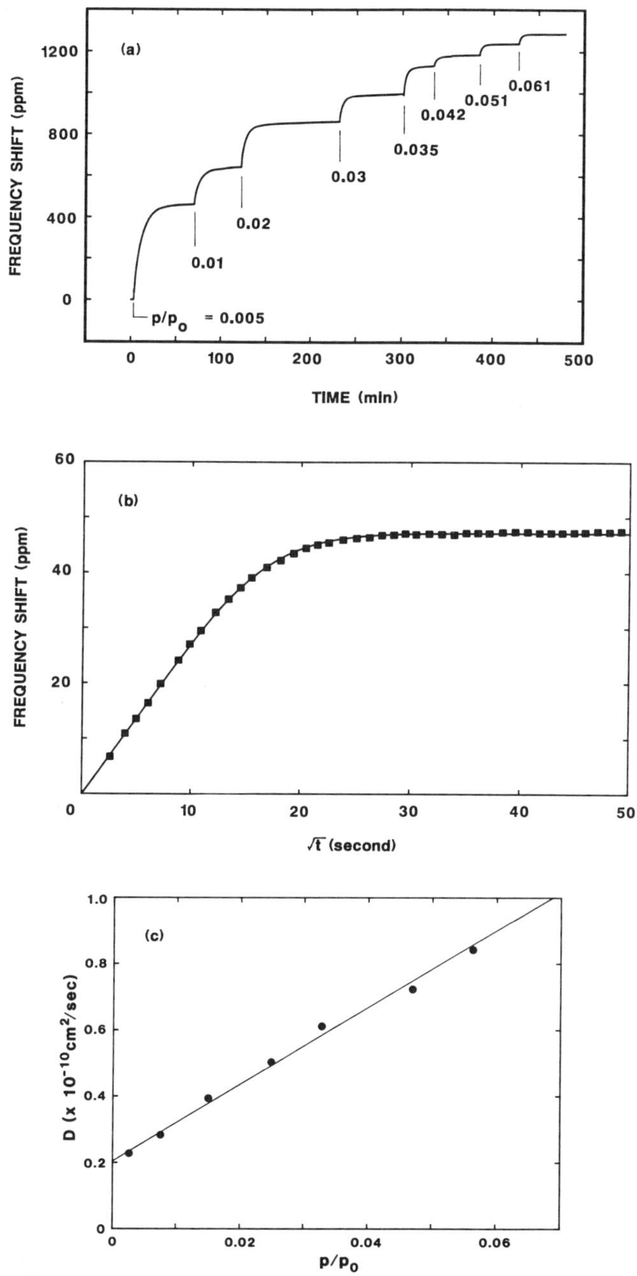

Figure 5. (a) SAW frequency transients during diffusion of methanol at various p/p_o values into a 1.8 μm polyimide film. (b) Fit of Equation 4 to the final frequency transient (p/p_o from 0.051 to 0.061). (c) D values calculated from the frequency transients as a function of the average value of p/p_o for each step.

organic species in other polymer systems that log(D) increases linearly with concentration in the polymer (25). For the polyimide film used in this study, frequency shifts appear to be dominated by elastic changes, making it impossible to quantify the amount absorbed. This means that these transients cannot be used to extract solubility information and, therefore, a direct comparison to these predictions is not possible. However, a similar effect of increasing D with concentration is seen. For a polystyrene film we have studied, mass loading does appear to dominate the frequency shift, allowing the extraction of both D and solubility values (i.e., using Equation 1). However, the very low D values (8.7×10^{-15} cm^2/sec for toluene; p/p_o from 0 to 0.05) made an investigation into the effect of concentration on D impractical (i.e., almost a full day would be required for each p/p_o step).

The main advantage of the SAW-based technique for monitoring the accumulation of species in a polymer sample is the ability to use a thin film. Since the time required to reach a set fraction of total saturation is proportional to the square of the diffusional length (20), the use of thin films dramatically reduces the time required for analysis over techniques (18) that use bulk polymer samples. The range of D values obtainable can be determined by considering various extremes. Using very thin films (tens of nm) and relatively long equilibration times (hours), D values below 10^{-15} cm^2/sec can be obtained. Alternatively, using thicker films (on the order of 1 μm) and shorter saturation times (tens of seconds), D values up to 10^{-9} cm^2/sec can be accommodated. This wide range of relatively low D values is useful for characterizing films which exhibit slow diffusional properties, such as organic molecules in polymers.

Conclusions

The results of this investigation illustrate the utility of SAW devices in characterizing the properties of thin films formed on the substrate of the device. Using the SAW device as an extremely sensitive microbalance, N_2 adsorption isotherms have been obtained directly on thin films. This has allowed the characterization of the surface area and pore size distribution in the film. In general, these measurements are not possible with conventional instrumentation due to the low total surface area present in thin film samples. As another application, SAW device frequency transients can be used to monitor diffusion of species in polymer films in real-time. The short diffusional length scale present in a thin film makes the time required for saturation of the film orders of magnitude shorter than would be required with bulk samples. This allows for a dramatic decrease in the time required for determining diffusion coefficients.

Acknowledgments

We gratefully acknowledge the helpful discussions and suggestions of Peter Green, Doug Adolf, and Dan Doughty and the technical assistance of Barb Lammie and Carol Ashley, all of Sandia National Laboratories. This work was performed at Sandia National Laboratories, supported by the U.S. Department of Energy under contract no. DE-AC04-76DP00789.

Literature Cited

1. Martin, S. J.; Frye, G. C.; Ricco, A. J.; Zipperian, T. E. Proc. 1987 IEEE Ultrasonics Symp., 1987, p 563.
2. Martin, S. J.; Schweizer, K. S.; Schwartz, S. S.; Gunshor, R. L. Proc. 1984 IEEE Ultrasonics Symp., 1984, p 207.
3. Venema, A.; Nieuwkoop, E.; Vellekoop, M. J.; Ghijsen, W. J., Barendsz, A. W.; Nieuwenhuizen, M. S. IEEE Trans. Ultrasonics, Ferroelectrics and Freq. Control 1987, UFFC-34, 148.
4. D'Amico, A.; Palma, A.; Verona, E. Sensors and Actuators 1982, 3, 31.
5. Bryant, A.; Lee, D. L.; Vetelino, J. F. Proc. 1981 IEEE Ultrasonics Symp., 1981, p 171.
6. Chuang, C. T.; White, R. M. Proc. 1982 IEEE Ultrasonics Symp., 1982, p 295.
7. Snow, A.; Wohltjen, H. Anal. Chem. 1984, 56, 1411.
8. Brace, J. G.; Sanfelippo, T. S.; Joshi, S. G. Sensors and Actuators 1988, 14, 47.
9. Grate, J. W.; Snow, A.; Ballantine, D. S.; Wohltjen, H.; Abraham, M. H.; McGill; R. A.; Sasson, P. Anal. Chem. 1988, 60, 869.
10. Ricco, A. J.; Frye G. C.; Martin, S. J. Langmuir 1988, 5, 273.
11. Martin, S. J.; Ricco, A. J.; Ginley, D. S.; Zipperian, T. E. IEEE Trans. on Ultrasonics, Ferroelectrics and Freq. Control 1987, UFFC-34, 142.
12. Ricco, A. J.; Martin, S. J.; Zipperian, T. E. Sensors & Actuators, 1985, 8, 319.
13. Wohltjen, H.; Dessy, R. Anal. Chem., 1979, 51, 1458.
14. Gregg, S. J.; Sing, K. S. W. Adsorption, Surface Area and Porosity; Academic Press: New York, 1982.
15. Brunauer, S.; Emmett, P. H.; Teller, E. J. Am. Chem. Soc. 1938, 60, 309.
16. Lowell, S.; Shields, J. E. Powder Surface Area and Porosity; Chapman & Hall: New York, 1984.
17. Adamson, A. W. Physical Chemistry of Surfaces, 4th ed.; John Wiley & Sons: New York, 1982; Chapter XVI.
18. Felder, R. M.; Huvard, G. S. In Polymers, Part C: Physical Properties; Fava, R. A., Ed.; Academic: New York, 1980; p 315.
19. Berens, A. R. J. Polymer Sci., Poly. Phys. Ed. 1979, 17, 1757.
20. Crank, J. The Mathematics of Diffusion; Clarendon Press: Oxford, 1975; Chapter 4.
21. Brinker, C. J.; Mukherjee, S. P. J. Mat. Sci. 1981, 16, 1980.
22. Brinker, C. J.; Hurd, A. J.; Ward, K. J. In Ultrastructure Processing of Advanced Ceramics; Mackenzie, J. D.; Ulrich, D. R., Eds.; Wiley: New York, 1988; p 223.
23. Born, M.; Wolf, E. Principles of Optics; Pergamon Press: New York, 1975; p 87.
24. Deopura, B. L.; Sengupta, A. K.; Verma, A. Polymer Communications 1983, 24, 287.
25. Fujita, H. In Diffusion in Polymers; Crank, J.; Park, G. S., Eds.; Academic: New York, 1968; p 75.

RECEIVED March 9, 1989

Chapter 15

Elastic Properties of Thin Polymer Films Investigated with Surface Acoustic Wave Devices

David S. Ballantine, Jr.[1,3], and Hank Wohltjen[2]

[1]Geo-Centers, Inc., 10903 Indian Head Highway, Fort Washington, MD 20744
[2]Microsensor Systems, Inc., P.O. Box 8, Springfield, VA 22150

The observed glass transition temperatures (T_g) of several thin polymer films on surface acoustic wave (SAW) devices are 50-60 °C higher than the T_g results reported using other methods such as DSC. The increase in the onset of T_g is the result of interaction of the high frequency SAW with the polymer film, consistent with the time-temperature superposition principle. The T_g were identified as localized minima in the frequency curves, or by changes in the slope of the curves, as the coated sensors were heated between 35-110 °C. Potential applications of SAWs for the characterization of polymer materials and the implications of these findings for the interpretation of SAW data are discussed.

Polymeric materials are being employed in an increasing number of novel applications. As stronger, more flexible and more durable materials are discovered the demand for these materials will continue to grow. The chemical and physical properties of these materials will determine the applications for which they may be employed. Thus, the rapid and reliable characterization of these properties will be crucial.

In the area of chemical sensors, thin polymer films are routinely used as coatings for the semi-selective sorption of chemical vapors. One sensor technology, the surface acoustic wave (SAW) device, has demonstrated excellent sensitivity as a vapor sensor when coated with films having appropriate solubility properties (1). To date, most sensor applications have utilized the extreme mass sensitivity of the devices. In this paper, we will examine the response mechanisms of the SAW sensor and demonstrate its sensitivity to changes in the elastic properties of the coating materials. Finally, we will discuss the significance of these results in terms of current sensor applications, and the advantages of the SAW for polymeric materials characterization.

[3]Current address: Department of Chemistry, Northern Illinois University, DeKalb, IL 60115

0097–6156/89/0403–0222$06.00/0

Background

Since its introduction in the early 1970's, the SAW device has seen increasing utilization as a chemical sensor due to its sensitivity, potentially rapid response, and low cost. A brief description of the operating principles of the SAW will aid in understanding the basis of this study and in the interpretation of the results. Typically, a SAW device consists of a set of two interdigital transducers which have been microlithographically fabricated on the surface of a piezoelectric substrate. The application of a time-varying electric potential to one transducer causes a mechanical deformation of the substrate, resulting in the generation of a surface acoustic wave. The second transducer converts the mechanical wave back into an electrical signal. The properties of this wave (amplitude, frequency, phase) are sensitive to perturbations occurring on or near the surface of the substrate. More detailed discussion of the SAW operation can be found in reference (2). The generation of different types of surface waves is possible (3); for the purposes of this work, when we refer to the surface acoustic wave we specifically mean Rayleigh-type surface waves.

The response of the SAW device is the combined result of changes in the mass loading, conductivity, or elastic properties of the surface film. Equations describing the effects of changes in these properties on the frequency of the device have been derived previously (2,4). For many sensor studies, non-conducting polymer films are employed. One equation, given below, describes the response behavior for a SAW device coated with a thin, lossless, isotropic, non-conducting film,

$$\Delta f = (k_1+k_2)\rho h f_o^2 - k_2 h f_o^2 (4\mu/V_R^2)[(\lambda+\mu)/(\lambda+2\mu)] \qquad (1)$$

where k_1 and k_2 are material constants for the quartz substrate, V_R is the Rayleigh wave velocity, h is the film thickness, ρ is the density, μ is the shear modulus, λ is the Lame constant, and f_o is the fundamental frequency of the device. Typical values for these parameters for ST-cut quartz substrates are given in Table I. The first half of the equation yields the frequency shift resulting from mass loading, while the second half describes the effect of changes in the elastic properties of the film on the resonant frequency.

To date, the selection of coatings for vapor sensor applications has been mostly empirical, requiring the screening of a large number of candidate materials to identify coatings with sufficient sensitivity to the vapor of interest. To address this problem, recent work has focused on characterizing the observed sensor responses in terms of solubility interactions (1,5). The sorption of a solute vapor into a solvent coating can be quantitatively defined as K, the partition coefficient. A modified version of Equation 1 has been used to predict the frequency response of coated SAWs to specific vapors utilizing K values calculated from gas-liquid chromatography (GLC) data (5). The inherent assumptions in that work are that (1) the polymer coating is a lossless film (that is, there is no significant attenuation of the surface wave resulting from interactions with the surface film), and (2) the contributions to the observed response from elastic properties of the film are negligible. To justify the second assumption, polymer films were selected that would be above their glass transition temperatures (T_g) at the operating temperature of the device.

TABLE I. TYPICAL PARAMETER VALUES*

parameter	description	value (units)
k_1	material constant	-8.7×10^{-8} m^2s/kg
k_2	" "	-3.9×10^{-8} " "
V_R	Rayleigh wave velocity	3158 m/s
	density of polymer film	1000 kg/m^3
h	film thickness	1×10^{-7} m
f_o	SAW resonant frequency	158 MHz
μ	shear modulus (glass)	10^{10} $dyne/cm^2$
	" " (rubber)	10^7 " "
$(\lambda + \mu / \lambda + 2\mu)$	----------------------	0.85 (typical value)

*Values for k_1, k_2, and V_R (for ST-quartz) are taken from reference (5).

In order to assess the validity of the second assumption, a brief discussion of the elastic properties of polymers is needed. The viscoelastic behavior of a polymer is depicted schematically in Figure 1. The parameter of interest in the case of SAWs is the shear modulus, denoted as G in Figure 1 and as μ in Equation 1. Simply stated, the modulus is a measure of the rigidity of the polymer. The regions of interest are the glassy region (where the polymer is a hard, rigid material) and the elastomeric region (where the polymer is a rubber). Rigid, glassy polymers typically have high modulus values on the order of 10^9 - 10^{10} $dyne/cm^2$. In this region, the polymer chains are locked into the lowest energy conformations and there is insufficient energy in the system to allow free rotation around the polymer backbone. As the temperature increases the polymer becomes an elastomer. In this region, there is sufficient energy in the system for free rotation to occur. This additional rotational freedom is manifested as a softening of the polymer, with a corresponding decrease in the modulus to 10^4 - 10^7 $dyne/cm^2$. The temperature at which this softening occurs is the T_g. Other changes occur at this temperature that can be monitored to identify the T_g experimentally. These include changes in specific volume of the polymer, index of refraction, gas diffusion coefficients, thermal expansion coefficients (measured by dilatometry), and specific heat (measured by differential scanning calorimetry (DSC) or by differential thermal analysis (DTA)). General discussions of the elastic properties of polymers can be found in references (6,7).

The T_g of a polymer increases as a function of the oscillating frequency of an applied stress. This phenomenon was first described by Williams, Landel and Ferry in 1955, and became the basis of the time-temperature superposition principle (8). Previous work demonstrated that the SAW could be used to characterize polymeric materials (9,10). Both thin films and bulk samples were characterized by monitoring changes in the amplitude of the surface wave as the temperature of the polymer samples was increased. The results indicated that, for thin polymer films, significant interaction of the polymer film with the high frequency surface wave can occur, resulting in an increase in the T_g of the polymer. This observation is consistent with the time-temperature superposition principle.

The work presented here was motivated by two factors. First, it had been assumed that polymer films used in previous sensor applications were above their T_g at the SAW operating temperature. If the T_g of a given polymer film increases significantly due to effects of the high frequency surface wave, then the elastic properties of the film must be taken into consideration when interpreting sensor responses. Second, since frequency measurements with the SAW device are inherently more sensitive than amplitude measurements, such measurements may prove useful in the area of materials characterization. The following experiments were performed to verify this potential and to investigate the possible effect of elastic properties on sensor responses.

Experimental

The polymer coatings studied are given in Table II, along with pertinent physical parameters. These coatings were selected based on availability, since they are among coatings previously used for chemical sensor studies at the Naval Research Laboratory (NRL). In addition, the T_g and melting temperature (T_m) are within the range that could be easily investigated using our experimental apparatus. Of these coatings, fluoropolyol (FPOL) and poly(ethylene maleate) (PEM) were provided by the Polymeric Materials Branch, Chemistry Division, of the NRL in Washington D.C. They are both linear polymers with no observed crystallinity. PEM is a polyester material with a repeating monomer unit of 35-50. FPOL is a highly viscous epoxy pre-polymer with

TABLE II. POLYMER COATINGS AND PHYSICAL PARAMETERS

polymer	T_g (°C)	T_m (°C)	n	ρ (g/cm^3)
fluoropolyol (FPOL)	10	---	8-10	1.653 (25 °C) 1.563 (90 °C)
poly(ethylene maleate) (PEM)	-10	---	35-50	1.353 (25 °C)
ethyl cellulose (ECEL)	43	157-165	---	1.14 (at T_m)
poly(caprolactone)	<-40	60	---	---

a repeating monomer of 8-10. Densities were measured using a float technique, and T_g were measured by DSC at the NRL Chemistry Division. Additional information regarding the structure and properties of these materials can be found in reference (1).

Poly(caprolactone) and ethyl cellulose were obtained from Aldrich, and were used as received. They are semi-crystalline materials and would be expected to exhibit localized changes in elastic properties at both T_g and the melting point, T_m. Physical parameters for these polymers were obtained from the Aldrich catalog. No value was available for the T_g of PCAP; the value in Table II is estimated using the relationship

$$T_m = K\ T_g \qquad (2)$$

where K = 1.4 - 2 (7).

The 158 MHz dual SAW devices were obtained from MicroSensor Systems, Inc., Fairfax. VA (MSI part #: SD-158A). They consisted of two individual delay lines manufactured on a single ST-cut quartz chip. Electrical connections to the transducers were established by wire bonding to a T-08 plug-in package. Additional information regarding the dimensions and geometry of the devices can be obtained from MSI.

Dilute solutions of the coating materials were prepared in chloroform. These solutions were then aspirated using an airbrush to generate a finely dispersed aerosol, which was deposited on one side (i.e. one delay line) of the dual SAW device. The film thicknesses were estimated using the first term from Equation 1, neglecting elastic effects. Estimated film thicknesses were on the order of 50 - 100 nm. The other delay line remained uncoated, acting as a reference to correct for frequency deviations due to fluctuations in ambient temperature and pressure.

The coated devices were then placed in a heating chamber consisting of a 3/4" brass Swagelok union fitting which was wrapped with heating tape. The temperature of the heating tape was controlled by a Variac, and the temperature of the chamber was monitored using a YSI Series 400 thermistor. This thermistor was incorporated in a circuit which converted the thermistor resistance into a voltage that could be read directly by the computer-controlled data acquisition system. Prior to reaching the sensor, the dry air stream was heated via a coiled tube located within the chamber. A constant flow rate of 40-50 ml/min was maintained during the experiments to minimize the effects of temperature gradients and localized heating near the sensor surface.

The frequencies of the individual delay lines were monitored independently, and the frequency difference between the two delay lines was output from an electronic mixer. These frequencies (and frequency differences) were recorded as the devices were first heated from 35 $^{\circ}$C to 110 $^{\circ}$C, and as they were subsequently cooled to room temperature. Each device was subjected to repeated heating/cooling cycles, with frequencies being recorded continuously as a function of temperature. The coatings were then removed by washing the SAW devices in chloroform in an ultrasonic cleaner. The bare devices were once again placed in the heating chamber and subjected to successive heating/cooling cycles. Frequency curves of the bare devices were recorded for use as blanks.

Results and Discussion

Once the effect of temperature on the frequency of a bare device was known, the effect of temperature on the coatings could be determined by subtracting the results for the bare devices from the frequency-temperature curves for the coated devices. The frequency shifts recorded during these temperature studies are reported in parts per million (ppm), which is defined as the observed frequency shift (in Hz) divided by the resonant frequency of the device (in MHz). For the devices used in this study, 1 ppm is equivalent to 158 Hz. In order for several related curves to be easily viewed on the same plot, individual curves are offset from one another by several hundred ppm.

Examples of the frequency-temperature curves for a bare device are given in Figure 2. Channel 1 and channel 2 refer to the two delay lines of the dual device. When used as a sensor, channel 1 would be coated and channel 2 would remain uncoated to act as a reference. The difference curve is the output from the mixer, and represents the frequency difference between the two delay lines.

It is worth noting that there is substantial frequency shift for both delay lines resulting from temperature changes, on the order of 3-6 ppm/^{o}C. This is the result of thermal stresses arising in the sensor package as well as from the temperature coefficient of the acoustic velocity of the ST-quartz. Even though the difference frequency of the dual device does correct for this temperature coefficient to a great extent, absolute correction is not possible since the stresses on the individual delay lines are not entirely reproducible. The net result is that there is a slight temperature-induced shift seen in the frequency difference output. For the most part, however, the frequency-temperature curves for the blank devices were reproducible, as indicated by the close match of the heating and cooling portions of the curves in Figure 2.

The frequency difference curves for the coated devices are given in Figure 3 through Figure 6. In the case of PEM and FPOL, curves are presented for two different coated devices. For ECEL and PCAP, only one device was run. For ECEL, results for successive heating/cooling cycles are given. For all sensors, the corresponding difference curves for the blank devices are presented for comparison. Several features of these coating curves are worth noting, and will be treated individually in the discussion that follows.

Slope of Frequency-Temperature Curves. For the blank curves in Figure 3 through Figure 6 the change in frequency with increased temperature is minimal. For the coated devices, however, there is a noticeable increase in the frequency difference curves for FPOL, PEM, and ECEL. For PCAP there is also an increase in the heating portion of the curve up to about 55-60 ^{o}C, after which the frequency difference decreases. The rate of frequency change is on the order of 2-4 ppm/^{o}C, which is slightly less than the temperature dependance observed for the uncoated individual delay lines. Although the exact cause of this gradual, reproducible effect has not been positively identified, it may be the combined result of thermal expansion of the films and elastic effects.

The frequency shift predicted from the first term of Equation 1 is proportional to the product of the density of the coating and the film thickness. Since density decreases with temperature and film thickness

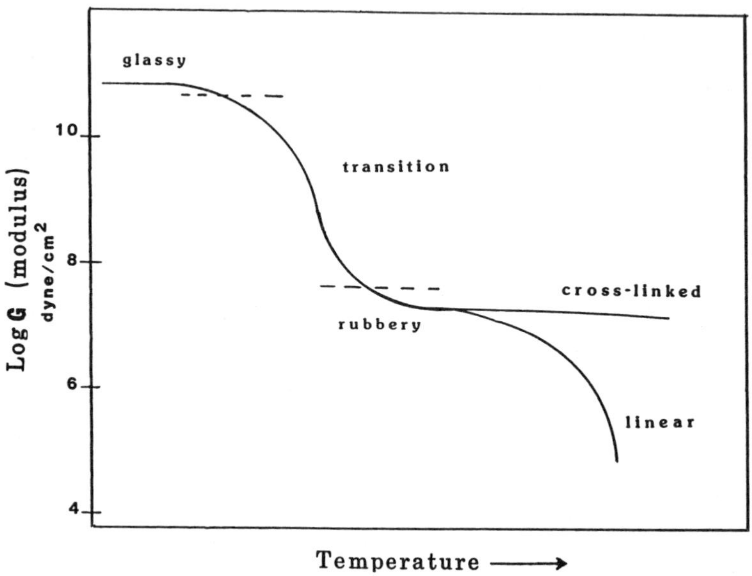

FIGURE 1. Idealized Modulus/Temperature Curves for Visco Elastic Polymers.

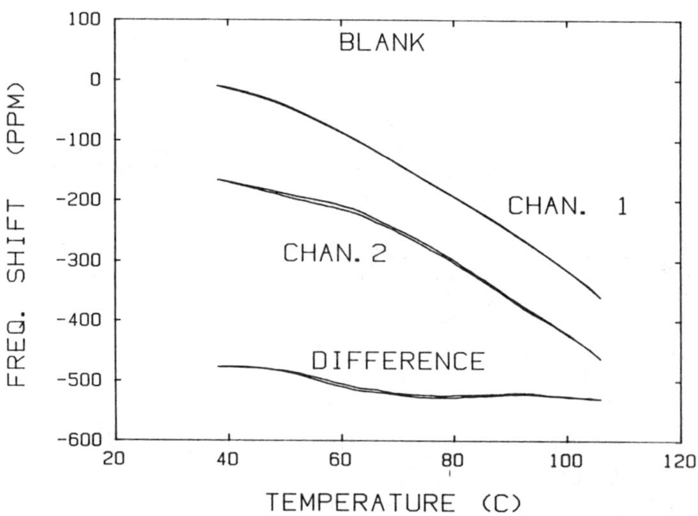

FIGURE 2. Frequency-Temperature Curves for Uncoated SAW Device.

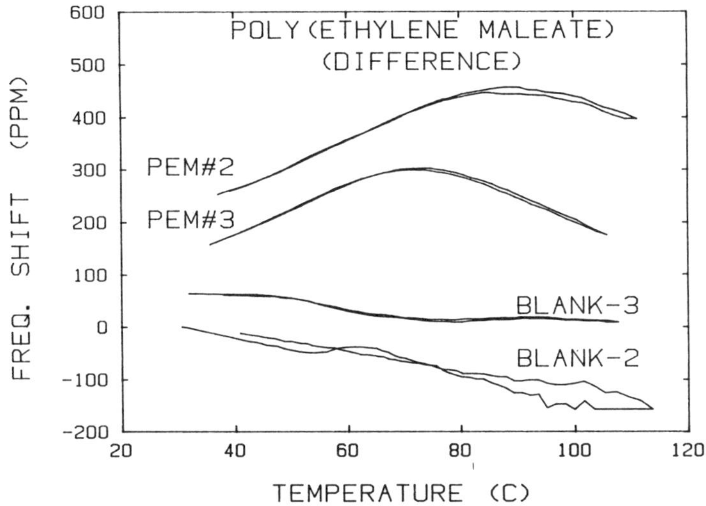

FIGURE 3. Frequency-Temperature Curves for two PEM-Coated SAW Devices, and Corresponding Blanks.

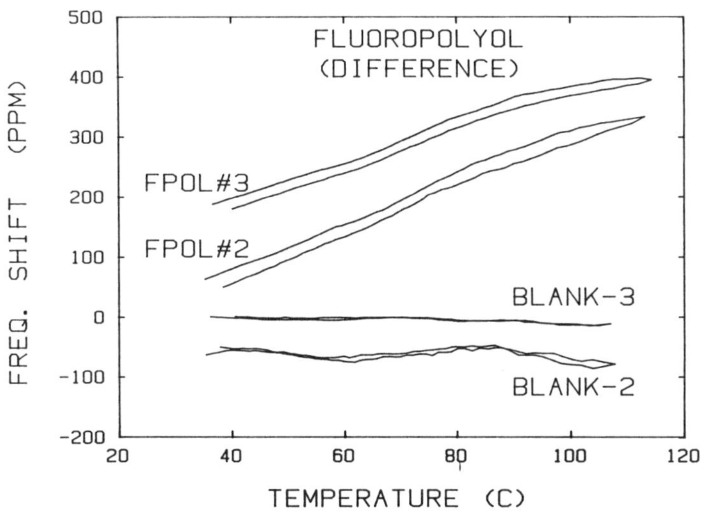

FIGURE 4. Frequency-Temperature Curves for Two FPOL-Coated SAW Devices, and Corresponding Blanks.

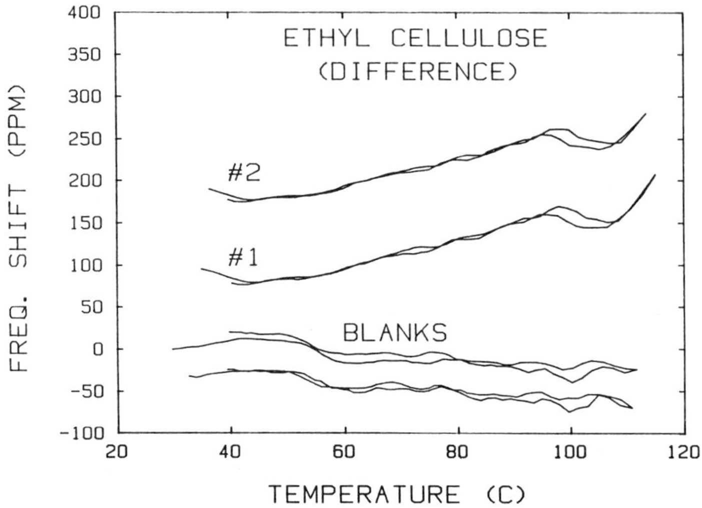

FIGURE 5. Frequency-Temperature Curves for Successive Heating/ Cooling Cycles for and ECEL-Coated SAW Device, and Blank.

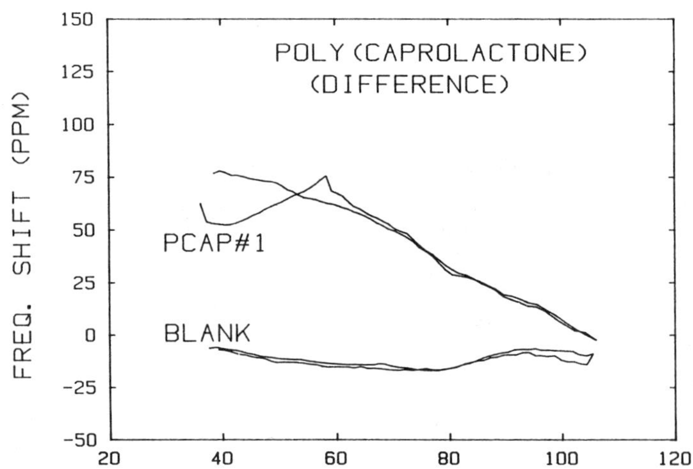

FIGURE 6. Frequency-Temperature Curve for PCAP-Coated Device and Corresponding Blank.

increases, the product of these factors remains essentially unchanged. Thus the mass per area of the film does not change significantly with temperature, and the frequency changes resulting from temperature effects on the first term in Equation 1 should be minimal. The second term deals with the effect of elastic properties on the SAW frequency. If the coating completely covers the active surface of the SAW device, the frequency of the device will be unaffected until the temperature approaches the T_g. If, however, the film does not completely cover the surface, increased temperature will result in increased surface coverage. The frequency of the device will decrease due to increased interaction with the film as it expands over the surface. Furthermore, the rate of frequency loss will be proportional to the rate of film expansion, i.e., to the thermal expansion coefficient.

As stated in the Experimental section, these coatings were applied using an airbrush technique. Visual inspection of these coatings under a microscope reveals that the films are not contiguous, but are dispersed on the surface as small islands. In this case, expansion of the films may be responsible for the gradual slope of the frequency-temperature curves.

This possibility is supported by the frequency-temperature curves for FPOL shown in Figure 7. The four curves shown represent the frequency-temperature curves for the individual delay line (channel 1), recorded during four successive heating/cooling cycles, with #1 being the first and #4 being the last. These curves have been corrected by subtracting the frequency curve for the blank device. It appears that there is a non-recoverable frequency shift during the course of a heating/cooling cycle. This effect decreases with successive cycles. FPOL is a viscous material and would be expected to flow as the temperature increased; over several cycles the flow would gradually diminish as the film spread to its limit. Beyond this point, further increases in surface coverage would be limited to thermal expansion only, and the heating and cooling portions of the curves would become superimposed.

Determination of T_g and T_m. In addition to frequency shifts due to expansion of the film noted in the preceding discussion, the frequency should decrease even further as the film approaches the T_g (or the T_m). To determine the magnitude of the expected shift, we refer again to Equation 1. Solving the equation for a rubbery film (where $\mu = 10^7$ dyne/cm^2) using the parameter values from Table I, yields an expected frequency shift from mass loading of -315 kHz, and a shift from the modulus contribution of only +3.3 Hz. For rubbery films the effects of elastic properties on observed frequency responses can be considered negligible. If, however, the film is glassy, with $\mu = 10^{10}$ dyne/cm^2, the effect of the modulus will be +3.3 kHz. A change in modulus at T_g from 10^{10} dyne/cm^2 to 10^7 dyne/cm^2 would produce a decrease in the observed frequency of 3.3 kHz. If the modulus change occurs over a 5 $^{\circ}$C temperature range, which is reasonable for many polymers, then a decrease of 4.2 ppm/$^{\circ}$C would be expected. (NOTE: 1 ppm = 158 Hz). This frequency shift is comparable to the frequency shifts from thermal expansion discussed previously, and should be observable.

The shear modulus is a complex term, and is the combined result of both loss and storage components. For a periodic or oscillatory stress, the polymer chain will tend to rearrange into conformations of

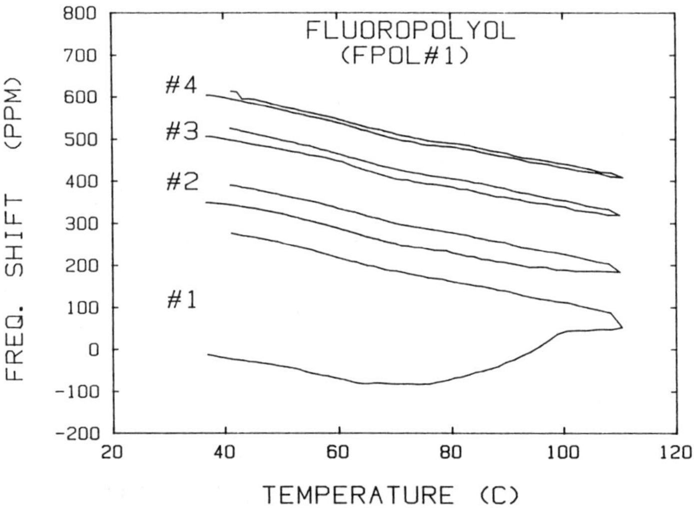

FIGURE 7. Frequency-Temperature Curves for Four Successive Heating/Cooling Cycles of a FPOL-Coated SAW Device.

lower energy to relieve this induced strain. The amount of energy dissipated during this deformation process is called the loss modulus, and is usually 90° out of phase with the strain (7). The loss modulus reaches a maximum value in the region of the T_g (7); this maximum energy loss will be manifested as a localized minimum in the frequency of a coated delay line as the polymer film approaches the T_g (or the T_m), according to Equation 1.

Increases in the T_g of polymers between 3-25 °C have been reported for each decade increase in the frequency (6, page 251), with increases of 5-10 °C representing a reasonable average. Thus, for 158 MHz SAW devices used during this study, the polymer films would be expected to exhibit increases in T_g of 25-80 °C. From the parameters given in Table II, the T_g for FPOL, PEM, and ECEL would be expected to fall within the temperature range of our experiments. For PCAP, the T_g may still be below the temperature range studied, although the T_m would be within the experimental range.

For PCAP and ECEL, there appear to be localized frequency difference maxima in the frequency-temperature curves. (NOTE: these localized maxima correspond to localized frequency minima in the frequency-temperature curves of the individual coated delay lines because of the practice of recording the frequency difference between the two delay lines of the dual device.) This maximum occurs between 55-60 °C for PCAP, which corresponds to the T_m given in Table II. The maximum is not seen in the cooling portion of the curve but reappears in successive heating cycles. For ECEL, the maximum occurs between 95-100 °C, which is 52-57 °C above the T_g listed in Table II, and well below the T_m. This maximum is seen in both the heating and cooling portions of the curve, and falls within the temperature range that would be expected for the T_g of ECEL on the SAW device. An additional point of interest is that, while the T_g is associated with a relaxation phenomenon and increases with frequency, the T_m represents a change of phase (from the crystalline to the liquid or amorphous state) which does not exhibit a frequency dependence.

The curves for PEM and FPOL are difficult to interpret. For PEM, the elevated T_g would be expected to fall between 15-70 °C based on the T_g in Table II. The curves in Figure 3 exhibit a maximum between 75-90 °C which is most likely not the T_g. It is possible that the T_g occurs in the lower end of the expected temperature range, and thus would fall outside the range of our experiments. For FPOL, the T_g would be expected to fall within the range of 35-90 °C. No frequency difference maxima are observed in this range. Close inspection of the curves in Figure 7 reveals an increase in the slope of the curve between 65-70 °C, which may correspond to the T_g. Since the thermal expansion coefficient increases significantly at the T_g, the rate of coating expansion can be observed as a frequency shift as discussed previously, and would be expected to increase in the region of the T_g.

There are other reasons why a readily identifiable frequency shift may not be seen at the T_g for the FPOL curves. As noted earlier, a SAW device coated with a glassy material having a modulus value of 10^{10} dyne/cm^2 would exhibit a frequency shift of the order of 3.3 kHz. If the glassy modulus is only 10^9 dyne/cm^2, however, the expected shift decreases to 0.33 kHz or less. If the transition occurs over a 5 degree temperature range, this is equivalent to a shift of only 0.42 ppm/°C, and is small compared to the frequency shifts observed for the bare devices as well as for shifts associated with thermal expansion of the coatings. In such a case, identification of T_g as a change in the slope of the curve may be a more definitive method.

Future Applications

These preliminary data demonstrate the sensitivity of the SAW device to the elastic properties of polymers. The SAW device has potential for use in the characterization of these materials. In addition to sensitivity, the SAW also represents a versatile and inexpensive alternative to existing methods. Some of the parameters which might be studied using the SAW are discussed below.

Thermal Expansion Coefficients. The thermal expansion of a lossy film should be observed as a steady decrease in frequency. The coated devices we studied exhibited a steady frequency shift with temperature. If this change in frequency is the result of film expansion, then the slope of this frequency curve should be proportional to the thermal expansion coefficient. The T_g of polymeric materials can be identified by the change in the thermal expansion coefficient at the T_g. To verify the effect of surface coating coverage on observed SAW frequency-temperature response behavior, the results for the airbrush-applied films studied here must be compared with results obtained from contiguous films of known thickness.

T_g and T_m. By taking advantage of the large modulus changes that occur at the T_g and T_m, the SAW can be utilized in the identification of these transition temperatures for polymeric materials. Our results have indicated, however, that there are limitations on the applicability of the SAW for these determinations. For example, if the modulus of the initial glassy material is less than 10^{10} dyne/cm^2, then the frequency shift observed at transition will be correspondingly diminished. Also, sharp transitions occurring over small temperature ranges will be more easily identified than gradual transitions spanning large temperature ranges. The use of contiguous films, rather than the dispersed droplet films studied here, should eliminate the gradual drift associated with thermal expansion of the coating and facilitate the identification of T_g. Also, for glassy materials with a modulus below 10^{10} dyne/cm^2, the T_g may still be determined using the thermal expansion technique described previously.

The identification of T_g and T_m for semi-crystalline materials poses a challenge. If there is a high degree of crystallinity, there may not be a sufficient amount of amorphous glassy polymer for a facile determination of T_g; values of T_m should still be obtainable as discussed previously.

Thermal Activation Energies. The relaxation that occurs as the polymer passes from the glassy state to the elastomeric state can be described using an Arrhenius type equation,

$$\tau = \tau_o \exp(\Delta H/kT) \tag{3}$$

where τ_o is the relaxation time at the reference temperature, T_o, and ΔH is the thermal activation energy (7). An external oscillating stress will induce conformational changes associated with the relaxation when the frequency (ω) corresponds to the relaxation frequency, or when $\omega = 1/\tau$. Thus, the thermal activation energy for the relaxation process can be determined by plotting the T_g vs frequency over a broad frequency range. The slope of the resulting curve will correspond to the ΔH for that relaxation. This method can be used not only for relaxations at T_g, but for secondary relaxations that occur below the T_g.

Elastic Modulus Determinations. The data presented here are limited due to a lack of information regarding the modulus values for polymers used during this work. The SAW could potentially be used for the determination of such values for glassy polymers. Before such determinations are possible, however, the SAW must be calibrated using well-characterized polymer systems. The mass sensitivity of the SAW can be calculated from the first term in Equation 1, and calibrated using rubbery films of known mass so that the modulus effects are negligible. The modulus of glassy materials can then be determined as the difference between the observed frequency shift for a given film and the expected shift due to the mass of the film.

Plasticization Effects. Many polymers undergo plasticization, or softening, when exposed to certain vapors. The effects of plasticization are seen as a decrease in the T_g and an expansion of the rubbery region. By determining the T_g of the polymer films in the presence and absence of a given vapor, the plasticization effects of that vapor can be studied. Also, the relative contributions of mass loading and elastic effects can be determined by monitoring the frequency and the amplitude of the surface wave simultaneously. The frequency will be perturbed by small mass loading effects as well as by changes in the elastic modulus of the film; the amplitude is also sensitive to elastic effects, but should exhibit little or no sensitivity to small increases in mass loading.

Summary

The SAW device has potential for use in the characterization of polymeric materials. It has several advantages over existing methods, including small sample requirements (1 μg or less), sensitivity, versatility, and low cost. It must be stressed that these SAW data were collected using very thin films. The ability to extrapolate from thin film data to determine the properties of bulk materials must be verified by studying well-characterized materials. In addition, uniform, contiguous films and/or thicker films should be studied to aid in the identification of transition temperatures.

In sensor applications, greater care should be taken in the selection and application of polymer films to minimize modulus effects. The effect of elastic properties of the coating materials must be taken into consideration in the interpretation of SAW sensor data. Alternatively, the sensor community might take advantage of the SAW sensitivity to elastic properties to devise more sensitive sensors, and to expand the applications of these sensors.

Acknowledgments

The work described here was supported by Navy contract N-00014-88-C-2247, and was performed at the Naval Research Laboratory. The authors wish to thank Dr. A. Snow (Code 6120), Dr. D. Venezky (Code 6170) from the Chemistry Division of NRL, and Dr. R. Ting of the NRL/USRD facility for insightful discussions.

Literature Cited

1. Ballantine, D. S., Jr.; Rose, S. L.; Grate, J. W.; Wohltjen, H. Anal. Chem. 1986, 58, 3058.
2. Wohltjen, H. Sens. Actuators 1984, 5, 307.
3. White, R. M. Proc. IEEE 1970, 58(8), 1238.
4. Ricco, A. J.; Martin, S. J.; Zipperian, T. E. Sens. Actuators 1985, 8, 319.
5. Grate, J. W.; Snow, A.; Ballantine, D. S., Jr.; Wohltjen, H.; Abraham, M. H.; McGill, A.; Sasson, P. Anal. Chem. 1988, 60, 869.
6. Ferry, J. D. In Visco-Elastic Properties of Polymers; J. Wiley & Sons: New York, 1961; Chapter 4.
7. Armeniades, C. D.; Baer, E. In Introduction to Polymer Science and Technology: An SPE Textbook; Kaufman, H. S.; Falcetta, J. J.; Eds.; J. Wiley & Sons: New York, 1977; Chapter 6.
8. Williams, M. L.; Landel, R. F.; Ferry, J. D. J. Am. Chem. Soc. 1955, 77, 3701.
9. Wohltjen, H.; Dessy, R. Anal. Chem. 1979, 51(9), 1470.
10. Groetsch, J. A.; Dessy, R. A. J. Appl. Poly. Sci. 1983 28, 161.

RECEIVED March 17, 1989

Chapter 16

Sensors Based on Biomolecules Immobilized on the Piezoelectric Quartz Crystal Microbalance

Detection of Glucose Using Hexokinase

Steven J. Lasky and Daniel A. Buttry[1]

Department of Chemistry, University of Wyoming, Laramie, WY 82071–3838

The fabrication of a biosensor based on the use of the QCM for glucose detection using hexokinase immobilized within a poly(acrylamide) matrix has been demonstrated. Very large frequency changes accompany the binding of glucose to the enzyme. The frequency changes are quite reproducible for a given crystal. The signal to noise level is in excess of 100 for a glucose concentration of 10 mM, a value close to that found in blood (ca. 7 mM). These frequency changes appear to be too large to arise solely from the mass change due the binding reaction. Two possible mechanisms for the production of such unexpectedly large frequency changes have been discussed, and experiments proposed to determine their applicability to the observations.

The importance of schemes for detection of glucose in liquids in real time is amply evidenced by the large literature on research into such sensors. Guilbault and coworkers (1-5) as well as a number of other research groups (6-8) have published a series of studies on the use of glucose oxidase in electrochemical sensors. In these sensors immobilized glucose oxidase catalyzes the reaction between glucose and molecular oxygen to produce hydrogen peroxide and the oxidized carbohydrate. Detection of glucose is accomplished indirectly by measurement of oxygen consumption or hydrogen peroxide appearance by reduction or oxidation, respectively, at Pt electrodes located near to the immobilized enzyme layer. More recently, a more direct method of detecting glucose has been realized by Heller and Degani, in which redox couples are attached directly to the glucose oxidase molecule so that facile electron transfer between the enzyme and an electrode surface occurs by virtue of electron mediation of these redox couples (9,10). In this scheme the detection of glucose relies on the direct measurement of the charge required to reoxidize the $FADH_2$ sites in the enzyme created by reduction of FAD sites by glucose. Oxygen would thus be an interferent in such a system because it can also reoxidize the $FADH_2$ sites in competition with the oxidation process mediated by the attached redox couples. Nevertheless, this type of methodology represents a promising new approach to the problem of glucose detection.

An entirely different approach than those based on electrochemical methods is

[1]Address correspondence to this author.

0097–6156/89/0403–0237$06.00/0

the subject of this contribution which describes a part of our efforts directed towards the development of a detector for glucose in liquids which is capable of real time operation. The method of detection is based on the quartz crystal microbalance (QCM), a mass sensitive piezoelectric device capable of mass measurements in the nanogram to milligram range. The principle of detection relies on the fact that the entire QCM device (including any materials attached to the surface of the QCM disk) is a composite resonator, the oscillation frequency of which depends sensitively on the mass of the entire device. To the extent that chemical reactions (such as binding of a substrate to an enzyme immobilized at the disk surface) can be used to induce mass changes of this composite resonator, then these chemical reactions can form the basis of an analytical method for detecting the substrate, with the QCM performing the function of transducing the chemical event into an electrical signal.

The use of piezoelectric devices as the active (transducing) component in detection systems has been an especially active research area in the past few years. Much of the early work was directed towards gas phase detectors (11) and will not be further discussed here. The use of these devices as mass detectors in liquids has only more recently been described. Bastiaans and coworkers studied the use of the QCM as a detector in liquid chromatography (12) and of SAW devices in an immunoassay application (13). They reported some difficulty in these applications due to instability in the oscillation frequency of the devices caused by viscous loading by the solution. Advances in oscillator design (14) have allowed for this problem to be overcome, so that the devices may be routinely used in the liquid environment. For example, our research group at the University of Wyoming has been actively involved in the application of this device to the study of mass changes at electrode surfaces during electrochemical events in redox polymer films (15-17). In addition, other groups have applied these devices to the detection of interfacial immunochemical reactions (18). These recent developments indicate that at this point in the development of this (and related) technique(s) the challenges are not so much instrumental as they are chemical. Methods for the stable immobilization of sufficient quantities of various types of biomolecules need to be further developed so that measurably large signals are generated.

The basis of our method for measuring glucose relies on binding of the glucose to hexokinase, an enzyme which catalyzes the phosphorylation of glucose to glucose-6-phosphate. The detection scheme is a simple one, in principle. The hexokinase can be immobilized onto the surface of a QCM via any of a number of immobilization methods. The appearance of glucose in the liquid near to the surface of the oscillating QCM disk with the immobilized enzyme will result in binding of glucose to the enzyme according the the equilibrium constant for this binding reaction. The binding process should induce a mass change of the composite resonator which is measured with standard frequency counting techniques. To the extent that the binding occurs in an amount proportional to the concentration of glucose in the liquid, the frequency change from binding provides a measure of this concentration. We show below that the signals generated by this binding process are far above the signal to noise ratio for the device, and are, indeed, related to glucose concentration.

Experimental

A schematic of the QCM instrument is shown in Figure 1. Two QCM crystals are used in the experiments, so that detection is done in a differential mode. One is termed the active crystal, which has hexokinase immobilized by entrappment in a poly(acrylamide) gel. The other is termed the reference crystal, which has bovine serum albumin (BSA) immobilized in exactly the same way. A similar differential mode technique, utilizing piezoelectric cystals, has been previously decribed using a

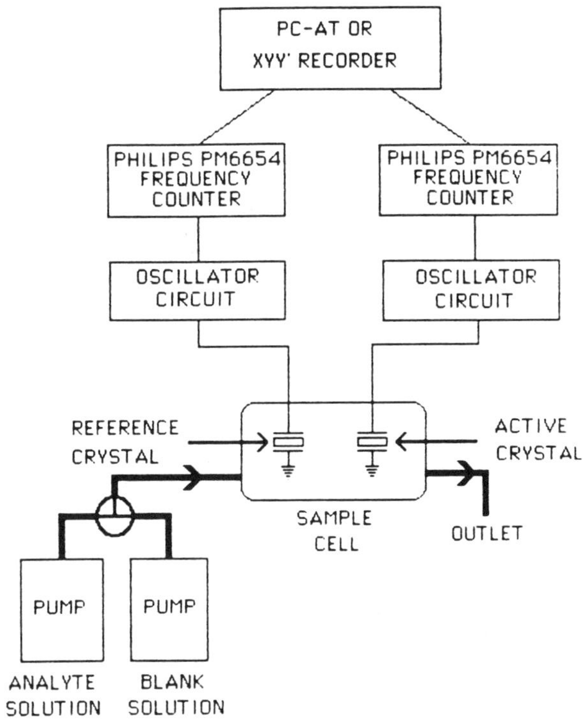

Figure 1. Schematic of the differential QCM glucose biosensor instrument.

SAW device (13). BSA is immobilized on the reference crystal to ensure that non-specific effects related to protein entrappment in the poly(acrylamide) gel do not influence the measurement. The purpose of the reference crystal is to allow for subtraction of the contribution to the signal which arises from the known dependence of the oscillation frequency of a QCM on the density and viscosity of the liquid at its surface (19). The cell is designed so that the flow is introduced between the parallel faces of the active and reference crystals. We assume that these changes are identical for both crystals. In addition, it will be seen below that this contribution is considerably smaller than that from the binding of glucose to the active crystal.

The two QCM crystals are each driven into oscillation by their own broadband oscillator circuit. The schematic of this circuit has been described (14). The oscillation frequency of each QCM oscillator circuit is measured with a Philips PM6654 frequency counter. These counters are equipped with analog outputs suitable for measurement on an XYY' recorder which has a time base on the x-axis. Thus, the oscillation frequencies of the active and reference crystals are recorded versus time. The crystals are mounted into a polypropylene cell in such a way that only the modified side of each crystal is exposed to the solution. Commercial peristaltic pumps (Isco Corp.) are used to flow solutions containing either pure buffer (30 mM TRIS buffer at pH 7.4) or various concentrations of glucose dissolved in this same buffer past the surfaces of the two crystals.

Enzyme immobilization is achieved using a poly(acrylamide) entrappment method similar to that previously described by Guilbault and Lubrano (2). A drop of the polymerization solution (containing the enzyme, acrylamide, N,N'-methylene-bis-acrylamide, and potassium persulfate) is placed onto the gold electrode on the face of the QCM disk, and the polymerization is initiated by addition of 2 μl of 3-dimethylaminopropionitrile. Films are then soaked in buffer for at least an hour before use.

The QCM crystals used in this study are 1 inch diameter, AT-cut quartz crystals with thicknesses adjusted so that they have resonant frequencies near 5 MHz (Valpey-Fisher). The central region of the crystal is sandwiched between two vapor deposited gold electrodes of ca. 300 nm thickness with a ca. 30 nm layer of Cr applied prior to the gold to promote adhesion to the quartz substrate. A diagramatical representation of the crystal is shown in Figure 2. The area of the central, circular pad (i.e. the area of the crystal which is sandwiched between the gold electrodes) is 0.28 cm^2. This is the piezoelectrically active area of the crystal, and only mass changes which occur there are sensed by the QCM.

The resonant frequency of a QCM in a liquid depends on the viscosity and density of the liquid. Since these quantities are temperature dependent it is necessary to control the temperature of the solution in the cell. This is accomplished by constructing the cell with feedthroughs for a circulating fluid which has its temperature controlled by a Neslab RTE-110 temperature control system. Temperature control to within 0.01 C is possible in this way. This corresponds to a frequency stability from temperature effects of less than 0.1 Hz (20). All experiments were performed at a temperature of 37 C.

Results

Figure 3 shows the response of the instrument to alternation of the solution between pure buffer and glucose in buffer. Large decreases in frequency are observed when the solution in the cell is changed from pure buffer to glucose in buffer. The frequency re-attains its original value when pure buffer is returned to the cell. The magnitude of the frequency decrease seen for this 10 mM glucose concentration is ca. 1200 Hz. These frequency changes are quite reproducible for a given crystal and glucose

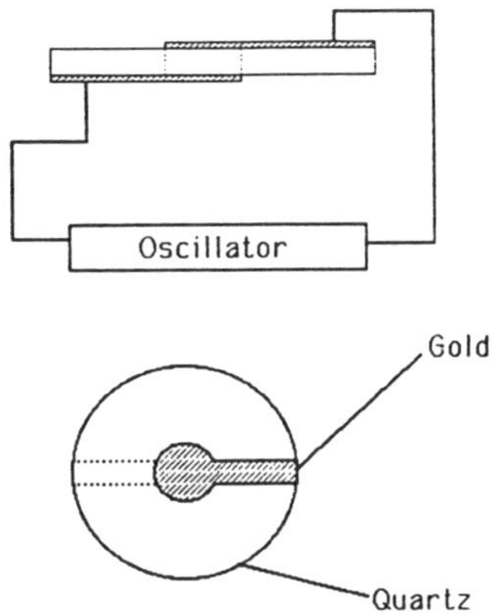

Figure 2. Bottom - Schematic diagram of a piezoelectric quartz crystal disk with the vapor deposited gold electrodes which sandwich the crystal in the central region. Top - Side view of crystal at bottom shown with connections to the oscillator circuit.

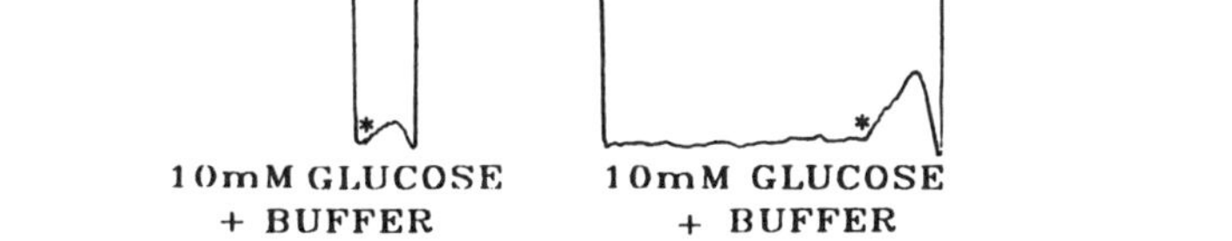

Figure 3. Data from the repeated exchange of the cell solution between pure buffer solution and glucose in buffer. The asterisks indicate the times at which the solution exchange was begun.

concentration. The time required for attainment of the final frequency change is much less than a minute, and may, in fact, be a consequence of the flow characteristics of the cell which have not been optimized for rapid exchange. The data exhibit small, but reproducible, variations in the resonant frequency just after the solutions are exchanged, especially when pure buffer is replacing the glucose solution. Their exact cause is unknown at this time. The oscillation frequency of the reference crystal is also seen to be affected by the change of solution from pure buffer to glucose in buffer. However, these changes are quite small in comparison to the frequency change for the crystal with immobilized hexokinase. A notable feature of these data is the remarkable stability of the oscillation frequencies of both the active and reference crystals. The noise level in the signal of the active crystal is less than 5 Hz, and that in the reference is considerably less than that. The origin of the smaller noise level in the reference crystal is not clear. It is worth pointing out here that the noise levels in other experiments with the QCM in our labs is on the order of 1 Hz, without the benefit of signal averaging.

Figure 4 shows a plot of the response of the crystal to several different concentrations of glucose. The frequency change is seen to be proportional to the concentration over the range shown, from 1 to 10 mM. Using different crystals prepared with a slightly different immobilization procedure we have demonstrated essentially linear behavior from 0.1 mM to 10 mM. The linearity of the plots is not perfect, nor is the slope of the plot identical for different crystals. In fact, the slope of the plot is observed to be strongly dependent on the details of the polymerization conditions, being especially sensitive to the amount of the crosslinking agent present. For example, a change in the ratio of N,N'-methylene-bis-acrylamide to acrylamide from 1:10 to 1:15 changes the magnitude of the frequency decrease for 10 mM glucose from 1200 Hz to 200 Hz, respectively. This unique and somewhat unexpected behavior will be reported in more detail in a future contribution.

Discussion

The frequency changes (Δf, in Hz) which arise from mass changes (m, in $\mu g\ cm^{-2}$) at QCM devices are usually described using the Sauerbrey equation, or some variation of it, which provides a linear relationship between these two quantities.

$$\Delta f = - 2\, f_o^2\, m / n\, (\rho_q\, \mu_q)^{1/2} \tag{1}$$

In this equation f_o is the resonant frequency of the composite resonator (i.e. quartz crystal, gold electrodes, and immobilized film in the solution) prior to exposure to glucose, n is the harmonic number of the shear mode excited in the QCM device, ρ_q is the density of quartz (2.648 g cm^{-3}), and μ_q is the shear modulus of quartz (2.947 x 10^{11} g cm^{-1} s^{-2}). For our experiments f_o is very near to 5 MHz and n is 1. The equation may be rewritten as:

$$\Delta f = - C_f\, m \tag{2}$$

where all the constants for a given crystal are collected into C_f, which has a value of 56.6 Hz μg^{-1} cm^2 for the present case. Thus, a mass change of 1 μg distributed uniformly over a 1 cm^2 area would give rise to a frequency change of 56.6 Hz in the resonant frequency of the 5 MHz crystal. Note that the negative sign in equations 1 and 2 indicate that the resonant frequency decreases when mass is added to the device.

The Sauerbrey equation is applicable under conditions in which the mass which induces the frequency change is rigidly attached to the surface of the QCM.

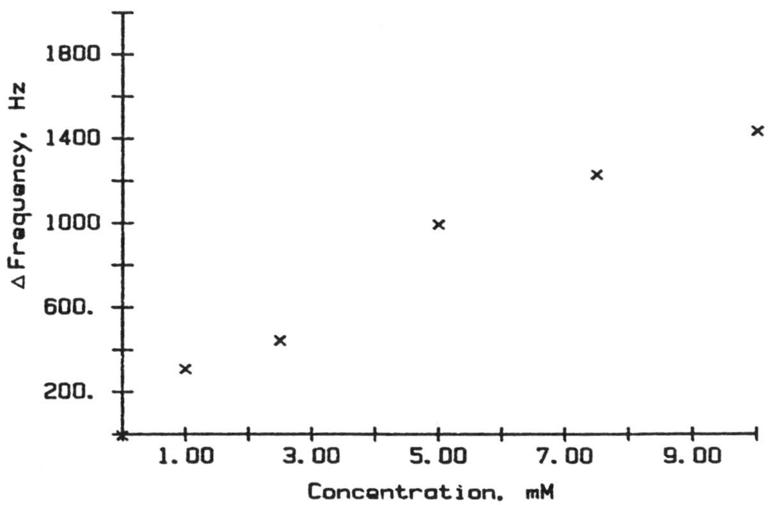

Figure 4. Plot of the frequency decrease observed for exposure of the differential QCM glucose biosensor to various concentrations of glucose.

When this is not the case, then more complex analyses are required to extract the true mass change from Δf. The detailed considerations which arise in such a case are beyond the scope of this contribution, but qualitative discussion is in order, as it relates to the origin and magnitude of the signal which is observed from exposure of the QCM device to glucose. First, however, we consider whether or not equation 2 is valid for the present case.

From the total amount of hexokinase which is immobilized onto the surface of the QCM, it is possible to use equation 2 to calculate the maximum possible value for Δf which could be observed from glucose binding. This maximum value would be that observed for the case in which the amount of bound glucose would start at zero (i.e. in the pure buffer solution) and end at 100% binding (i.e. in a glucose solution which was concentrated enough to induce saturation of the enzyme). Thus, it is assumed that each and every hexokinase molecule can bind one molecule of glucose, and that the masses of all bound glucose molecules are transduced into a change of the resonant frequency of the QCM device according to equation 2. For the data shown in Figure 3 a total mass of 4.9×10^{-5} g of hexokinase was immobilized onto the 0.28 cm^2 gold pad (the piezoelectrically active area) at the center of the QCM disk. Given a molar mass for hexokinase of ca. 50,000 g mol^{-1}, one calculates a maximum frequency decrease for binding of glucose at saturation of 36 Hz. Note also that the glucose concentration in Figure 3 should not be sufficient to saturate the enzyme, because an increase in glucose concentration leads to even larger values of Δf. These considerations strongly indicate that the frequency changes observed in Figure 3 are not the result of mass binding and linear mass-frequency response by the device. Thus, it appears that equation 2 is not valid for the case at hand.

There are at least two possible explanations for the very large observed values of Δf from glucose binding. One is that the mass-frequency correlation really is linear, but that the binding of glucose to hexokinase causes the incorporation of additional material (perhaps solvent) into the poly(acrylamide) matrix. Changes in solvent content could be caused by a change in the free volume within the film. Such a coupling between free volume changes and solvent transport has recently been demonstrated (17). A reasonable speculation would be that the changes in free volume within the film (if, indeed, they do occur) could be related to the striking conformational changes induced in hexokinase by the binding of glucose (21). However, the amount of solvent transport into the film induced by such changes would need to be very large. An excellent test of such an explanation is to measure the value of Δf for glucose binding in a solution of H_2O and then in a solution of D_2O. If significant solvent transport occurs during binding, then Δf should be larger in the D_2O experiment by an amount related to that fraction of the total mass change which arises from solvent transport (17). We are currently performing such experiments.

A second possible explanation for the larger than expected values of Δf is that the QCM is sensing changes in the viscoelastic properties of the poly(acrylamide)/enzyme film which are induced by the binding reaction. It is known that the frequency of a QCM device is related to the viscoelastic properties of any medium present at its surface. In fact, this has been used to advantage in the study of such properties for polymeric samples of various types (22). In this regard, the conformational change in hexokinase induced by glucose binding may well be changing the viscoelastic properties of the poly(acrylamide) matrix in such a way as to cause the large decreases in Δf. Quantitative treatment of this problem is quite complex, and will undoubtedly require measurement of the frequency dependence of the mechanical impedance of the QCM/film composite resonator, which is a useful indicator of such effects (23).

Conclusions

This study has demonstrated the utility of chemically derivatized QCM devices for the detection of analytes in solutions. The signal to noise level for the measurement of glucose concentrations in liquids is excellent, and shows that, even though the QCM devices are not intrinsically as sensitive as SAW devices, they can be competitive under the proper conditions. It is worth noting in this regard that the signal levels are so large that correction for the frequency change due to changes in density and viscosity of the solution is not required, as judged from the much smaller frequency changes which occur at the BSA reference crystal during solution exchange in the cell. Thus, differential measurements need not be required unless other factors come into play (such as large fluctuations in temperature).

The response of this QCM glucose biosensor is roughly linear with glucose concentration in the range found in blood (ca. 7 mM). Also, the response time is quite short, so that real time detection is possible. These results are especially significant in terms of possible clinical applications. In future communications the response of this biosensor to possible interferents (e.g. other sugars, such as fructose) and the cofactors for the enzymatic reaction (i.e. ATP and Mg^{2+}) will be reported. Preliminary results from such experiments indicate that the trend in selectivity of the sensor towards other monosaccharides follows their known affinities for hexokinase.

The origin of the large values of Δf observed for the binding of glucose to the sensor are not entirely clear at this time. Two possible mechanisms have been discussed. Further experimentation will be necessary before the matter is resolved. In either case, since conformational changes are a reasonably general phenomenon in the binding of enzymes to substrates, it should prove possible to expand the possible applications of this QCM methodology to other biorecognition systems. Also, to the extent that larger direct mass changes may be induced by achieving larger surface coverages of the active biological component of the sensor, extension to a very wide variety of systems should occur in the future.

Acknowledgments

We are grateful for the generous support of this work by a grant from the Whitaker Foundation. Also, initial funding for support of this work was provided by the Biomedical Research Support Grant S07RR07157-11, which was awarded to the University of Wyoming by the Biomedical Research Grant Program, Division of Research Resources, National Institutes of Health.

Literature Cited

1. G. G. Guilbault and G. L. Lubrano, Anal. Chim. Acta, 1972, 60, 254.
2. G. G. Guilbault and G. L. Lubrano, Anal. Chim. Acta, 1973, 64, 439.
3. G. G. Guilbault and G. L. Lubrano, Anal. Chim. Acta, 1978, 97, 229.
4. G. Nagy, L. H. von Storp, and G. G. Guilbault, Anal. Chim. Acta, 1973, 66, 443.
5. M. Nanjo and G. G. Guilbault, Anal. Chim. Acta., 1974, 73, 367.
6. S. J. Updike and G. P. Hicks, Nature (London), 1967, 214, 986.
7. D. L. Williams, A. R. Doig, Jr., and A. Korosi, Anal. Chem., 1970, 42, 118.
8. T. Yao, Anal. Chim. Acta, 1983, 148, 27.
9. Y. Degani and A. Heller, J. Phys. Chem., 1987, 91, 1285.
10. Y. Degani and A. Heller, J. Am. Chem. Soc., 1988, 110, 2615.
11. J. F. Alder and J. J. McCallum, Analyst, 1983, 108, 1169.
12. P. L. Konash and G. J. Bastiaans, Anal. Chem., 1980, 52, 1929.

13. J. E. Roederer and G. J. Bastiaans, Anal. Chem., 1983, 55, 2333.
14. O. R. Melroy, K. K. Kanazawa, J. G. Gordon, and D. A. Buttry, Langmuir, 1986, 2, 697.
15. D. O. Orata and D. A. Buttry, J. Am. Chem. Soc., 1987, 109, 3574.
16. P. T. Varineau and D. A. Buttry, J. Phys. Chem., 1987, 91, 1292.
17. S. J. Lasky and D. A. Buttry, J. Am. Chem. Soc., 1988, 110, 6258.
18. M. Thompson, C. L. Arthur, and G. K. Dhaliwal, Anal. Chem., 1986, 58, 1206.
19. K. K. Kanazawa and J. G. Gordon, Anal. Chem., 1985, 57, 1770.
20. J. J. Donohue and D. A. Buttry, unpublished results.
21. W. F. Anderson and T. A. Steitz, J. Molec. Biol., 1975, 92, 279.
22. see, for example, "Viscoelastic Properties of Polymers", J. D. Ferry, Wiley and Sons: New York, 1961.
23. M. O'Donnell, L. J. Busse, and J. G. Miller, In "Methopds of Experimental Physics"; L. Marton and C. Marton, Eds.; Academic Press: New York, 1981; Vol. 19, p. 29.

RECEIVED March 9, 1989

SENSORS BASED ON OPTICAL MEASUREMENTS

Introduction: Optical Sensors

The chapters in the optical sensor section of this book cover current critical issues in development of chemical sensors based on spectroscopy. Primary emphasis is on the chemistry of sensing, i.e., the formulation of solid phase indicators that interact with analyte to produce an optically detectable signal. The major theme of the first three chapters is the development and characterization of porous polymers to serve as indicator substrates. The second three chapters describe systems based on the use of biological molecules including enzymes, antibodies and other receptors, which provide a high degree of selectivity. The technology of waveguide measurements is also an important component of the field. Various aspects of this technology are considered in the last three papers. A theme which appears throughout is the use of new indicators, which expand upon the number of analytes amenable to fiber optic sensing.

Polymers as Indicator Substrates

Relatively few analytes can be detected directly by spectroscopy. It is necessary to add reagents that interact with the analyte to form a product which is optically detectable. In laboratory methods, these reagents are generally in solution. However, reagents used for in situ optical sensing generally have to be formulated in a solid phase. The solid phase has to be permeable to analyte but may be designed to exclude other substances that might otherwise interfere. Furthermore, there should be a convenient method for formulating the polymeric indicator substrate and coupling it to fiber optics. The chapter by Walt et al. (1) in this section describes systems based on polyacrylamide. The tip of a fiber is chemically modified so that it covalently bonds directly to polyacrylamide. A particularly attractive feature of this approach is that the sensors are prepared by a convenient dip-coating procedure which can be used to simultaneously prepare a large number of sensors. Polyacrylamide is a versatile substrate which can be coupled to a variety of indicator systems.

The chapter by Seitz et al. (2) describes the use of poly(vinyl alcohol) (PVOH) as an indicator substrate. Indicator is covalently bonded to PVOH prior to crosslinking with glutaraldehyde. The rate of the crosslinking reaction is controlled so that a known volume of substrate can be transferred to a fiber optic bundle as a liquid and allowed to form a solid gel in place. The procedure allows for independent control of the amounts of indicator and substrate.

Cellulose acetate is investigated as a substrate for indicator immobilization in the chapter by Porter et al. (3). Base hydrolysis can be used to enhance the permeability of cellulose acetate films to analyte, increasing the surface area available for indicator immobilization as well as shortening sensor response times. Techniques developed for the textile industry may be used to couple dyes to cellulose acetate.

An alternative approach is to use a polymer that slowly hydrolyzes as a delivery system to provide a slow but continuous supply of indicator. The use of ethylene-vinyl acetate copolymer for this purpose is described in the chapter by Walt et al. (1). This is a highly significant development because it allows irreversible reactions to be used for continuous sensing for extended periods of time, greatly expanding the range of possible applications.

Selective Indicator Systems based on Biomolecules

Many sensing applications require a degree of selectivity that can be only achieved using biomolecules. The coupling of biomolecules with sensors is well established using other types of sensors. There is considerable current interest in extending this to optical sensors.

The Chapter by Arnold (4) describes various ways of preparing sensors based on biocatalysts, i.e. enzymes, as the molecular recognition element. The measured parameter in these sensors is the steady state concentration of an optically detectable product or substrate. Systems have been developed in which the product or substrate is directly detected spectroscopically. Alternatively, substrate or product can be detected indirectly through a second indicator. The chapters by Walt et al. (1), Arnold (4) and Giuliani (5) describe enzyme-based indicator systems coupled to optical detection of changes in pH, ammonia and reduced viologen concentration respectively.

Antibodies provide a powerful alternative approach to achieving high selectivity. The chapter by Sepaniak et al. (6) discusses the design of antibody-based fiber optic chemical sensors using fluorescence. Naturally fluorescent antigens may be detected directly from the increase in observed fluorescence as antigen binds to antibody immobilized at the tip of an optical fiber. Non-fluorescent antigens may be detected indirectly by their ability to competitively displace fluorophor labelled antigen from immobilized antibody. Reversible sensors are possible in principle but have very slow response times due to slow antigen/antibody dissociation kinetics. The chapter by Sepaniak et al. (6) describes a new system for adding and removing reagents from the tip of an optical fiber so that in situ measurements can be repeated without have to remove the fiber from the sample.

In addition to antibodies, other types of biological receptors may also be used for sensing. For example, the possibility of sensing based on acetylcholine receptors is considered in the chapter by Krull et al. (7).

New Indicators for Fiber Optic Sensing

The chapters in this book include descriptions of several new indicator reagents for optical sensors. Some involve the immobilization of chromogenic reagents that have previously been characterized in solution. For example, the chapter by Porter et al. (3) describes immobilization of several colorimetric pH indicators. Other approaches are more creative. The chapter by Walt et al. (1) describes indicators prepared by covalently coupling an absorbing indicator to a fluorophor. Interaction of the analyte with the absorbing indicator is detected indirectly via its effect on the observed fluorescence intensity. This approach addresses the problem of making single fiber absorbance measurements without interference from stray light. The same chapter also describes an indicator for gasoline vapors based on fluorescence enhancement accompanying changes in the microenvironment of a polymer-bound fluorophor when the polymer is exposed to the vapors.

The chapter by Krull et al. (7) describes a quite different approach to the problem of optical transduction. The molecular recognition element is a receptor embedded in a lipid membrane. Fluorescent probes are used to sense perturbations in the membrane accompanying analyte binding by the receptor. This is a potentially generic approach to transduction, applicable to any molecular recognition element that can be incorporated into a membrane.

Measurement Technology

Development of instrumental techniques for measurements through fibers is an important component of optical sensor research. The chapter by Angel et al. (8) shows how remote absorbance measurements can be implemented and demonstrates that, with an indicating reaction for halogenated organics, the absorbance measurement improves on both the repeatability and sensitivity of fluorescence measurement. The same chapter also describes the use of copper and gold colloids as substrates for surface enhanced Raman spectroscpy. Data is presented documenting the possibility of sensitive Raman measurements in the near infrared region of the spectrum where optical fibers have the greatest transmittances.

The chapter by Giuliani (5) describes a completely different approach to optical sensing instrumentation. Rather than coupling a sampling site to a sophisticated spectrometer in a protected location, the idea is to build a miniature spectrometer which can be taken directly into the field. Using modern optical technology, e.g. light emitting diodes as sources, and indicator coated capillary tubes, response has been demonstrated for several different analytes.

The chapter by Bright et al. (9) demonstrates that phase-resolved fluorescence measurements may be implemented remotely through optical fibers. This is significant since phase-resolved techniques may be used to greatly enhance the amount of information available from a fluorescent indicator.

Literature Cited

1. Walt, D.R.; Munkholm, C.; Yuan, P.; Luo, S.; Barnard, S.; this Volume.
2. Wangbai, M.; Zhujun, Z.; Seitz. W.R.; this Volume.
3. Stole, S.M.; Jones, T.P.; Lai-Kwan, C.; Porter, M.D.; this Volume.
4. Arnold, M.A.; this Volume.
5. Giuliani, J.F.; this Volume.
6. Sepaniak, M.J.; Tromberg, B.J.; Alarie, J.P.; Bowyer, J.R.; Hoyt, A.M.; Vodinh, T.; this Volume.
7. Krull, U.J.; Brown, R.S.; Dyne, K.; Hougham, B.D.; Vandenberg, E.T.; this Volume.
8. Angel, S.M.; Nugent, M.; Langry, K.; Kulp, T.J.; this Volume.
9. Litwiler, K.S.; Bright, F.V.; this Volume.

W. Rudolf Seitz
Department of Chemistry
University of New Hampshire
Durham, NH 03824

RECEIVED March 29, 1989

Chapter 17

Design, Preparation, and Applications of Fiber-Optic Chemical Sensors for Continuous Monitoring

David R. Walt, Christiane Munkholm, Ping Yuan, Shufang Luo, and Steven Barnard

Max Tishler Laboratory for Organic Chemistry, Department of Chemistry, Tufts University, Medford, MA 02155

Fiber optic sensors now exist to measure a variety of analytes. When used for *in situ* and *in vivo* applications, the sensor reagents must be attached directly to the optical fiber tip. This paper describes technology for covalently attaching indicator reagents to the fiber sensing tip. The method serves to increase the fiber surface area and results in an amplified signal. The individual processing steps of the technique are described, as are the preparation of sensors measuring pH, CO_2, and penicillin. New indicator systems based on energy transfer are also described. A physiological pH sensor based on energy transfer from a pH insensitive fluorophore to a pH sensitive absorber was prepared successfully. Finally, a sensor based on irreversible reagents are described. This sensor utilizes a polymeric release system to deliver fresh reagent over a long period of time.

The advent of optical fibers has initiated a revolution in telecommunications technology and is producing a subsequent and possibly equal impact on chemical sensor technology (1,2). Optical fibers, also known as lightguides or optical waveguides, permit the low loss transmission of light through the fiber core by the phenomenon of total internal reflection. Fibers can also be used in a bidirectional mode in which light propagates through a fiber and returns *via* the same or a second fiber, thereby allowing spectroscopy to be performed at a distance (3). The optical fiber functions as a conduit for light and can be used to monitor changes in absorption, reflectance, chemiluminescence, and fluorescence in the sample.

Since optical fibers can be many meters in length, are flexible, and have diameters typically 125 - 1000 μm, it is feasible to perform continuous spectroscopy in previously inaccessible or remote sites. Sensors based on fiber optic technology also provide some interesting advantages over

0097–6156/89/0403–0252$06.25/0

electrochemical sensors. Their sturdy and simple construction permits placement in harsh environments. They are immune to electromagnetic interference, and require no reference electrode. The low cost of optical fibers permits the sensors to be disposable for many applications. Among the potential applications are the in situ toxic waste sites monitoring, in vivo blood gas monitoring and in situ process control monitoring. The very significant potential for in vitro clinical diagnostic applications is also being actively pursued.

Fiber optic sensors are classified as intrinsic or extrinsic sensors. With an intrinsic sensor the optical fiber itself acts as an optical component and is modulated directly by the change in a physical parameter, thus altering the transmitted light. It is possible to detect very subtle influences by coiling a long length of fiber in the modulating area. Such intrinsic responses initiated research activities in the 1970s and produced the first generation of optical sensors (4). Intrinsic sensors exist for the measurement of temperature, magnetic fields, acoustics, strain and electrical current as well as other physical parameters.

An extrinsic sensor is used for specific chemical detection and requires the association of an optical transducer with the fiber. The transducer must induce an optical signal change (absorption, fluorescence or reflectance) in response to the selective detection of an analyte in a complex mixture. The preferred mode is fluorescence due to its inherent sensitivity, the easy separation between exciting and emitted light, and the technical expediency of coupling laser excitation to optical fibers. Since fluorescence intensity is directly proportional to incident light power, lasers can produce intense fluorescence signals. In using fluorescence, a laser or other light source such as a Xenon arc lamp or LED, is coupled to one end of the fiber and the excitation light propagates to the distal tip of the fiber, where it interacts with a specific component of the analyte solution. A portion of the isotropic emission from a fluorophore returns through the same optical fiber. (Figure 1). Appropriate filters or monochromaters are used to separate the scattered excitation light as well as stray light from that of the emission signal before it enters the photon counting device such as a photomultiplier or photodiode. In optical sensors the fiber is completely passive and simply serves as a light guide. The physical parameter or analyte being measured causes a change in the transducer which then interacts with the excitation radiation and thus modulates the returning light to the detector. An analyte dependent fluorescence intensity change is a typical transduction mechanism. pH sensors based on this mechanism use a fluorescent pH indicator, whose fluorescence depends on the acidity of the solution. The operating principle of our first pH sensor used the pH dependent fluorescence of fluorescein (5). The fluorescein is immobilized on a polymer fixed to the end of a fiber. The excitation and emission wavelengths for fluorescein are 488 and 520nm respectively (Figures 2A and 2B).

Direct physical contact between the sensor reagent or transducer and the fiber is not a requirement. The indicator can also be in a sample that is viewed through a window via the optical fiber, such as in a flow-injection-analysis sensing scheme used for process control (6). Although such applications will be an important part of optical sensor technology, the more demanding approach is the preparation of extrinsic sensors with the reagent phase attached directly to the fiber tip, a requirement for in vivo and in situ applications. The inventive work of sensor chemists now focuses

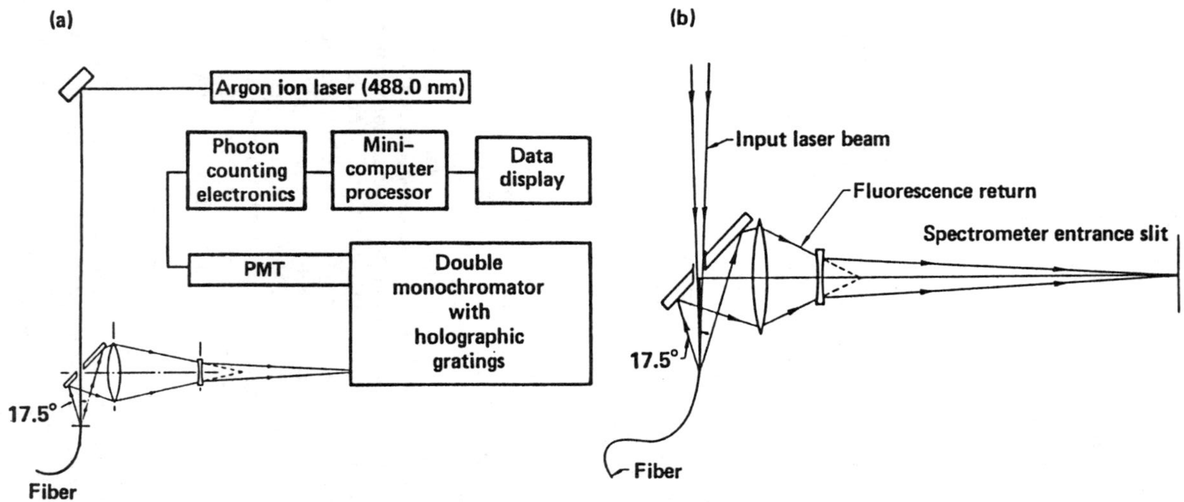

Figure 1. Schematic (a) of the overall fiber optic spectrometer and (b) magnified view of the optical coupler. (Reproduced from ref. 5. Copyright 1986 American Chemical Society.)

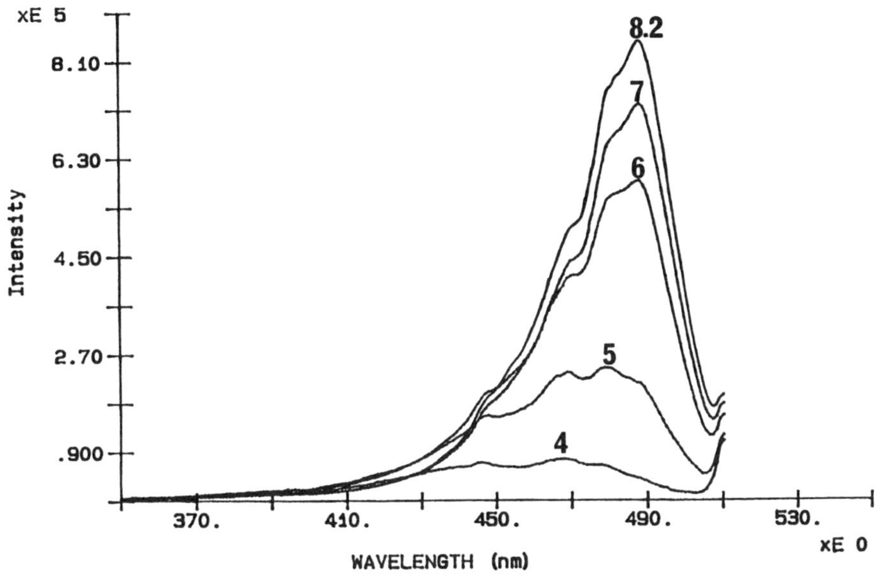

Figure 2A. Excitation spectra of fluorescein at different pH values; emission was monitored at 520 nm.

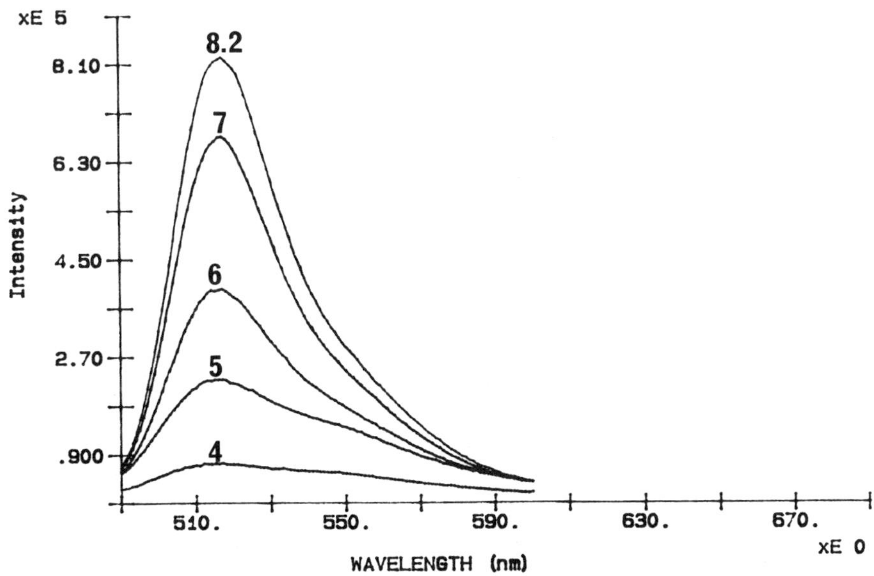

Figure 2B. Emission spectra of fluorescein at different pH values; excitation was at 488 nm.

on two concerns: the design of methods for fixing the transducing element to the fiber tip, and the identification of spectroscopic indicating systems that can be linked to changing concentrations of analytes.

Methods for Attaching Sensing Reagents to the Fiber Tip

General. A variety of fabrication schemes for coupling the transducer to the fiber tip have been employed. Chemically selective reagents have been contained within analyte permeable membranes (7) or tubing (8,9) that cover the end of the fiber. Sensors have been prepared by attaching a glass bead on the tip of a fiber with the reagents adsorbed or bound to the bead (10). A variety of polymers and resins have been used with the sensor material adsorbed to their surface (11,12). The bare optical fiber is then placed in contact with the material and fixed by sealing with a membrane.

The appropriateness of a particular sensor preparation depends on the intended application, however, all of the previous methods suffer from at least one of the following limitations: 1) lengthy response times due to the diffusional limitations of membranes, 2) sensor reagent leaching from the support matrix, 3) small signal levels due to low reagent concentration in the acceptance cone of the optical fiber, 4) enlarged dimensions of the fiber tip due to the reagent fixation method, and 5) the cumbersome task of assembling individual sensors. We have developed a technique for sensor preparation that overcomes these deficiencies and also greatly advances the real-time performance of fiber optic sensors.

Amplification Technology. Direct attachment of an indicator monolayer to the fiber surface produces a sensor with very low signal levels. Our technique for fiber optic sensor preparation is based on the covalent attachment of a polymer to the distal surface of the fiber tip (5). Similar to the polymer modifications performed on electrodes (13,14) the polymer serves to increase the surface area in the sensing region. The indicator reagent, a fluorescent dye or biomolecule, is copolymerized with the fiber into the polymer (Figure 3). Since this procedure greatly increases the amount of sensing reagent immobilized on the fiber tip, the signal intensities are significantly amplified. Furthermore, since no membrane is used to contain the sensing reagent, the polymer is in direct contact with the analyte solution, resulting in a reversible sensor response that depends only on passive diffusion within the polymer layer. Because this layer is only a few microns thick, the sensor responds very rapidly.

We have prepared a variety of sensors with this methodology, thus demonstrating its generic potential (12,15-17). Before describing the preparation of particular sensors, the individual process elements will be discussed. We have used glass fibers exclusively, however the same process elements could be applied, in principle, to plastic fibers with some modification of the experimental details.

Surface Activation. Since the sensor reagents are to be bonded directly to the fiber tip it is first necessary to activate the tip surface. A great variety of silanizing agents exist for derivatizing glass and it is therefore possible to convert an unreactive glass surface to a functionalized surface. The choice of functional group will depend on the subsequent chemistry used to attach the polymer to the surface. Since our initial sensor preparation was based on a vinyl polymerization it was necessary to

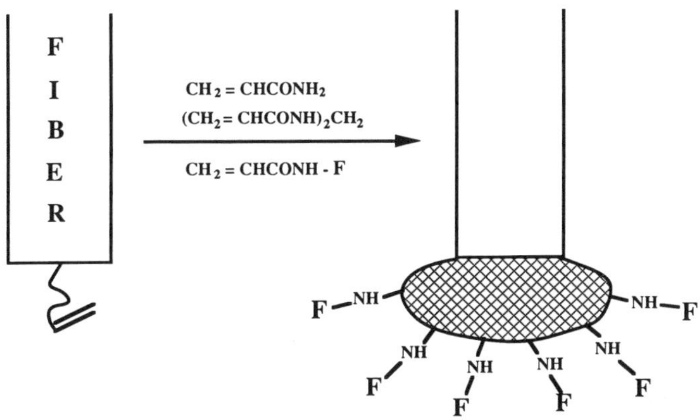

F —— Fluorescein

Figure 3. Copolymerization of fluorescein molecules into the polymer attached directly to the fiber tip.

functionalize the glass tip with vinyl groups, thus enabling the tip to be incorporated covalently into the polymeric sensor layer.

Surface activation can be accomplished by solution or vapor phase silanization or by deposition of a gaseous plasma. A useful silanization reagent is γ-methacryloxypropyltrimethoxysilane which, in one step, functionalizes the glass surface with a vinyl group. Another commmonly used silanization reagent is aminopropyltriethoxysilane, which converts the glass to an amino surface (18). Subsequent treatment with acryloyl chloride converts the amino surface to a vinyl surface. Vapor phase silanization with the same reagent can be accomplished by suspending the fibers over a refluxing solution of the silanizing reagent (19). Vapor deposition is preferred over the solution reaction as it allows more complete monolayer coverage, with less formation of polysiloxane aggregates. However, plasma deposition is the superior method for homogenous surface coverage and given access to the apparatus, is the method of choice (20). This method requires containing the fibers inside a plasma deposition chamber into which monomer vapors are introduced. Passing an electric charge or radio-frequency field through the vapor forms a highly reactive plasma which then reacts with the glass surface. An alkylamine surface can be deposited by reacting glass first with a hexane plasma followed by an ammonia plasma.

Indicator Derivatization. Since the sensor indicator reagents are copolymerized with the immobilized polymer it is necessary to functionalize these reagents with polymerizable groups. It is important that derivatization of these molecules does not compromise their luminescent, optical, or chemical properties. A variety of derivatized molecules exist, such as fluoresceinisothiocyanate (FITC) and fluoresceinamine, and many of these may be functionalized for copolymerization and immobilization. In our first pH sensor preparation, fluoresceinamine was activated for polymerization with acryloyl chloride. Although fluoresceinamine has a very quenched fluorescence, once acylated it acquires the highly fluorescent properties of fluorescein (21). The acrylic double bond then enables the dye molecule to be copolymerized with a vinyl polymer. FITC is also easy to immobilize through reaction with residual amine groups on proteins such as bovine serum albumin. The pH sensitive dye hydroxypyrenetrisulfonic acid (HPTS) can be covalently attached through reaction of its sulfonate groups (22). Derivatization of HPTS requires protection and deprotection of the pH sensitive hydroxyl group and immobilization is therefore not as straightforward as with fluoresceinamine. Numerous other fluorescent and absorbing dyes exist, many with functional groups that lend themselves to facile incorporation into a polymer matrix. Other dyes, such as HPTS, require carefully planned synthetic schemes to preserve their chemical properties.

When the immobilized sensing reagent also contains a bioreceptor, such as an enzyme or an antibody, the device is regarded as a biosensor (23). Such sensors hold great promise as they exploit the inherent ability of the biomolecule to selectively and sensitively recognize a particular chemical species in a complex matrix. Enzyme-based sensors produce a signal due to a selective enzyme-catalyzed chemical reaction of an analyte and form a product that is detected by a transduction element in the sensor. The analyte concentration is thus monitored either through the rate of product formation or the steady-state product concentration.

Polymerization. The key aspect of sensor preparation is copolymerization of the polymer and indicating reagent to form a sensing layer on the fiber tip. This polymerization is accomplished by placing the activated fiber tips into a solution containing monomer, crosslinker, functionalized indicating reagent and polymer initiator. Addition of light or heat may be required to initiate the reaction. As polymerization proceeds, the activated fiber tip is incorporated in the copolymer.

Hydrogels are very suitable polymer matrices for most sensors since a hydrophilic microenvironment is necessary for rapid exchange of analytes between the bulk solution and indicating layer. The first application of the amplification technique was a pH sensor constructed of acrylamide, methylene bisacrylamide as a crosslinker, and acryloylfluorescein (5). We have also constructed pH sensors with acryloylfluorescein immobilized in hydroxyethyl methacrylate (HEMA). This sensor has a slower response time,(ca. 20 sec), than the acrylamide pH sensor but exhibits a greater dynamic range. The contrast in the performance of these two pH sensors suggests that the polymer matrix can alter the behavior of the transducer significantly; such microenvironmental considerations must be taken into account when designing a sensor.

We have also prepared pH sensors using bovine serum albumin (BSA) and HPTS with glutaraldehyde crosslinking on an amino silanized fiber. This preparation is less reliable than the first method due to unpredictable loss of the dye's pH sensitivity following reaction with glutaraldehyde. Furthermore, it is difficult to apply this preparation to the fiber tips. Given the extensive applications of the BSA-glutaraldehyde method to enzyme and protein immobilization, it offers a useful means of covalently immobilizing protein to optical fibers.

Optimization of Amplification Technology. Although this method has been established as a general one for preparing a variety of different sensors, a number of factors have emerged as potentially crucial to the successful preparation of sensors.

a. Although the technique permits the batch preparation of sensors, it is not possible to make a batch of fibers with identical properties. Fiber to fiber variation is due to the inherent difficulty of removing the fiber tips from a fully polymerized gel, which leads to varying amounts of material on the fibers. We have been able to mitigate this limitation by withdrawing the fiber tips during the gelation phase, and allowing them to remain suspended inside the reactor until polymerization is complete.

b. Since the polymerization would initiate ideally at the fiber tip, rather than the solution, we have experimented with initiation catalyzed by light conducted to the fiber tip. Preliminary work indicates that this technique is promising.

c. The pH of the solution can be a factor in the successful copolymerization of a dye in a polymer. We have found that at certain pH values some fluorescence indicators are immobilized in a pH insensitive form, suggesting that the pH of the solution is affecting irreversibly the structure of the dye in the polymer.

d. The concentration of the dye or biomolecule can be varied to increase sensor sensitivity. However, certain dyes may form hydrophobic aggregates which could lead to concentration quenching effects. The dye's physical and chemical properties will determine its optimum concentration in the polymer.

e. The kinetics of the heterogenous polymerization is also a factor in the effective incorporation of sensor reagent. Although we have conducted no direct investigation of this aspect of sensor preparation, we have observed that certain polymers form at rates that exclude the sensor reagent from the copolymer. In addition, the concentration of the dye or biomolecule in the reaction mixture can also affect the rate of polymerization. At very high concentrations of dye, the polymerization may be completely inhibited.

f. As mentioned previously, the polymer microenvironment may significantly influence the behavior of the transducer. This effect is not necessarily detrimental; the choice of polymer may sensitize the dye, as was observed in the case of fluorescein immobilized in HEMA.

g. The concentration and chemical nature of the cross-linker are parameters that should be included in any sensor optimization scheme.

h. Chemical derivatization of dyes and biomolecules must be performed to avoid loss of their desirable characteristics.

i. Purification of the derivatized dyes and biomolecules before polymerization may improve the reproducibility and sensitivity of the sensor. Purification of acryloylfluorescein greatly enhances the success of the pH sensor preparation.

Examples of Sensors Prepared with Amplification Technology

The variety of fiber optic sensors prepared with amplification technology demonstrate the generality of the method.

pH sensors. A pH sensor based on fluorescence intensity changes was made by incorporating acryloylfluorescein into an acrylamide-methylenebis(acrylamide) copolymer attached to a glass fiber modified with γ-methacryloxypropyltrimethoxysilane (5). The sensor gives very rapid responses (ca. 10 sec) and reversible measurements (Figure 4) over the pH range of 4.0 - 7.0, with a signal-to-noise ratio ca. 275/1. The overall dimension of the sensor tip is ca. 220 μm, and the sensor reagent is permanently bound to the fiber tip.

Gas sensors

CO_2 Sensor. A carbon dioxide sensor based on the fluorescence change due to pH modulation of dissolved carbon dioxide was prepared (16). This sensor is comprised of a covalently attached pH sensing layer to which a second, hydrophobic siloxane polymer membrane is applied. The membrane permits the passage of carbon dioxide into the pH sensing region. The sensor was prepared with varying compositions of the pH sensing layer and produced sensors with different dynamic response ranges. The three compositions used for pH layers were: 1) fluorescein covalently immobilized with acrylamide, 2) fluorescein covalently immobilized with HEMA, and 3) HPTS adsorbed on acrylamide. The second polymer membrane coating was prepared from a 4% solution of dimethylsiloxane-bisphenol A carbonate copolymer in a mixture of dichloromethane and n-hexane (1:1), pipetted onto the pH fiber tip and allowed to dry at room temperature for 24 hr. The integrity of the membrane was checked by testing for fluorescence changes in different pH buffers. Any changes in fluorescence indicated that the membrane was not pinhole free and the process was repeated as necessary. The sensor responds reversibly to physiologically significant concentrations of carbon dioxide. The response is 70% complete in 30 sec and reaches equilibrium in 60 sec, with a similar recovery rate.

Gasoline Sensor. A sensor that detects the concentration of gasoline vapor in air has been developed. The transducing mechanism is based on the principle of fluorescence enhancement. The dye immobilized on the tip of the sensor is environmentally sensitive, and provides fluorescent signal enhancement in a nonpolar environment. When gasoline vapor is absorbed by the polymer, it decreases the polarity of the dye microenvironment, resulting in a fluorescence intensity increase with a concomitant blue shift of the emission maximum wavelength from 625nm to 590nm. This result may be explained by assuming that the dye is more polar in the excited state than in the ground state. As the microenvironment of the dye becomes solvated with gasoline vapor, the local polarity of the indicating layer decreases. The excited state of the dye becomes destabilized, increasing the energy difference between the excited and ground state, resulting in the shift to shorter wavelengths. The increase in fluorescence intensity is the result of a shorter excited state lifetime in the nonpolar environment.

Figure 5 shows the results obtained with a gasoline vapor concentration of 166ppb. The sensor has a dynamic range of over 5 orders of magnitude, and has provided a reversible and continuous response for over three weeks.

Biosensors

a. Penicillin, Acrylamide-FITC. Enzyme-based optical biosensors have been prepared with amplification technology by inclusion of the enzyme penicillinase with the components of a pH sensor. Penicillin can be detected according to the following enzyme catalyzed conversion.

$$\text{Penicillin G} \xrightarrow{\text{Penicillinase}} \text{Penicilloate} + H^+$$

The pH change caused by the action of penicillinase on penicillin allows concentrations of penicillin to be measured as a function of a pH change in the microenvironment of the sensor, detected by a pH sensitive fluorescent dye such as fluorescein or HPTS contained in the sensing layer. Monitoring a non-specific property such as pH can reduce the selectivity advantage of bioreceptor reactions because the sensor is also sensitive to pH changes in the bulk medium.

Enzyme immobilization is a well developed area of chemistry and numerous methods are available (18). We have prepared a variety of enzyme based biosensors using the following immobilization procedures: 1) entrapment in crosslinked polyacrylamide (17), 2) glutaraldehyde crosslinking to BSA (17), 3) copolymerization of an enzyme with acrylamide after activation of the enzyme's amine groups with acryloyl chloride (24).

b. Penicillin, Avidin-biotin. Surface amplification has also been used to develop a general method to immobilize enzymes and utilizes the interaction between the glycoprotein avidin and the vitamin biotin. This sensor operates on the same principle as the penicillin sensor made by

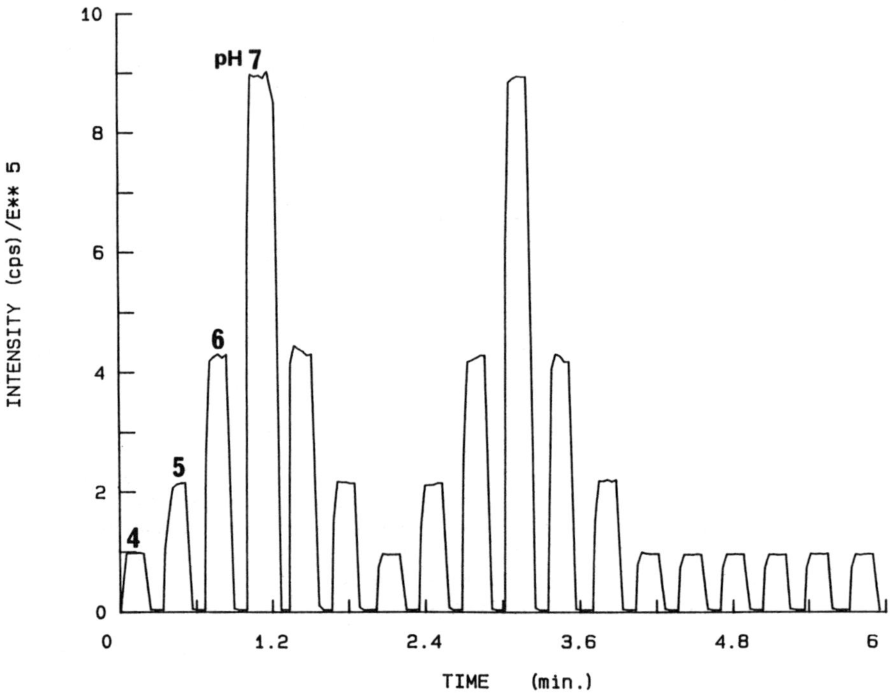

Figure 4. Data produced by pH fiber optic sensor based on fluorescein. Signal was produced with laser excitation at 488 nm; observation was at 530 nm.

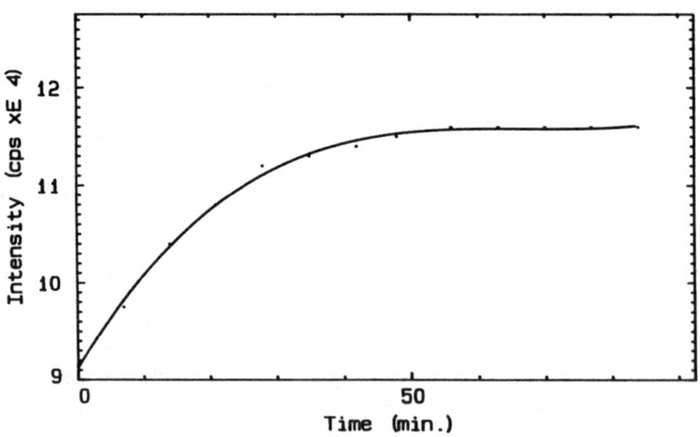

Figure 5. Time Spectrum of the lowest detectable concentration of gasoline (166 ppb). Excitation was at 570nm and emission at 600nm.

coimmobilization as described above. Penicillinase catalyzes conversion of penicillin to penicilloic acid producing a microenvironmental pH change at the fiber tip. This pH change is sensed through the fluorescence change of fluorescein attached to the enzyme. The detection limit was found to be $1x10^{-4}$ M when measured in pH 7.0 0.005M phosphate buffer kept at constant ionic strength. This sensor has demonstrated its reversibility, sensitivity and stability over a three-month period with steady state response times of 60-90 seconds.

To prepare this sensor the fiber is first silanized with γ-methacryloxypropyltrimethoxysilane and then copolymerized with acrylamide, cross-linkers and monomers containing free amino groups. The amino-containing fiber reacts further with succinimidyl-derivatized biotin and then with avidin, furnishing a biotin-avidin complex on the fiber. Because of avidin's multiple valency, a biotin and fluorescein-labeled enzyme could then be coupled to the fiber (Figure 6). This unique noncovalent binding of avidin and biotin is best known for its extremely high binding constant and stability towards pH changes, temperature and denaturing agents.

Fluorescence Energy Transfer Based Sensors

In the direct fluorescence measurement based sensors, there are some limitations imposed on the preparation and application of more sensors. These include the stringent criteria placed on the dye's properties:

1. High quantum yield for fluorescence.
2. Excitation and emission wavelengths commensurate with the light source and detection systems.
3. Sensitivity to a particular chemical species in the desired concentration range.
4. Selectivity for the species of interest in a heterogeneous sample matrix.

These criteria limit the applicability of fluorescent dye-based sensors to only a handful of analytes. If a fluorescence-based method does not exist or is incompatible with an optical sensor, an absorption sensor is the only alternative. There exist a large number of different absorbing dyes that are sensitive, selective and absorb in convenient regions of the spectrum. Several absorbance-based fiber-optic sensors have been described (8,25,26). Unfortunately, a major drawback with absorption spectroscopy is its inherent insensitivity. We decided to explore the possibility of using other optical techniques that can be used with indirect methods. A successful example has been presented which combines the sensitivity of the fluorescence measurement with the specificity of an absorbing dye and is based on an energy transfer mechanism (15).

Energy transfer fibers are based on the principle that, upon excitation, a donor molecule will transfer a portion of its energy to an acceptor molecule if there is overlap between the donor's emission and the acceptor's absorption spectrum. This transfer occurs without the emission of a photon and is primarily the result of a dipole-dipole interaction between the donor and acceptor. The efficiency of singlet dipole-dipole energy transfer is predicted by Förster theory (27).

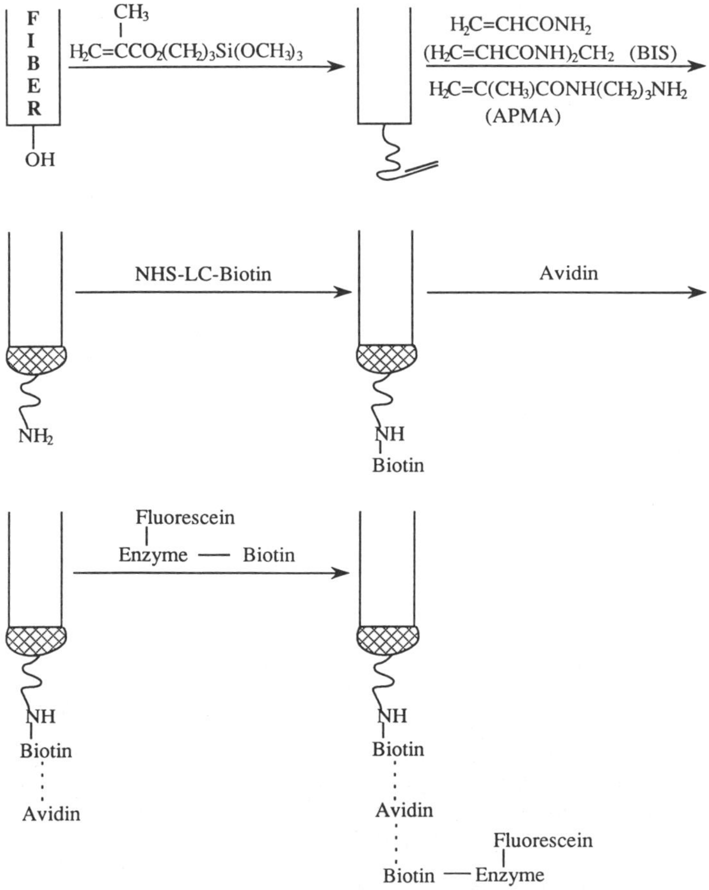

Figure 6. General reaction procedures for preparing an enzyme optical sensor based on avidin-biotin interaction.

$$\mathrm{Eff} = \frac{R_0^6}{R_0^6 + R^6} \qquad [1]$$

where R is the distance between the two chromophores and R_0 is the distance at which energy transfer is 50%. R is dependent on the concentration of absorber while R_0 depends primarily on the spectral overlap integral. Förster proposed an expression for the relationship between R and acceptor concentration

$$R = \left(\frac{3000}{4\pi N[A]}\right)^{1/3} \qquad [2]$$

where N is Avogadro's number and [A] is the concentration of acceptor. Alternatively, we have

$$E = \frac{1}{1 + \left(\frac{[A]_0}{[A]}\right)^2} \qquad [3]$$

A pH fiber-optic chemical sensor based on energy transfer has been prepared successfully. This sensor utilizes a polymeric bichromophore and employs a fluorescent donor, eosin, and a non-fluorescent pH sensitive acceptor, phenol red. As pH increases, phenol red's absorption increases in the spectral region that overlaps with the emission spectrum of eosin (Fig.7). Since the extent of energy transfer is proportional to the spectral overlap integral, the efficiency of energy transfer increases as the pH increases and is detected as a decrease in eosin's fluorescence.

The preparation of this sensor was accomplished by covalently attaching both dye moieties to the end of an optical fiber employing the same polymerization chemistry used in the direct fluorescence measurement fiber described above. This intimate mixing of the two dyes leads to efficient energy transfer since the effect is inversely proportional to the sixth power of the distance between the donor and acceptor.

The results of a fiber prepared in this manner are shown in Figure 8. As predicted, the fluorescence intensity decreases as the pH increases, because as the pH increases, phenol red increases its spectral overlap integral with eosin and becomes more efficient at accepting energy from eosin. The sensor has proven to be highly sensitive for reversible pH measurements in the physiological range. The titration curve made by this sensor is virtually identical to a titration curve for phenol red indicating that the absorbance characteristics of phenol red are being detected indirectly as a fluorescent signal.

This approach provides a new sensor design and a greater flexibility in the choice of dyes utilized. Therefore, the applicability of fiber optic sensors for analysis should be enhanced dramatically. However, the polymer-based energy transfer systems still suffer from several limitations. First, in order to obtain sufficiently small intramolecular distances between donor and acceptor moieties, the prepolymerized concentrations of the dyes must be high. Oftentimes, high dye concentrations are precluded by limited solubility. Second, in polymers, the distance between dyes is only an

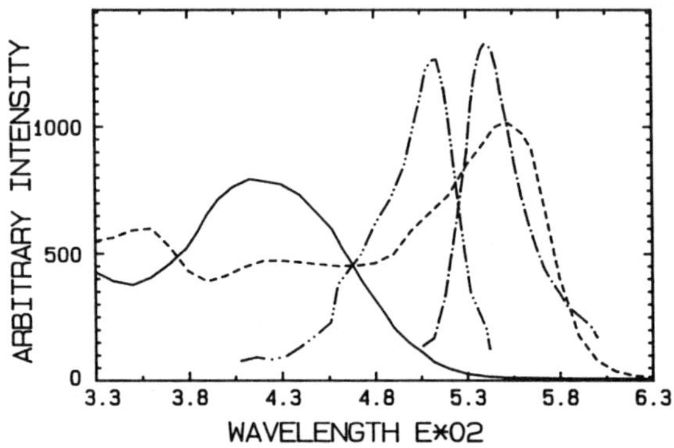

Figure 7. Phenol red pH 4.0 - absorption ———— ; Phenol red pH 8.0 - absorption - - - - - ; Eosin - excitation -··-··- ; Eosin - emission -·-·-.

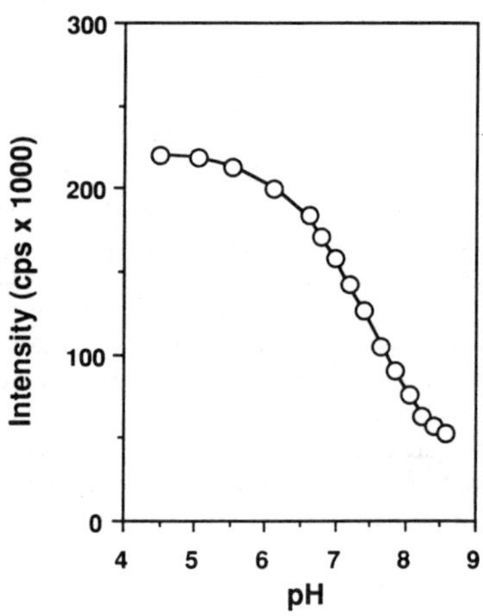

Figure 8. pH response of phenol red-eosin fiber. (Reproduced from ref. 15. Copyright 1987 American Chemical Society.)

average and is poorly defined. Third, the chemical incorporation and distribution of dyes in the polymer cannot be controlled. Finally the polymer cannot be characterized adequately.

Sensors Based on Irreversible Chemistry

The sensors described above have all been based on the reversibility of the indicating scheme. However, many colorimetric or fluorometric techniques are irreversible due to the formation of a tight binding complex of reagent and analyte or the formation of an irreversibly colored adduct. To solve these problems, a long lasting, continuous reagent delivery sensor is required. We have successfully employed a polymeric delivery system to construct a long lasting irreversible pH fiber optic sensor. This polymeric system is able to deliver fresh sensing reagent over a few months (28).

Polymeric delivery systems are under rapid development (29-31). They have been used for the sustained release of macromolecules such as polypeptide hormones, polysaccharides, antigens, antibodies and enzymes. Based on the work of Langer et al, the best long-term release results were obtained with ethylene-vinyl acetate copolymer (EVA). We have employed EVA as a polymer matrix throughout our experiments. The polymer was dissolved in an appropriate solvent such as methylene chloride. The powdered form of a sensing reagent was added to the polymer solution and mixed completely. The resulting mixture is cast in a mold and dried .

Selection of the sensing reagent depends on the availability of suitable instrumentation. Two approaches have been explored to make a pH sensor. Both approaches use a ratiometric method to compensate for variations in indicator release rate. First, a single pH sensitive dye, 8-hydroxypyrene-1,3,6-trisulfonic acid (HPTS) which has dual excitation wavelengths was utilized (22,32). The ratios of fluorescence emission intensities at 515 nm resulting from excitation at 405 nm (acid form) and 450 nm (base form) were used to measure pH values from 5.5 to 8.0 as shown in Figure 9. The excitation light source of this system employs a high pressure Xenon arc lamp which gives a continuous spectrum from 190 nm to 750 nm and a double monochromator for selecting any specified excitation wavelength. The advantages of ratio measurements lie in their internal calibration removing sensitivity to fluctuations in source intensity, amount of reagent in the sensing region at the time of measurement and photobleaching.

In cases where dual excitation or dual emission reagents are not available or where only a single wavelength excitation source such as a 488nm argon ion laser is available, a mixture of two reagents could be used. In this case, one reagent responds to changes of analyte concentrations while the other serves as an internal reference. However, it should be emphasized that the variations in the release rate of the two reagents could cause significant changes in the accuracy of the ratio measurement. Therefore the two reagents must be mixed thoroughly before they are entrapped into the EVA copolymer. In our second approach, HPTS and sulforodamine 640 (SR-640) were first dissolved and mixed in water to ensure a uniform distribution of the dyes. Water was then removed by lyophilization and the two-dye powder was incorporated into the EVA copolymer. Both dyes were excited by an argon ion laser at 488 nm and their emission ratios at 530 nm and 610 nm were used to measure pH values between 5.5 and 8.0 as shown in Figure 10.

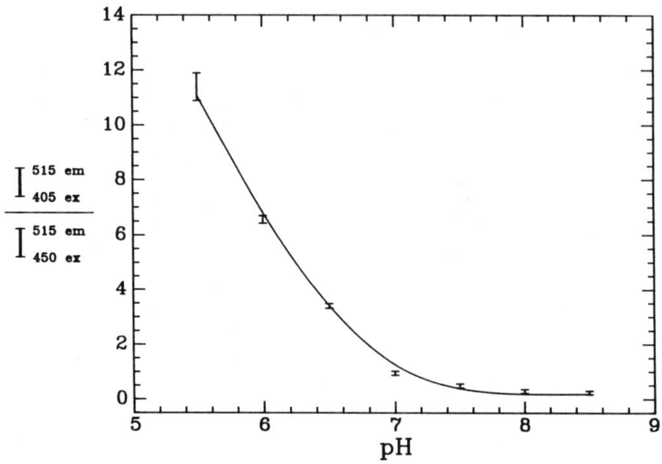

Figure 9. Results of pH fiber-optic sensor based on HPTS. Signals were produced with excitation at 405 and 450 nm; observation was at 515 nm. (Reproduced from ref. 28. Copyright 1989 American Chemical Society.)

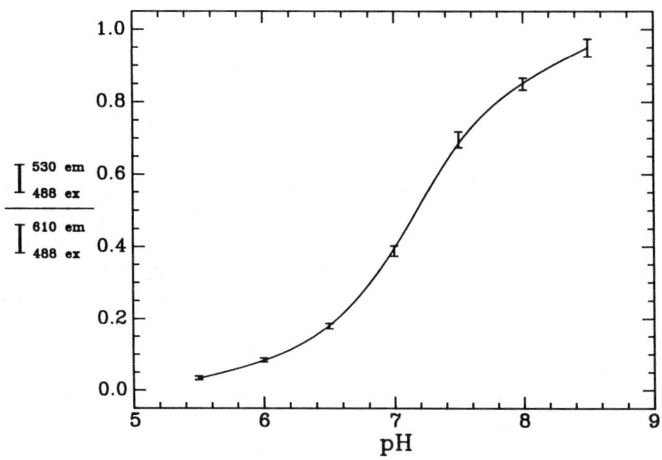

Figure 10. Results of pH fiber-optic sensor based on two dyes: HPTS and SR-640. Signals were produced with laser excitation at 488 nm. (Reproduced from ref. 28. Copyright 1989 American Chemical Society.)

To construct a sensor based on the polymeric delivery system, both the longevity and the response time of the sensor must be considered. The sensor is composed of three parts: a polymer reservoir, a sensing region with many tiny holes and a fiber with its tip protruding through the polymer reservoir and positioned at the top of the sensing region (Figure 11, part C, D and E). The dye-containing polymer was placed in a reservoir of sufficient size to meet the required longevity of the sensor. The response time per pH unit change of a sensor is a function of its design and the condition of the solution being measured. Here we define the response time as the time required to reach 95% of the steady state signal. As determined experimentally (Table 1), stirring the analyte solution greatly enhanced the rate for establishing equilibrium between the analyte solution and the solution inside the sensor. Shortening the sensing region by half from 6 mm to 3 mm decreased the response time by 33%. In addition, increasing the hole diameter in the sensing region from 0.25 mm to 0.36 mm decreased response time by a factor of three. Under the best conditions (3 mm sensing length, 0.36 mm hole diameter, stirring), the response time is about 10 minutes which is longer than other pH sensors with immobilized reagents on the fiber tip. However it is acceptable when sensors of this type are used in remote sensing of groundwater or hazardous environments for extended periods of time.

Sensors based on reagent delivery with controlled release polymers have opened up the possiblity of designing sensors based on analytical reactions that consume reagent. Many cation or anion determinations use chelating reagents that form irreversible adducts and can potentially be used with this type of sensor. Also, this type of sensor may find application to fluorescence immunoassay because most antigen-antibody binding constants are sufficiently high restricting the development of reversible fiber-optic immunosensors. Although such sensors necessarily have finite lifetimes, these lifetimes can be quite long if the rate of reagent consumption is small relative to the total amount of reagent available. Furthermore, sensors based on using ratio measurements do not require steady-state mass transfer to obtain a constant signal.

TABLE 1. Response Time of a HPTS Sensor (mins/pH unit)

sensing length / hole diameter	3 mm		6 mm	
	a	b	a	b
0.25 mm	35	160	50	300
0.36 mm	10	60	15	90

a. data shown were measured with stirring.
b. data shown were measured without stirring.

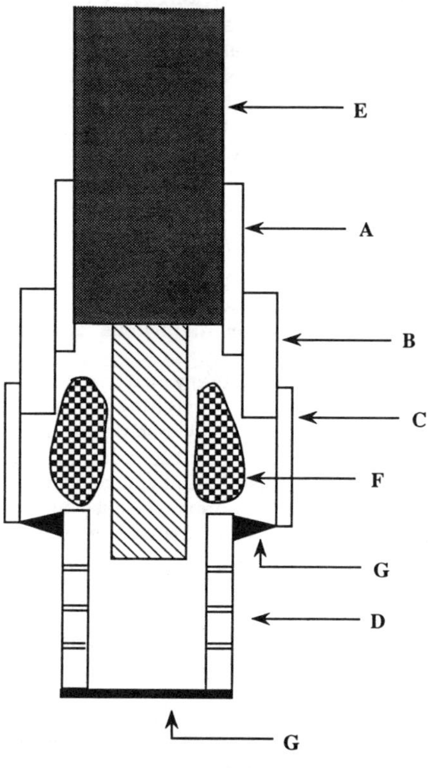

A. Teflon capillary tube (I.D. = 1/32", O.D. = 1/16")
B. Tygon tube (I.D. = 1/16", O.D. = 3/16")
C. Teflon tube (I.D. = 3/16", O.D. = 1/4")
D. Teflon tube (I.D. = 1/16", O.D. = 1/4") with tiny holes (0.25 - 0.36 mm)
E. Optical Fiber
F. Polymer matrix
G. Parafilm

Figure 11. Configuration of a fiber-optic sensor based on reagent delivery with controlled-release polymers.

Summary

As described in the foregoing article, fiber optic sensors are achieving a place in the analytical sensor arena. The successful development of a particular fiber optic sensor involves the suitable selection of an indicating chemistry accompanied by a means to couple this indicating chemistry to the sensor tip. In designing a sensor it is possible to take advantage of a wide variety of sensing mechanisms. First, one can choose a direct measurement in which the change in optical signal, such as fluorescence intensity, is measured directly by the indicating reagent. In the absence of a suitable fluorescent indicating material it is possible to couple a colorimetric indicator with a fluorescent indicator using energy transfer as the transducing mechanism. This approach enables one to choose from the plethora of available colorimetric chemistries without sacrificing the sensitivity inherent in the fluorescence measurement. For biologically active materials it is possible to couple the indicating dye with an enzyme to generate a selective sensor based on the enzyme's specificity. Finally in those systems where reversible indicators do not exist, it is possible to use a slow release polymer to deliver fresh sensing reagent to the distal tip of the fiber on a continuous basis. This method, however, suffers from longer response times but enables the analytical chemist to expand the available indicating chemistries suitable for application to fiber optic sensors.

The field of optical sensing was initiated only a decade ago. With the rapid advances being made in sensor design coupled with the advances being made on indicating chemistries, the prospect for widespread use of optical sensors for analysis is promising. Developments in the fields of optics, optical fiber materials, indicating dyes, and of new sensing schemes will enhance the utility of these devices.

Acknowledgments

The authors are deeply indebted to Dr. Fred Milanovich of Lawrence Livermore National Laboratory for developing the instrumentation used in this work. Financial support for this work was provided by the Environmental Protection Agency to Tuft's Center for Environmental Management.

Literature Cited

1. Seitz, W.R. C.R.C. Reviews in Analytical Chemistry 1988, 19, 135.
2. Wolfbeis, O.S. Trends in Analytical Chemistry 1985, 4, 184.
3. Angel, S.M. Spectroscopy 1987, 2, 38.
4. Fluitman, J.; Popma, Th. Sensors and Actuators 1986, 10, 25.
5. Munkholm, C.; Walt, D.R.; Milanovich, F.P.; Klainer, S.M. Anal. Chem. 1986, 58, 1427.
6. Ruzika, J.; Hansen, E.H. Analytica Chimica Acta 1985, 173, 3.
7. Scheggi, A.M.; Baldini, F. Optica Acta 1986, 33, 1587.
8. Peterson, J.I.; Goldstein, S.R.; Fitzgerald, R.V. Anal. Chem. 1980, 52, 864.
9. Saari, L.A.; Seitz, W.R.; Anal. Chem. 1983, 55, 667.
10. Fuh, M-R.S.; Burgess, L.W.; Hirschfeld, T.; Christian, G.D.; Wang, F. Analyst 1987, 112, 1159.
11. Zhujun, Z.; Seitz, W.R. Analytica Chimica Acta 1984, 109, 15.

12. Kirkbright, G.F.; Narayanaswamy, R.; Welti, N.A. Analyst 1984, 109, 15.
13. Itaya, K.; Bard, A.J. Anal. Chem. 1978, 50, 1487.
14. Elliot, C.M.; Murray, R.W. Anal. Chem. 1976, 48, 1247.
15. Jordan, D.; Walt, D.R.; Milanovich, F.P. Anal. Chem. 1987, 59, 437.
16. Munkholm, C.; Walt, D.R.; Milanovich, F.P. Talanta 1988, 35, 109.
17. Kulp, T.J.; Camins, I.; Angel, S.M.; Munkholm, C.; Walt, D.R. Anal. Chem. 1987, 59, 2849.
18. Methods in Enzymology Mosbach, K., Ed; Academic Press, New York, 1976, 44, 139.
19. Haller, I. J. Am. Chem. Soc. 1978, 100, 8050.
20. Yasuda, H. J. Poly. Sci. 1977, 15, 81.
21. Munkholm, C.; Parkinson, D-R.; Walt, D.R. submitted for publication.
22. Offenbacher, H.; Wolfbeis, O.S.; Furlinger, E.; Sensors and Actuators 1986, 9, 73.
23. Arnold, M. A.; Meyerhoff, M. E.; Critical Reviews in Analytical Chemistry, 1988, 20, 149.
24. Walt, D.R.; Luo, S.; Munkholm, C. Proceedings of SPIE, Optical Fibers in Medicine III January 1988, 906
25. Coleman, J. T.; Eastham, J. F.; Sepaniak, M. J. Anal. Chem. 1984, 56, 2246
26. Jones, T. P.; Porter, M. D. Anal. Chem. 1988, 60, 404.
27. Förster, T. Discuss. Faraday Soc. 1959, 27, 7.
28. Luo, S.; Walt, D. R. Anal. Chem., 1989, 61, 174.
29. Langer, R. Methods in Enzymology 1981, 73, 57.
30. Langer, R. Chemtech, 1982, February, 98.
31. Rhine, W.D.; Hsieh, O.S.T.; Langer, R. J. Pharm. Sci. 1980, 69, 256.
32. Zhang, Z.; Seitz, W.R. Anal. Chim. Acta 1984, 160, 47.

RECEIVED March 24, 1989

Chapter 18

Poly(vinyl alcohol)-Based Indicators for Optical pH and Mg(II) Sensing

Ma Wangbai[1], Zhang Zhujun[1], and W. Rudolf Seitz

Department of Chemistry, University of New Hampshire, Durham, NH 03824

A fiber optic pH sensor has been prepared by covalently coupling fluoresceinamine to poly(vinyl alcohol) (PVOH) via cyanuric chloride. A drop of aqueous PVOH/indicator conjugate is then crosslinked with glutaraldehyde in situ on the common end of a bifurcated fiber optic bundle. Fluorescence intensity excited at 480 nm and observed at 520 nm increases as the pH is varied from 3 to 8. Fluorescence intensity excited as 420 nm and observed at 520 nm is constant over the same range and can serve as a reference to compensate for instrumental fluctuations. A Mg(II) sensor is prepared by coimmobilizing eriochrome black T (EBT) and fluoresceinamine on the poly(vinyl alcohol). At pH 9.6, increasing the Mg(II) concentration from 0 to 20 micromolar leads to a decrease in fluoresceinamine emission intensity. This occurs because the absorption spectrum of Mg(II)-EBT overlaps the fluoresceinamine emission, causing quenching via fluorescence energy transfer.

Poly(vinyl alcohol) (PVOH) crosslinked with glutaraldehyde offers important advantages as a substrate for immobilizing indicators for fiber optic chemical sensors (1). Sensor preparation involves the following steps: (a) indicator is covalently coupled to PVOH using cyanuric chloride as a linking reagent; (b) the indicator/PVOH conjugate is dissolved in water and the amount of indicator is determined by spectrophotometry; (c) if necessary, the immobilization is repeated or the indicator/PVOH conjugate is diluted with underivatized PVOH to adjust the indicator level to a desired value; (d) glutaraldehyde and HCl are added to initiate crosslinking at a relatively slow rate so that solid doesn't start to form for several minutes; and (e) an acurately known volume of indicator/PVOH/glutaraldehyde is transferred to the end of a fiber optic bundle and allowed to gel in situ. This is the first reported procedure for preparing fiber optic chemical sensors that allows for reproducible

[1]Current address: Department of Chemistry, Shaanxi Normal University, Xian, People's Republic of China

0097-6156/89/0403-0273$06.00/0

and independent control of both the amount of indicator and the amount of substrate.

An earlier report focussed on the properties of the PVOH gel itself and demonstrated how the amount of indicator and substrate could be systematically controlled (1). Here we describe the response characteristics of sensors for pH and Mg(II) prepared using PVOH as the indicator substrate. The pH sensor is based on immobilized fluoresceinamine, an indicator used in previous optical pH sensors (2-4). The Mg(II) sensor is based on eriochrome black T (EBT) coimmobilized with fluoresceinamine. Because the absorption spectrum of the Mg(II)-EBT complex overlaps the emission spectrum of immobilized fluoresceinamine, the presence of Mg(II) leads to a decrease in fluoresceinamine fluorescence intensity because there is energy transfer to the Mg(II)-EBT complex. Energy transfer based sensors provide a way of coupling a color change to a fluorescence measurement (5).

Several of the experiments reported here involve manipulation of the amount of immobilized indicator and serve to illustrate the advantages of using the PVOH based procedures.

Experimental

Apparatus. The fiber optic photometer is similar to the instrument used in earlier work (2). It includes a tungsten-halogen source, a photomultiplier detector, interference filters for wavelength selection and a bifurcated fiber optic bundle 3 mm in diameter at the common end. In this study, excitation filters had peak transmittances at 420 and 480 nm and the emission filter had peak transmittance at 520 nm. The bandwiths at half-maximum transmittance were 9.4, 8.2 and 8.6 nm for the 420, 480 and 520 nm filters, respectively. An automated filter wheel allowed for rapid switching from one filter to another.

Fluorescence excitation and emission spectra were recorded with an SLM 8000 spectrofluorometer. They are not corrected for instrumental effects. Absorption was measured with a Spectronic 200 spectrophotometer. An Orion 501 pH meter was used to determine pH.

Reagents. Poly(vinyl alcohol) (100 % hydrolyzed) with average molecular weight 14,000, cyanuric chloride (97%), fluoresceinamine isomer I and eriochrome black T (indicator grade) were purchased from Aldrich Chemical Co. Glutaraldehyde solution (50% (w/w) was obtained from Fisher Scientific.

Magnesium standard solutions were prepared by dissolving 100.0 mg of magnesium metal in 2 mL of 1.0 M HCl and diluting to 100.0 mLs.

Hydrochloric acid or potassium hydroxide were added to buffers to vary pH. Acetate, phosphate and tris(hydroxymethyl)aminomethane buffers were used for the pH ranges 1.40 to 6.00, 6.00 to 8.00 and 8.00 to 10.00 respectively.

Mg(II) measurements were all made in pH 9.6 ammonia buffer.

Procedures. The first step in indicator immobilization is to react poly(vinyl alcohol) (PVOH) with cyanuric chloride in acetone/water as described earlier (1). The PVOH/cyanuric chloride conjugate is then reacted with 10 mLs of a solution of indicator in acetone. For pH sensing, the indicator solution contained 100 mg of fluoresceinamine. For Mg(II) sensing, the indicator solution typically contained 100 mg of eriochrome black T and 5 mg fluoresceinamine. After 30 minutes, the product is separated and washed first with acetone and then with water until no further unreacted indicator can be observed in the washings. The product was dried and stored as a solid. The amount of immobilized indicator was determined by spectrophotometry using standard curves prepared from unreacted indicator.

The first step in preparing a sensor is to dissolve 100 mg of PVOH/indicator conjugate in 2.0 mL of water. Five minutes in a water bath at 80°C is required to get the PVOH to dissolve. After cooling to room temperature, 0.5 mL of PVOH/indicator solution is combined with 0.050 mLs each of 2% aqueous glutaraldehyde and 4 M HCl. This mixture can be manipulated as a liquid for about five minutes until sufficient crosslinking takes place to cause the solution to gel. During this interval a micropipet is used to precisely transfer 3 microliters to the common end of the bifurcated fiber optic bundle so that the gel forms in situ.

Results and Discussion

pH Sensor. Figure 1 shows excitation and emission spectra for fluoresceinamine covalently coupled to a PVOH gel. Spectra are very similar to the spectra observed for fluorescein. This is expected since the immobilization ties up the amino group of fluoresceinamine so that it does not influence the acid-base properties of the conjugate. This is consistent with previous studies of pH sensors based on covalently immobilized fluoresceinamine (2-4).

Fluorescence intensity is much lower at pH 1.4 than at pH 9.1. The intensity at 520 nm at pH 1.4 is only 7% of the intensity at pH 9.1. In figure 1, the spectra at pH 1.4 have been magnified 10X so that they can be shown on the same scale as the spectra at pH 9.1.

Figure 2 shows emission intensity at 520 nm vs pH for excitation at 420 and 480 nm. The band excited at 480 nm increases in intensity as the pH goes from about 3.5 to 7.0. This is similar to that observed for previously reported pH sensors based on immobilized fluoresceinamine and is due to the formation of the highly fluorescent base form of fluorescein.

The band excited at 420 nm decreases in intensity as the pH increases from 2 to 3. This band is due to the protonated from of fluorescein which has a pKa around 2.2 (7). However, above pH 3 up to pH 9 the intensity of this band is not affected by pH. We do not understand why this band stays constant since the protonated form of fluorescein should not be present at high pH. It may be due to a different form of fluorescein, e.g. fluorescein covalently coupled to PVOH through a hydroxy group as well as the amino group. While the origin of the band excited at 420 nm is uncertain, it does provide a reference signal so that pH can be related to an intensity ratio which is insensitive to drift and other parameters that affect absolute intensity measurements. Previous fluorescein-based pH sensors have not been based on intensity ratio measurements (2-4).

The response time and the sensitivity of the pH sensor depend on the amount of substrate and indicator, respectively. As described earlier, the procedure for preparing PVOH based sensors permits independent control of these parameters (1).

Mg(II) Sensor. Figure 3 shows excitation and emission spectra of

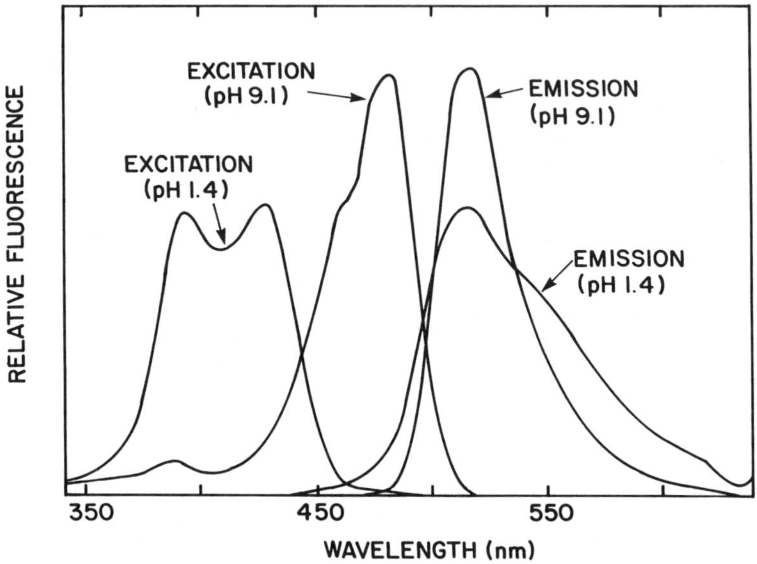

Figure 1. Excitation and emission spectra for immobilized fluoresceinamine at pH 1.4 and 9.1. The intensity of the pH 1.4 spectra has been increased 10X to be on the same scale as the pH 9.1 spectra.

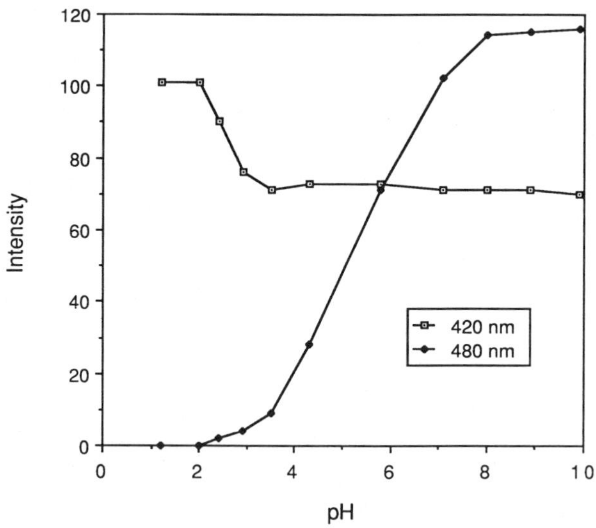

Figure 2. Relative emission intensity at 520 nm vs. pH for fluorescence excited at 420 and 480 nm.

immobilized fluoresceinamine and absorption spectra of eriochrome black T (EBT) in both the free form and complexed to Mg(II). Because the Mg(II)-EBT absorption overlaps fluoresceinamine emission, Mg(II)-EBT can quench emission either by an inner filter effect or by long range energy transfer. Free EBT absorbs at longer wavelengths and, therefore, is not an efficient quencher of fluorescein emission. Thus, it is possible to base sensing on the decrease in fluoreseinamine emission intensity as a function of Mg(II) concentration.

Figure 4 shows the reactions involved in immobilizing EBT and forming the Mg(II)-EBT complex. Although immobilization ties up a hydroxy group ortho to the nitrogen-nitrogen double bond, it does not interfere with the ability of EBT to complex with Mg(II), as evidenced by the fact that the intensity ratio decreases with Mg(II) concentration as expected. Furthermore, the sensitivity to Mg(II) is in the range expected from the conditional formation constants for Mg(II)-EBT complexes at pH 9.6 (8). Our observations in combination with the earlier report of 3:1 Mg(II)-EBT complexes strongly indicate that only ne of the ortho hydroxy groups is involved in complexation.

Figure 4 shows cyanuric chloride reacting with the hydroxy group on the same naphthalene ring as the sulfonate and the nitro groups. Because these electron withdrawing groups should enhance the acidity of the hydroxy group and make it more reactive with cyanuric chloride, we believe that the structure shown in figure 4 is the major product. However, the other hydroxy group can also react with cyanuric chloride. The actual distribuion of products is not known.

The complexation reaction involves displacement of a proton from the EBT. As a result, response to Mg(II) will be inherently pH dependent. As pH decreases, the sensor will be less sensitive because a larger Mg(II) concentration will be required to displace the proton and form the complex. To avoid this effect, all measurements were made in an ammonia/ammonium buffer with a pH of 9.6. This buffer serves not only to control pH but also as a secondary ligand to tie up metal ions such as Cu(II) that might otherwise interfere in the measurement.

The emission band excited at 420 nm was found to be insensitive to quenching. Although this emission also overlaps Mg(II)-EBT absorption, it is due to a less efficient fluorophor, as evidenced by the observation that emission intensity is much lower for 420 nm excitation than for 480 nm excitation. This suggests that the fluorophor excited at 420 nm has a shorter fluorescence lifetime than the fluorophor excited at 480 nm and, therefore, is less subject to quenching by long range energy transfer. The band excited at 420 nm thus can serve as a reference intensity to compensate for instrumental fluctuations.

Figure 5 shows that the ratio of fluorescence intensities excited at 480 and 420 nm decreases as a function of added Mg(II), confirming that the Mg(II)-EBT is quenching fluorescence from the base form of immobilized fluoresceinamine. The immobilized indicator phase contained 0.07 mg fluoresceinamine and 4.28 mg EBT per gram of PVOH.

The data in figure 5 do not fit a model which assumes a 1:1 complex. Instead, relative intensity decreases more rapidly with increasing Mg(II) concentration than expected for the 1:1 model. This suggests that 2:1 and possibly even 3:1 complexes are forming. Higher order complexes have been shown to form in solution (8,9).

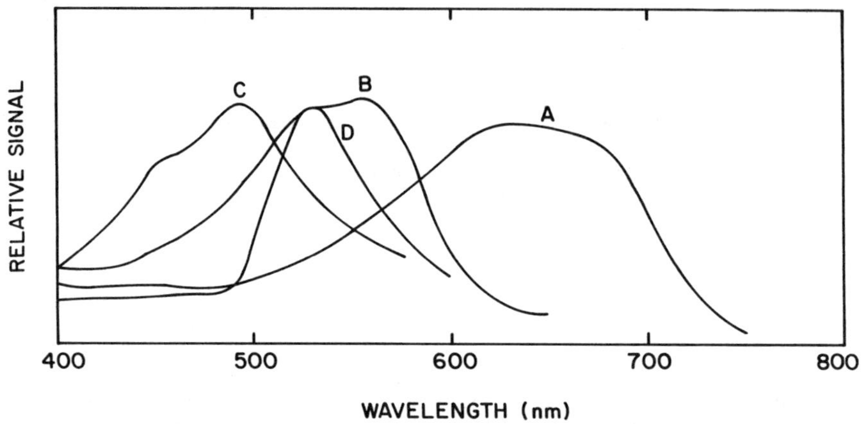

Figure 3. Absorption spectra of immobilized EBT (A) and Mg(II)-EBT (B). Excitation (C) and emission spectrum (D) for immobilized fluoresceinamine. All spectra at pH 9.6.

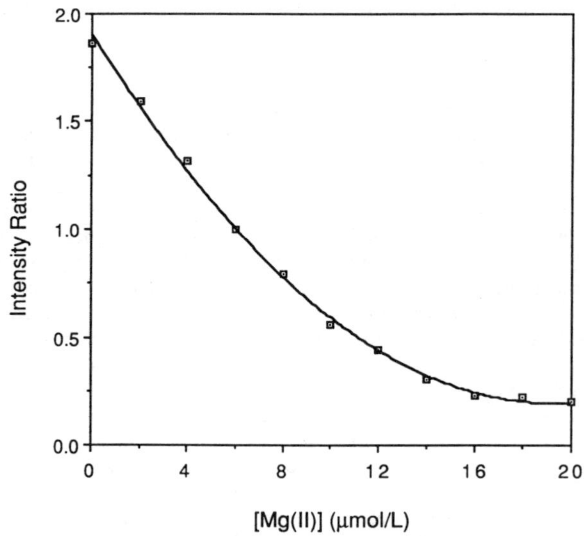

Figure 4. Reactions involved in covalently coupling EBT to PVOH.

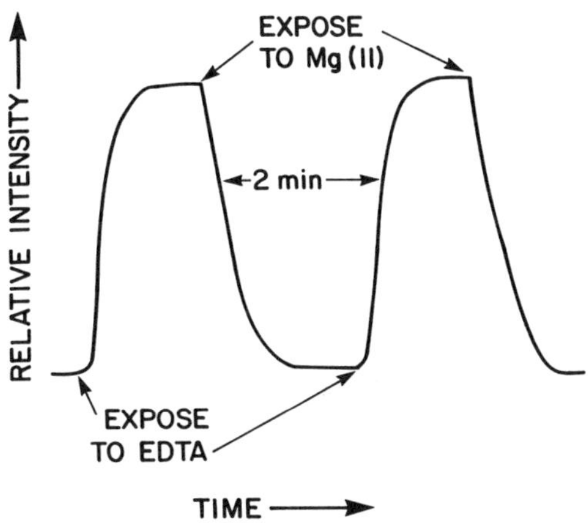

Figure 5. Decrease in the ratio of intensity excited at 480 nm to intensity excited at 420 nm as a function of added Mg(II).

Figure 6 shows typical raw data. The response time is about two minutes when the sensor is exposed to a solution of Mg(II). If the sensor is reexposed to a Mg(II)-free blank, fluorescence intensity returns to its original value. Because this is slow, sensors were exposed to a solution of EDTA to accelerate response. In the presence of EDTA, intensity returns to its original value in about one minute.

Quenching can occur by several mechanisms, including long range energy transfer, and absorption of the emitted fluorescence, commonly known as an "inner filter" effect. The effect of conditions on various quenching mechanisms has been examined (6). Several lines of evidence indicate that long range fluorescence energy transfer is the dominant quenching mechanism in the Mg(II) sensor. First, the absorbance of the PVOH/EBT phase is too low to cause an inner filter large enough to account for the response curve in figure 4. Additional evidence is shown in figure 7. Added Mg(II) causes a

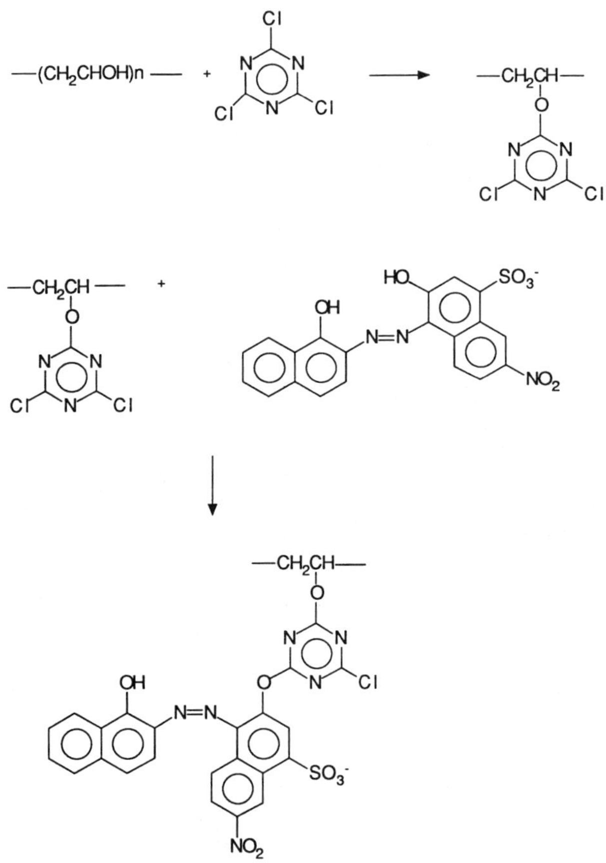

Figure 6. Raw data showing changes in intensity with time as the sensor is sequentially inserted into solutions containing Mg(II) and EDTA, both at pH 9.6.

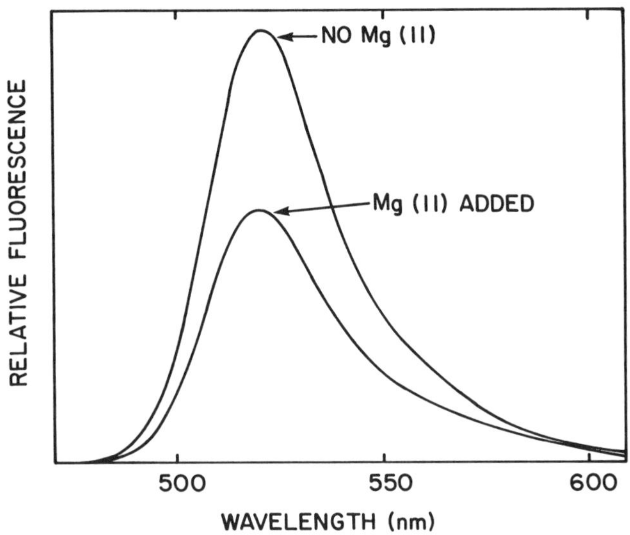

Figure 7. Indicator emission spectrum in the presence and absence of Mg(II).

decrease in fluorescence intensity but does not lead to a change in spectral distribution. If the inner filter effect were an important quenching mechanism, one would expect a change in spectral distribution because the inner filter effect is not equal at all wavelengths.

To further confirm that response involves energy transfer, a sensor was prepared by mixing two separate preparations of PVOH/indicator conjugate, one coupled only to fluoresceinamine and the other coupled only to EBT, prior to crosslinking. When this sensor was exposed to Mg(II), intensity decreased by 10%. A sensor with identical amounts of fluoresceinamine and EBT coimmobilized to PVOH showed a 42% decrease when exposed to the same amount of Mg(II). The observation that sensitivity is considerably greater when fluoresceinamine and EBT are coimmobilized indicates that distance between EBT and fluoresceinamine is an important parameter influencing response. This distance is a critical parameter determining the efficiency of fluorescence energy transfer but would not have any influence on the extent of inner filter effects.

Figure 8 shows that the extent of quenching increases with the amount of immobilized EBT as expected. In contrast, increases in the amount of immobilized fluoresceinamine cause an increase in absolute intensity values but have very little effect on the relative extent of quenching. The response time was not significantly affected by changes in the amounts of the two indicators.

The sensor shows selectivity expected for EBT. A four-fold excess of calcium does not interfere with sensor response. Potential interferences from transition metal ions can be eliminated by including masking agents in the buffer required for pH control.

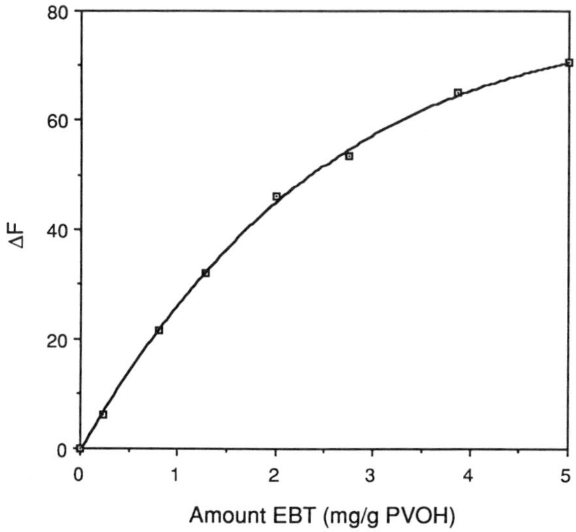

Figure 8. Effect of the amount of EBT on sensor response. The amount of immobilized fluoresceinamine was 0.07 mg/g PVOH.

Acknowledgments

Partial support for this research was provided by NSF Grant CHE-8502061.

Literature Cited

1. Zhujun, Z.; Zhang, Y.; Wangbai, M.; Russell, R.; Zhakhsher, Z.M.; Grant, C.L.; Seitz, W.R.; Sundberg, D.C.: Anal. Chem. 1989, 61, 202-205
2. Saari, L.A.; Seitz, W.R. Anal. Chem. 1983, 54, 821-823
3. Munkholm, C.; Walt, D.R.; Milanovich, F.P.; Klainer, S.M. Anal. Chem. 1986, 58, 1427-1430
4. Milanovich, F.P.; Hirschfeld, T.B.; Wang, F.T.; Klainer, S.M. Proc. SPIE 1984, 494, 18-24
5. Jordan, D.M.; Walt, D.R.; Milanovich, F.P. Anal. Chem. 1987, 59, 437-439
6. Yuan, P.; Walt, D.R. Anal. Chem. 1987, 59, 2391-2394
7. Diehl, H.; Horchak-Morris, N. Talanta, 1987, 34, 739-741
8. Schwarzenbach, G.; Biedermann, W. Helv. Chim. Acta 1948, 31, 678-687
9. Young, A.; Sweet, T.R. Anal. Chem. 1955, 27, 418-420

RECEIVED April 28, 1989

Chapter 19

Construction and Characterization of Optical Sensors with Porous Polymer Film

Scott M. Stole, Thomas P. Jones, Lai-Kwan Chau, and Marc D. Porter[1]

Ames Laboratory–U.S. Department of Energy and Department of Chemistry, Iowa State University, Ames, IA 50011

The range and scope of the applicability of direct dyes as immobilized indicators for the construction of optical pH sensors have been examined. The sensors are fabricated by the immobilization of these indicators at a porous cellulosic film, exhibiting a rapid response time. The rapid response results from the porous microstructure of a cellulose acetate film, which is manipulated by the duration of a base hydrolysis. To gain insights into the optimization of the microstructure of the film, a multifaceted characterization was undertaken. The change in permeability of films coated on a glassy carbon electrode was examined by cyclic voltammetry with electroactive probes of various sizes and shapes. The change in the composition of the film was assessed with infrared external reflection spectroscopy by monitoring the hydrolytic removal of the acetate functional group from the backbone of the cellulosic polymer. In addition to a discussion of the results of these structural studies, the properties of a variety of new pH sensors fabricated from the immobilization of various direct dyes are described. This includes the performance characteristics of a two-component pH sensor - a first step towards the development of an optical device that exhibits a response over a broad pH range.

Interest in the properties, structure, chemical modification, and application of thin films of organic polymers has grown enormously in recent years. The impetus for this rebirth derives from the relevance of such materials to adhesion (1), microelectronics (2), lubrication (3), and biocompatibility (specifically the interface between living and nonliving components as occurs with joint replacements or artificial hearts (4,5)). These materials have also gained increasing importance in the construction of sensors that are based on electrochemical and

[1]Address correspondence to this author.

0097–6156/89/0403–0283$06.00/0

spectrochemical detection schemes. These latter applications exploit the chemical and physical properties of thin films to control the selectivity and sensitivity of an analysis. For the development and construction of electrochemical sensors, strategies have focused on manipulating the composition and porosity of "redox" and ion exchange polymer films. Applications include the mediation of the heterogeneous electron transfer of large biomolecules with redox polymers (6) and the preconcentration of metal ions at ion exchange polymers (7). An alternative scheme utilizes the porosity of cellulosic (8,9) and other polymeric materials (10,11) to control the selectivity and to enhance the stability of an electroanalysis. Such approaches have been applied to the electroanalysis of H_2O_2 (8) and low molecular weight organic compounds (9) in the presence of macromolecules which may adsorb and subsequently poison an electrode surface.

Strategies for the design and incorporation of polymeric films as components of optical sensors have, to a large degree, paralleled those of their electrochemical analogs. As is the case with electrochemical sensors (12), the selectivity and sensitivity of a chemical analysis with optical sensors are governed by a complex mixture of chemical and physical properties and interactions, as well as by instrumentation performance characteristics. Of particular practical and fundamental importance are questions that concern changes in the reactivity of an immobilized species with respect to its solution analog. For instance, how does the mode of attachment alter the reactivity of the immobilized species? What other intermolecular interactions, such as those that are involved in solvation, require a close examination? To what extent does the molecular architecture of the electrical double layer influence reactivity? Answers to such questions, which are also relevant to a substantial number of surface and material science problems, promise to enhance our ability to design sensors with specific performance characteristics.

To assess specifically the influence of immobilization on the reactivity of colorimetric reagents used for optical sensors, it is instructive to examine the entries in Table I. This Table, although not exhaustive, lists several characteristics of sensors that have recently been developed for the determination of pH, metal ions (M^{n+}), and halide ions (A^-), in aqueous solutions. Reviews of optical sensors for gas-phase chemical analysis are available elsewhere (28). The listing includes the scheme for immobilization, the reactivity of the solution and immobilized forms of the ligand, the mode of optical analysis, and the response time. Ideally, to facilitate the predictive design of optical sensors, the mode of attachment should exert a minimal influence on the reactivity of an immobilized reagent. However, as shown in Table I, a direct translation of reactivity from the solution structure to the surface structure is rarely observed. For example, the pH sensor based on immobilized Congo Red has an acid strength of almost two orders of magnitude greater than the solution form of the indicator. Several of the metallochromic indicators exhibit differences that are even more dramatic; in fact, several completely lose their chelating ability upon immobilization. Possible insights into these differences can be found by an examination of the structure of the immobilized indicator and its interactions with the support. For example, the metallochromic azo reagent, calcichrome (2,8,8′-trihydroxy-1,1′-azonaphthalene-3,6,3′,6′-tetrasulfonic acid) contains four sulfonic acid functional groups (30), each of which may participate in its electrostatic immobilization at an anion exchange polymer film (24). An examination of a molecular model suggests that the ability of calcichrome to adopt a conformation that favors the formation of a metal complex with calcium (II) would be greatly hindered if the chelate were tethered to the polymeric film by more than one electrostatic linkage. This suggests that metallochromic reagents which are tethered through a single linkage may more readily undergo a structural rearrangement to achieve a conformation that is favorable to the formation of a metal complex, whereas immobilization through multiple linkages may severely limit the rearrangement requisite for complex formation. Furthermore, the effects on the formation of a metal complex after immobilization should be minimal for those ligands, such as quinolin-8-ol-5-sulfonate (27), which are naturally fixed in a configuration requisite for complexation. Several other factors also merit consideration, including electronic induction effects and the ionic characteristics of the polymeric support. With the former, the mode of attachment may sufficiently alter the electron density at a chelating group, translating into a change in the formation constant, K_f, for a complex. For the latter, the ionic charge at a polymer film may influence the partitioning or diffusion of an ionic species into the film by a charge (Donnan) effect (31,32).

Table I also lists response time as an important performance characteristic of optical sensors. This characteristic is particularly relevant in defining the range and scope of the utility of the sensor; that is, the faster the response, the more widespread its applicability. At present, the response of most sensors is limited by the slow diffusional mass transport of analyte through an impermeable polymeric support. Recent studies have shown, however, that response times can be dramatically decreased by the construction of sensors from porous polymeric materials that minimize barriers to the mass transport of an analyte (22).

As the previous discussions illustrate, the considerations relevant to the optimization of the performance characteristics of optical sensors represent a complex mixture of fundamental and practical issues. Further, as research progresses, it will become increasingly important to develop detailed molecular level descriptions of the microstructure of these sensors. Such descriptions will provide a direct correlation between the interfacial structure, reactivity, and performance, enhancing the ability to design sensors with new and/or improved performance characteristics. Thus, future progress will demand an increasingly interdisciplinary research strategy, drawing from specialties such as synthesis, theory, instrumentation, and materials characterization.

In a recent communication (22), we described the fabrication and preliminary evaluation of a chemically modified cellulose acetate film as an optical sensor for the selective determination of pH. This sensor was fabricated by the immobilization of Congo Red at a porous cellulosic film. Advantageous features of this design

Table I. Characteristics of Optical Sensors for the Chemical Analysis of Aqueous Solutions (H^+, M^{n+}, and A^-)

Sensor Composition (Indicator/Support)	Analyte	Reactivity[a] Solution Analog	Reactivity[a] Immobilized	Optical Mode of Analysis	Estimated[b] Response Time (min.)	Mode of Immobilization	Reference
Phenol red/ acrylamide	H^+	7.9	7.6	reflectance	0.7	adsorption	13
Bromophenol blue/ XAD[c]	H^+	4.1	3.7	reflectance	—	adsorption	14
Bromocresol purple/ XAD	H^+	6.1	5.8	reflectance	—	adsorption	14
Bromothymol blue/ XAD	H+	7.1	7.7	reflectance	1.1	adsorption	14,15
Thymol blue/XAD	H^+	8.9	9.6	reflectance	—	adsorption	14
Phenolphthalein/XAD	H^+	9.6	10.2	reflectance	—	adsorption	14
Thymolphthalein/XAD	H^+	9.3	11.0	reflectance	—	adsorption	14
HPTS[d]/SG[e]	H^+	7.3	6.8	fluorescence	0.8	covalent	16
HCC[f]/SG	H^+	7.0	6.9	fluorescence	0.8	covalent	16
Fluoresceinamine/SG/ cellulose acetate	H^+	—	—	fluorescence	0.3	covalent	17
Fluoresceinamine/ acrylamide- methyl acrylamide	H^+	—	—	fluorescence	≤0.15	covalent	18
HOPSA/anion exchange[g]	H^+	7.3	7.6	fluorescence fluorescence	 0.4	electrostatic	19
Phenol red/ acrylamide- methylacrylamide	H^+	7.9	~7.4	fluorescence	0.03	covalent	20
Fluoresceine isocyanate/glass	H^+	6.5	5.9	fluorescence	0.15	covalent	21
Congo red/cellulose acetate	H^+	—	0.8 2.8	transmission	0.005	adsorption	22

Direct Blue 8/ cellulose acetate	H^+	10.3	10.6	transmission	—	adsorption	23
Direct Orange 8/ cellulose acetate	H^+	—	3.1 11.8	transmission	—	adsorption	23
Calcichrome/anion exchange	H^+	7.2 11.5	5.3 10.7	reflectance	—	electrostatic	24
Morin/cellulose	Be(II)	5.4[h]	4.0	fluorescence	1.3	covalent	25
	Al(III)	6.3[h]	4.5	fluorescence	0.7		26
Quinolin-8-ol-5-sulfonate/anion exchange	Mg(II)	4.0[h]	4.9	fluorescence	~1	electrostatic	27
	Al(III)	—	6.5	fluorescence	~1		27
	Zn(II)	7.0[h]	8.0	fluorescence	~1		27
	Cd(II)	—	7.1	fluorescence	~1		27
Calcichrome/anion exchange	Ca(II)	3.9	1.5	reflectance	0.06	electrostatic	24
Beryllon II/anion exchange	Mg(II)	i	j	reflectance	—	electrostatic	24
Congo Red/anion exchange	Hg(II)	i	j	reflectance	—	electrostatic	24
Zincon/anion exchange	Zn(II)	i	j	reflectance	—	electrostatic	24
Fast Sulphon Black F/anion exchange	Cu(II)	i	j	reflectance	—	electrostatic	24
Alizann Complexone Ce(III)/XAD	F^-	4.6	3.7	reflectance	~12	adsorption	28

a)Reactivity indicators for hydrogen ion given as pK_a. Those for metal ion indicators given as log K_f. b) Estimated from the literature as the time required for the optical change to reach 63% of its equilibration value, ie. 1-1/e. c) XAD = styrene-divinylbenzene co-polymer. d) HPTS = 1-hydroxypyrene-3,6,8-trisulfonate. e) SG = sintered glass. f) HCC = 7-hydroxycoumarin-3-carboxylic acid. g) HOPSA = 8-hydroxyl-1,3,6-pyrenetrisulfonic acid. h) Calculated for dissolved 1:1 metal-ligand complex from formation constants at an ionic strength of 0.1 (31). i) Observable reaction. j) No observable reaction.

included: 1) a rapid response time (<1.3 s), 2) a large dynamic range (> 4 pH units), 3) a selective response, and 4) ease of fabrication. The rapid response resulted from the porous microstructure of the polymeric support, which was achieved by a carefully controlled base hydrolysis. This porous microstructure minimized the barriers to the mass transport of the analyte to the immobilized indicator. The large dynamic range resulted from both the polyprotic acid-base reactivity of Congo Red and the high optical absorptivity of its various ionic forms. Additionally, this sensor was easily fabricated by the immersion of the hydrolyzed film into an electrolytic dye bath which contained Congo Red. Furthermore, this sensor exhibited a response that was selective only to hydrogen ion. This was a consequence of the immobilization chemistry which apparently prohibits the amine groups of Congo Red from participating in the formation of a metal complex.

As part of our on-going efforts to develop novel design and fabrication schemes to enhance sensor performance, we have continued to explore the utility of porous polymeric films as support materials for immobilized colorimetric reagents. In this paper, we report the results from a multifaceted characterization of the porosity and composition of thin films of cellulose acetate as a function of the duration of the base hydrolysis. This characterization was undertaken to correlate the microstructure and composition of the film with changes in its permeability, thereby providing insights for the optimization and the extension of the applicability of this material as a support for sensor fabrication. The change in permeability was characterized by cyclic voltammetry (CV) with electroactive probes of various sizes and shapes. This was facilitated by coating the film onto a glassy carbon electrode. The change in film composition was examined with infrared external reflection spectroscopy (IR-ERS). These measurements focused on monitoring the hydrolytic removal of the acetate functional groups from the backbone of the cellulosic polymer. In addition to these structural studies, the results from an exploration of the range and scope of dyes similar to Congo Red, viz. other direct dyes, as immobilized colorimetric reagents for optical sensors will be described. A goal of this latter effort was to identify and to test dyes that could be incorporated as part of a multi-component sensor which would exhibit a response over a broad pH range, i.e. something akin to "optically transparent litmus paper".

EXPERIMENTAL

Preparation of Cellulose Acetate Films. Cellulose acetate films (Aldrich, Inc., Milwaukee, WI) were cast onto either GC-20 electrodes (Tokai Carbon Co., Tokyo, Japan) or microscope slides by spin-coating techniques. This polymer has a 39.8% acetyl content (by weight) which represents an average of 2.45 acetyl groups per glucosidic unit. Sheets of GC were cut to provide plates that were approximately 2.5 by 2.5 cm. The plates were successively polished with slurries of 600-grit silicon powder and 1 μm, 0.3 μm, and 0.05 μm alumina (Buehler Ltd., Evanston, IL) until a mirrorlike finish was obtained. Between each polishing step, the substrates were washed in an ultrasonic bath with deionized water (Millipore Corp., Bedford, MA). Glass substrates were carefully degreased with acetone.

Films were prepared by flooding a surface with a solution of the polymer and spinning at 1000-4000 rpm. Thicknesses were controlled by the variation of the concentration of cellulose acetate in cyclohexanone (1-10% w/v) and by the alteration of rotation rate. After casting, the films were stored under a watch glass and allowed to dry for 24 h.

Film thicknesses were measured prior to hydrolysis with an Alphastep surface profiler (Tencor Instruments, Mountainview, CA). A knife edge was used to remove several portions of the film from the substrate with the resulting step used for the thickness measurements. Thicknesses are reported as the average from at least three different locations on the film; the uncertainty is reported as the range of these measurements.

Base-Hydrolysis of Cellulose Acetate Films. The cellulose acetate films were hydrolyzed by immersion in 0.070 M KOH at 25°C. Immersion times varied from intervals of a few minutes up to 24 h. Upon removal from the hydrolysis bath, the samples were immersed in a cold water bath to quench the reaction. A 24 h hydrolysis resulted in ~47% decrease in the mass of the film and a ~53% decrease in thickness.

Dye Immobilization. The acid-base indicators were immobilized at the hydrolyzed cellulose acetate films via a conventional dye bath recipe (33). This consisted of immersing a hydrolyzed film into an electrolytic dye bath of approximately 2 mM Na_2SO_4 and 2 mM K_2CO_3 for 10 min. Dye concentrations were 0.1-1.0 mM. In addition to Congo Red (Direct Red 28), several other dyes were tested including the Direct dyes: Blue 8, Brown 2, Orange 8, Red 1, Red 2, Black 38, Red 13, Yellow 4, Yellow 12, Yellow 44, Violet 1, Red 75, Red 126, Blue 78, Orange 6, and Green 6. A chemical and structural description of these and other direct dyes can be found in reference 34, along with their IUPAC names. Based on the length of this listing, further details regarding the structure of these dyes will be described only when germane to the discussion.

Colorimetric indicators such as Congo Red are known as direct dyes (33,35). This terminology derives from the ability of these dyes to adsorb strongly to cellulose-based polymers simply by immersing the polymer into a hot dye solution at a high ionic strength. The factors that govern the formation and adsorption strength of such dyes are a complex mixture of chemical, physical, and structural effects. For Congo Red, it is apparent that the chemical interactions between its amine groups and the hydroxyl groups of the cellulosic support play an important role in the formation of the immobilized structures (36). The relative importance of the molecular planarity (flatness) and solubility, both of which have been found to contribute at differing degrees to the binding strength of such dyes, remains the subject of extensive controversy (35,37).

Instrumentation. Infrared spectra were obtained with a Nicolet 740 Fourier transform IR spectrometer (Nicolet Inst., Madison, WI). Reflection measurements employed a modified, variable angle reflection accessory (Harrick Sci., Ossining, NY) set to an angle of incidence of 60°. The fundamental considerations for performing IR-ERS measurements at materials with a low IR reflectivity, such as GC, have been previously described (38). An aluminum wire grid polarizer on KRS-5 (Cambridge Physical Sciences) was placed immediately before the accessory to provide selection of p-polarized light. Spectra were obtained in either a reflection or a transmission mode with a liquid nitrogen-cooled, narrow-band HgCdTe detector. After Happ-Genzel apodization, the spectral resolution was nominally 4 cm^{-1}. The spectrometer and sample chamber were purged with boiloff from liquid N_2. Spectra are presented as the ratio of 1024 sample to 1024 reference scans. For the IR-ER spectra, the y-axis is defined as $-\log(R/R_o)$, where R is the reflectivity of the coated electrode and R_o is that for a bare electrode. All samples were carefully rinsed and dried under a stream of dry N_2 gas before placement in the spectrometer.

The UV-VIS data were obtained with a DMS-200 spectrometer (Varian Instruments Co., Palo Alto, CA). A flow cell (22), configured in a conventional transmission mode, was mounted to the base plate of the spectrometer. At a solution flow rate of 1.0 $mL\text{-}s^{-1}$, the time required to effect a complete change (99.9%) in the optical signal was 1.22 ± 0.18 s.

Cyclic voltammetric measurements were made with a BAS CV-27 potentiostat (Bioanalytical Systems, West Lafayette, IN). A Ag/AgCl electrode (saturated with KCl) served as the reference.

Reagents. All reagents, except Congo Red, were used as received. The Congo Red was purified by recrystallization in absolute ethanol, and was determined to be 97% pure by potentiometric titration. The pH of the solutions was controlled with HCl or NaOH or with citrate/phosphate buffers. The ionic strength of the solutions was maintained with 0.5 M KCl.

RESULTS AND DISCUSSION

Characterization of Microstructure and Permselectivity of Cellulose Acetate Films at Glassy Carbon.

In this section, the results from a structural characterization of thin films of cellulose acetate at GC are presented. Cyclic voltammetric measurements were used to characterize the change in permeability as a function of hydrolysis time. The corresponding compositional changes of the film were examined with IR-ERS. A goal of this study was to correlate the changes in the permeability of the film with those of its composition, providing insights into optimizing this material as a support for sensor fabrication. A second goal was to determine the hydrolysis time required to remove exhaustively the acetate groups from the film. Earlier studies (23) had indicated that the stability of the response of these sensors in strongly acidic solutions was dependent on the duration of the base hydrolysis. Since an acid hydrolysis represents another pathway for the removal of an acetate moiety from the film, it was apparent that the exhaustive depletion of this moiety was requisite for the attainment of a stable sensor response.

Electrochemical Characterization. Films of impermeable, electrically insulating materials represent a substantial barrier to heterogeneous electron-transfer (39,40). Ideally, electron-transfer through a pinhole-free film of such a material should proceed through a highly nonadiabatic pathway with the kinetics exhibiting an exponential dependence on the separation between the electron donor and electron acceptor. Cellulose acetate, however, is well known for its permselective properties, which have led to its widespread application as a dialysis membrane (41,42). Recent studies have also shown that a controlled hydrolysis of such films regulates their permselectivity based on the size and shape of a solute (41,43). By employing electrochemical probes of different sizes and shapes, Wang and Hutchins (9) have characterized the change in porosity of these films as a function of hydrolysis time.

In our studies, the permeability of cellulose acetate films was monitored as a function of hydrolysis time with cyclic voltammetry (CV). The electrochemical probes were 1,4-hydroquinone (HQ), the cofactor nicotinamide adenine dinucleotide (β-NADH), and $Fe(CH)_6^{4-}$. Table II provides estimates of the molecular dimensions for each of these probes.

Table II. Estimated Molecular Dimensions of the Electrochemical Probes

Probe Species	Dimensions (Å)[a]			Molecular Volume (Å^3)	Molecular Weight
	Length	Width	Thickness		
Hydroquinone	8.4	6.6	3.4	190	110.1
$Fe(CN)_6^{4-}$	9.4	9.4	9.4	830	212.0
β-NADH	18.2	10.7	5.4	1050	709.4

a) Dimensions estimated from tabulations of Van der Waals and covalent radii (44) where the length equals the *a* axis, width the *b* axis, and thickness the *c* axis (45).

Figure 1A shows the normalized electrochemical charge, Q'_{ox}, as a function of hydrolysis time for a 170 ± 10 nm film of cellulose acetate at GC. The ordinate, Q'_{ox}, in Figure 1A equals $K'(Q_{ox}/n)$, where K' equals the ratio of the concentrations of β-NADH or $Fe(CN)_6^{4-}$ to that of HQ, and n is the number of electrons generated by the oxidation reaction. The n value for HQ and β-NADH equals two, whereas that for $Fe(CN)_6^{4-}$ equals one. The change in Q'_{ox} in Figure 1A verifies the size discrimination of mass transport through the hydrolyzed films. At short hydrolysis times (<40 min) only the

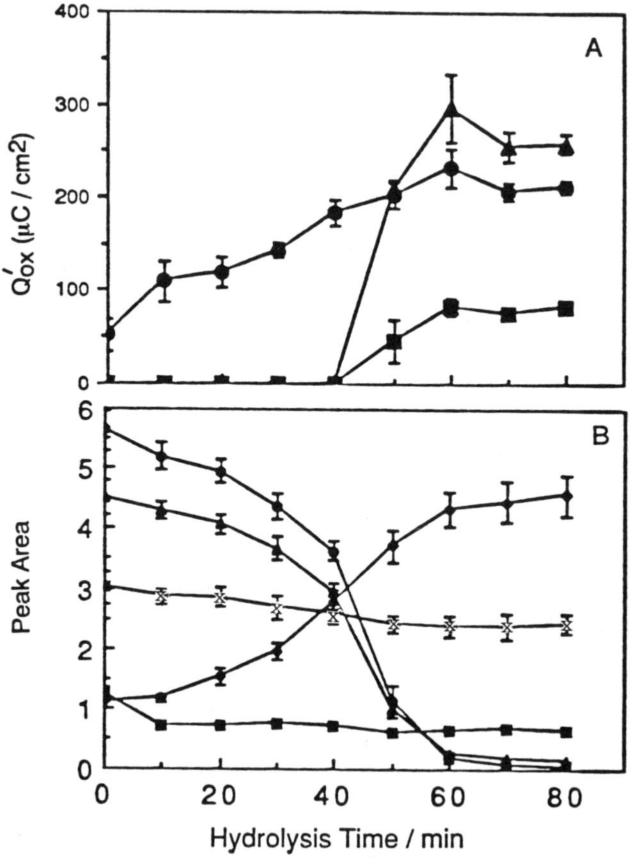

Figure 1. Dependence of (A) Q'_{ox} and (B) the integrated absorbances for the marker bands on base hydrolysis time. In A: (●) hydroquinone, (■) β-NADH, (▲) $Fe(CN)_6^{4-}$. In B: (●) $\gamma(C{=}O)$, (▲) $\gamma(c\text{-}O)_{(ACETATE)}$, (X) $\gamma(C\text{-}O)_{(RING)}$, (◆) $\gamma(OH)$, (■) $\gamma(C\text{-}H)_{(STRETCH)}$.

smaller HQ is capable of diffusing through the film and undergoing oxidation at the electrode. After 40 min of hydrolysis, however, the microstructure of the film has changed such that the larger probes, β-NADH and $Fe(CN)_6^{4-}$, can readily diffuse through the film. These responses remain relatively constant after a hydrolysis time of 60 min, indicating that the electrolysis of these probes is independent of any further changes in the porous microstructure of the film. More importantly, these results indicate that a hydrolysis time of ~60 min is needed to optimize the transport of an analyte through a cellulose acetate polymer film of this thickness. Increases in thickness require a correspondingly longer hydrolysis to achieve similar transport characteristics.

Infrared External Reflection Spectroscopic Characterization. To examine the compositional evolution of these films as a function of hydrolysis, IR-ERS measurements were performed. Figure 2 shows the IR spectra between 1850-850 cm^{-1} for two films of cellulose acetate. Figure 2A is an IR transmission spectrum for a 1.85 ± 0.01 μm film on silicon. Based on previous assignments (46-48), the bands at 1730 cm^{-1} and 1234 cm^{-1} are attributed to the υ(C=O) and υ(C-O) of the acetate functional group. The band on the low-energy side of the υ(C=O) results from hydrogen bonding interactions between C=O and the hydroxyl groups on the polymeric backbone. The broad band at 1052 cm^{-1} is primarily a result of the υ(C-O) stretch of the ether linkage between glucosidic units; the υ(C-O) of the ether linkage within a glucosidic unit, which also appears in this region, is largely masked by this band.

Figures 2B-E show the IR-ER spectra for a 170 ± 10 nm cellulose acetate film at GC as a function of hydrolysis time. Figure 2B is the spectrum of the film prior to hydrolysis. Figures 2C-E are IR-ER spectra for the film at hydrolysis times of 40, 50, and 80 min, respectively. Optical dispersion effects, which are enhanced due to the low IR reflectivity of materials such as GC, cause the observed band-shape distortions in the reflection spectra (38). For short immersion times (≤40 min), the changes in the absorbances for the acetate marker bands (1731 and 1234 cm^{-1}) suggest a reaction mechanism that is consistent with the hydrolysis of an ester. For example, at an immersion time of 40 min the integrated absorbances of the acetate markers decrease by about 35% whereas that for the marker for the polymeric backbone (1052 cm^{-1}) decreases only by ~14%. Although not shown in the figure, the decrease in the acetate composition of the film is accompanied by an increase in the number of hydroxyl groups of the film. These data indicate that at short immersion times the composition of the film evolves from that of cellulose acetate to one which is more "cellulosic" in nature.

In addition to the continual, and eventually exhaustive, loss of the acetate groups at longer immersion times, the gradual decrease in the integrated absorbance for the marker of the backbone indicates that the film is slowly degraded by the cleavage of the ether linkages between glucosidic units. This reaction apparently proceeds primarily by the stepwise removal of a terminal glucosidic group, progressively shortening the average length of the polymeric backbone (49). A competing reaction, which corresponds to a conversion to a glucometasaccharinic acid derivative, slowly "caps" the end group, preventing the complete degradation of the film.

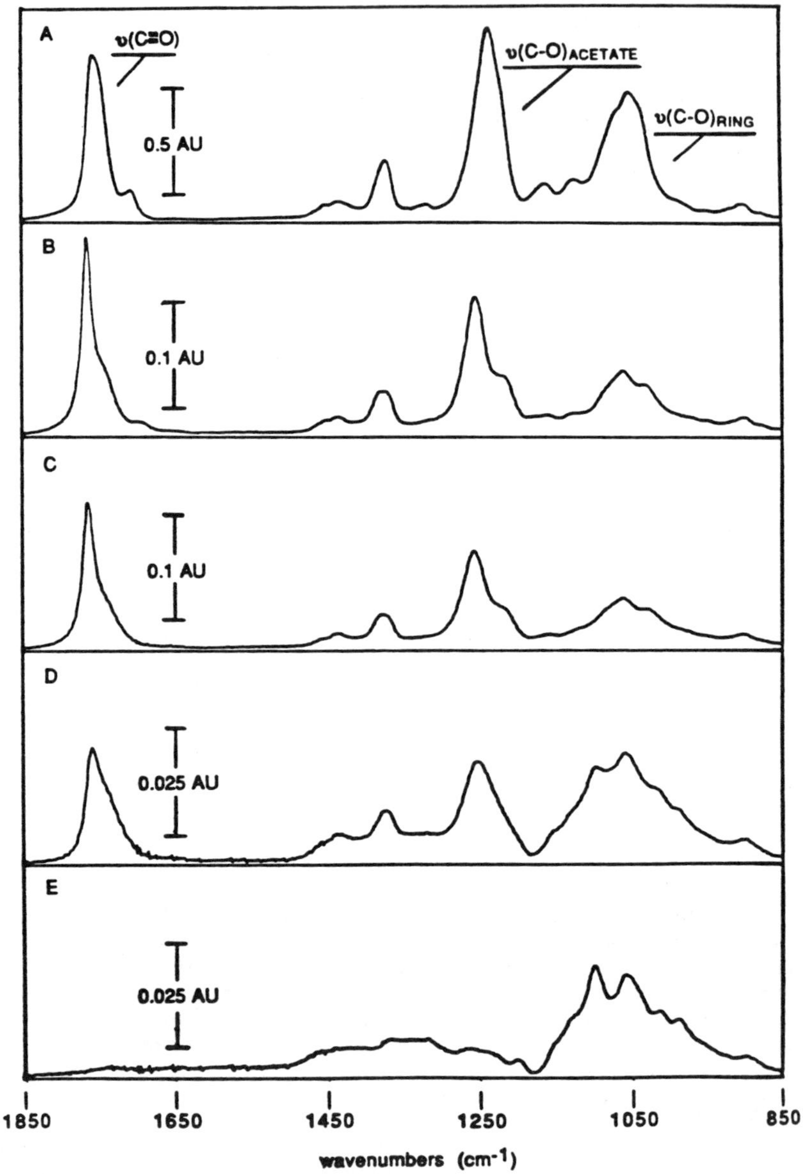

Figure 2. Infrared spectra for cellulose acetate films at silicon and glassy carbon. (A) Transmission spectrum for a 1.85 ± 0.01 μm film at silicon; (B-E) IR-ERS spectra for a 170 ± 10 nm film at glassy carbon as a function of base hydrolysis: (B) 0 min.; (C) 40 min.; (D) 50 min.; and (E) 80 min. hydrolysis. Reflection spectra were acquired at an angle of incidence of 60° with p-polarized light.

Figures 1A and 1B provide a respective comparison between the electrochemical permeability data and the molecular composition information from the IR-ER spectra. Figure 1B shows the change in the integrated absorption strengths, A_t, for the compositional marker bands as a function of hydrolysis time. A comparison of the data in these figures provides insight into the relationship between the permeability and the molecular composition of the film. At the hydrolysis times (~50 min) in which the larger probes gain access to the electrode, the IR-ER spectrum shows that the film has lost almost 79% of its initial acetate composition. This indicates that the loss of the acetate groups induces a change in the microstructure of the polymer film. These "bulky" acetyl groups limit the average distance of closest approach between adjacent polymeric chains (49). However, the gradual removal of these groups from the film allows an increase in intermolecular hydrogen bonding interactions. This decreases the average separation between neighboring chains and increases the free volume of the film thus inducing small pores and channels in the microstructure of the film.

Based on the estimated molecular dimensions of the probe molecules, together with the change in the composition of the film, a possible model which depicts the structural evolution of cellulose acetate is shown in a cross-sectional view in Figure 3B. Figure 3A represents the reaction for the hydrolysis of the cellulose acetate. This model is based on considerations of the changes in the electrochemical response of the larger probe molecules, β-NADH and $Fe(CN)_6^{4-}$. The electrochemical responses for the smaller probe, HQ, may result, in part, from a partitioning into the film (9), limiting the utilization of these data in developing a simple description of the microstructure of the film. Structure a (Figure 3B) depicts a few segments of the polymer chains in a small region of a cellulose acetate film prior to hydrolysis. Cellulose acetate exhibits varying degrees of crystallinity, which is a strong function of preparation parameters such as solvent, rotation rate, and extent of acetylation (50). For simplicity, structure a shows a relatively well-ordered region of the film with an average separation between neighboring chains of d_1. After a 40 min hydrolysis, the microstructure evolves into that resembling structure b. In structure b, the average distance between neighboring chains, d_2, is less than d_1. This is a result of the partial removal of the bulky acetyl groups and the subsequent increase in intermolecular hydrogen bonding interactions. The decrease in d_2 increases the free volume of the film by opening small channels and pinholes. Based on the estimated molecular dimensions of the electrochemical probes, the average diameter, d_3, of these channels, must be comparable or greater than that of β-NADH (see Table II).

Scope of Sensor Fabrication for Immobilized Direct Dyes at Porous Cellulosic Films.

In this section, the results from a survey of the utility of direct dyes as immobilized pH indicators at hydrolyzed cellulose acetate are presented. A goal of this effort was to identify those direct dyes that could serve as components in the construction of an optical sensor that would exhibit a response over an extended pH range. The characteristics of such a sensor, fabricated by the co-adsorption of Congo Red and Direct Blue 8, are described. In addition, differences between the reactivity of the solution and immobilized forms of these reagents are discussed.

Survey Data. A collection of direct dyes were tested regarding their utility as an optical pH sensor at hydrolyzed cellulose acetate. Each of the dyes was composed of at least one azo group which was linked to a variety of aromatic moieties (phenyl, diphenyl, and naphthyl) of differing substitution. The solution forms of these dyes exhibited a variety of color changes between pH values of 2-12. After immobilization, however, several of the dyes (the last eight in the listing in the Experimental Section) failed to exhibit an optical response to a change in pH. Further, the acid-base reactivities of the functioning immobilized dyes differed in varying degrees from those of their solution analogs (see below).

The optical properties of two of these new pH sensors as functions of pH are shown in Figure 4. The thickness of the cellulose acetate film was 2.9 ± 0.1 μm prior to a 24 h base hydrolysis. After hydrolysis, the thickness was 1.3 ± 0.1 μm. Figures 4A and 4B show the respective responses for Direct Orange 8 (I)

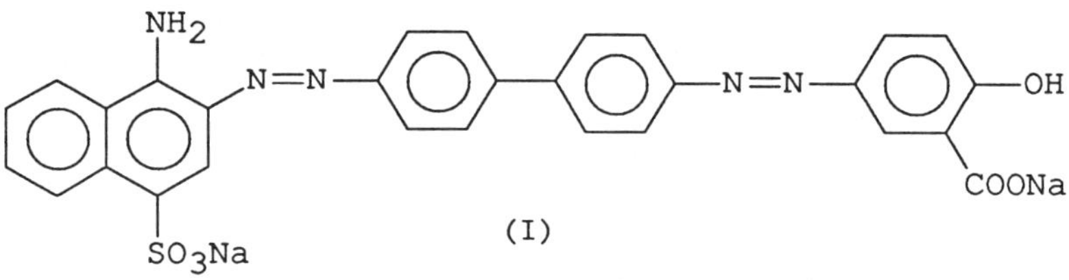

(I)

and Direct Blue 8 (II)

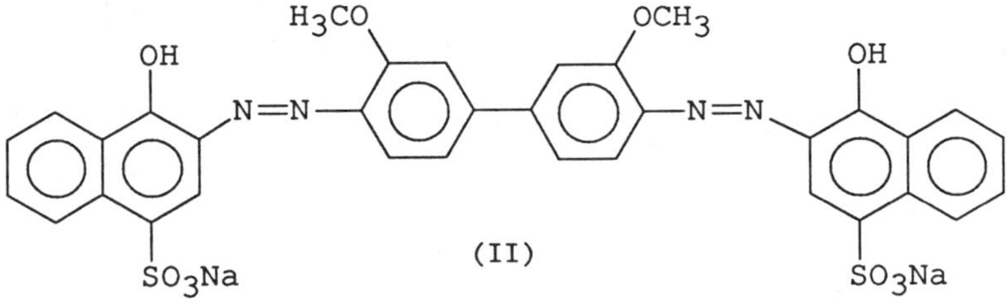

(II)

after immobilization at hydrolyzed cellulose acetate. As the spectra in Figure 4A indicate, the Direct Orange 8 pH sensor behaves as a polyprotic acid, undergoing changes in its optical properties as a result of a proton dissociation in low pH solutions and in high pH solutions. The dissociation in acid solutions corresponds to the deprotonation of an azo group, whereas the dissociation in basic solutions results from the deprotonation of the hydroxyl group. The

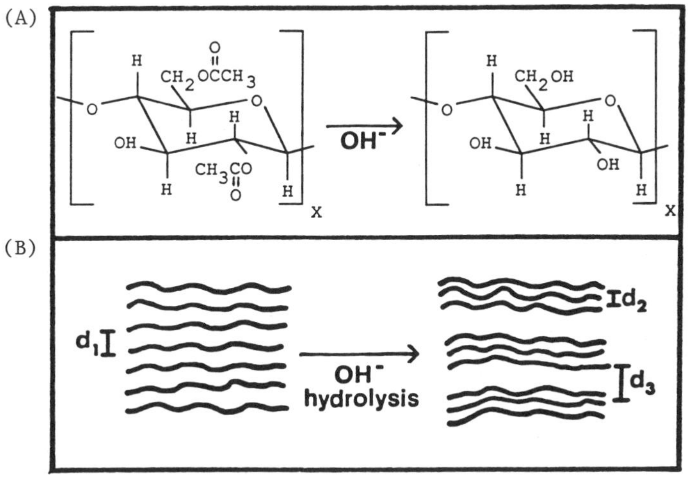

Figure 3. Reaction scheme (A) and representation of the evolution of the microstructure (B) of base-hydrolyzed cellulose acetate film. See text for details.

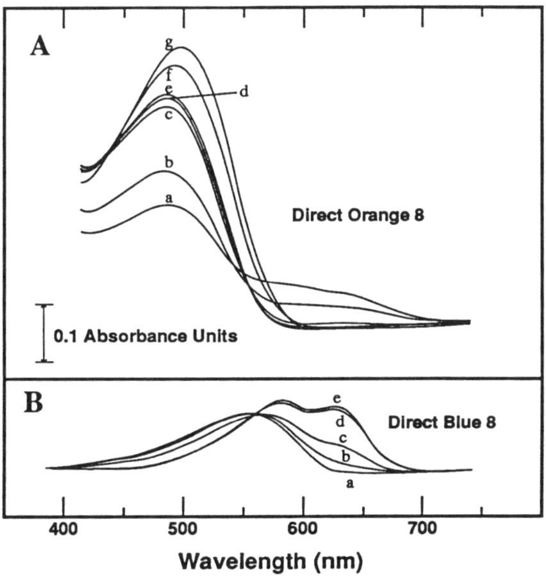

Figure 4. Absorption spectra of pH sensors constructed from (A) Direct Orange 8 and (B) Direct Blue 8 immobilized at hydrolyzed cellulose acetate. For (A) pH values: (a) 2.0 (b) 3.0 (c) 4.0 (d) 5.0 to 10.0 (no change over this range) (e) 11.0 (f) 12.0 (g) 12.9.; for (B) pH values: (a) 12.8 (b) 11.5 (c) 10.5 (d) 9.5 (e) 8.5.

pK_a values for these two deprotonation steps, estimated from spectrophotometric data obtained at 0.5 pH increments, are 3.4 and 11.8, respectively. A quantitative comparison of the reactivity differences between the solution and immobilized forms of the dye was hindered by aggregation and by susceptibility to phototropic isomerization in solution.

The optical response of a Direct Blue 8 pH sensor is shown in Figure 4B. This sensor undergoes a single proton dissociation step with a pK_a of 10.6, which corresponds to the deprotonation of a hydroxyl group. The presence of the dimethoxy groups at the 3 and 3′ positions on the biphenyl groups sterically hinders the protonation of the azo species. Similar effects have been observed for other dyes, such as Direct Red 2 (23). As with the Congo Red pH sensor, these new sensors also exhibit a rapid (~2 sec) response time.

Multiple-Component pH Sensor. A purpose of the previously described survey was to identify immobilized dyes that could be incorporated into a multiple-component pH sensor. The goal was to co-adsorb several different dyes at a hydrolyzed cellulose acetate film for the construction of a pH sensor which would exhibit a broad response, i.e., something akin to "optically transparent litmus paper." An example of an as yet incomplete version of such a sensor is given in Figure 5. Efforts to identify a direct dye which, after immobilization, exhibits a response over the physiological pH range have to date proven unsuccessful. This figure shows the optical response of a pH sensor that has been constructed by the co-adsorption of Congo Red (III) and Direct Blue 8 at a hydrolyzed

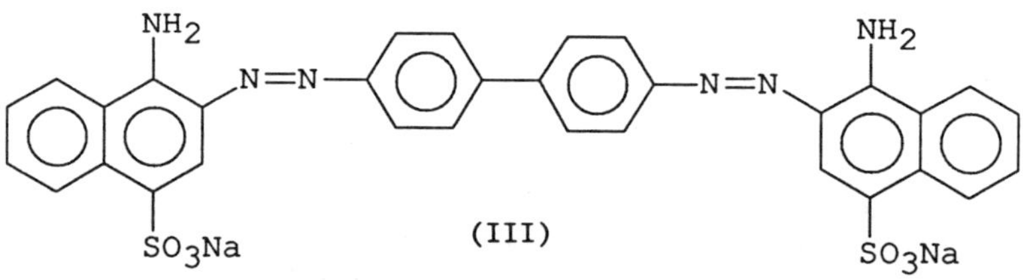

(III)

cellulose acetate film, representing a summation of the absorbances for the various ionic forms of the two dyes. Spectrum a in Figure 5 (pH 0.1) results from the absorbances of the protonated forms of Congo Red and Direct Blue 8. Congo Red, which is not completely protonated at this pH (22), has an absorbance maximum at 577 nm, whereas the absorbance maxima for Direct Blue 8 are at 583 and 628 nm. Thus, two absorbance maxima appear at 589 and 634 nm. Exposure to solutions with a pH less than zero results in the rapid deterioration of the film via the extensive cleavage of the glucosidic ether linkages. Spectra b and c at respective pH values of 6.0 and 8.0 correspond essentially to the sum of the absorbances for the completely deprotonated Congo Red and the protonated Direct Blue 8. These spectra yield absorbance maxima at 540, 582, and 634 nm. The sum of the absorbances for the completely deprotonated forms of the two dyes give rise to spectrum d at pH 12.0. This

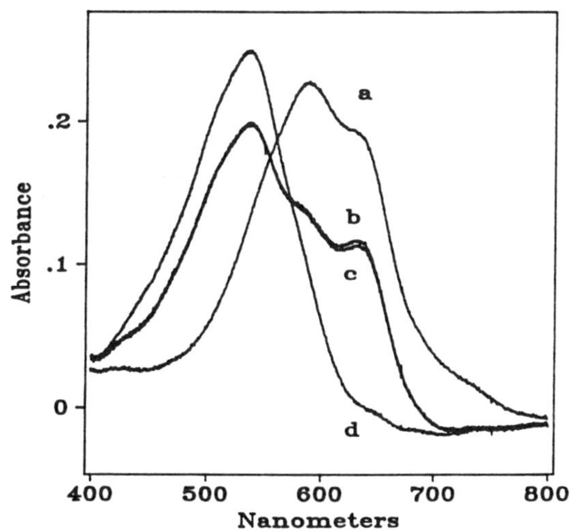

Figure 5. Absorbance spectra of a two-component pH sensor constructed from the co-immobilization of Congo Red and Direct Blue 8 at hydrolyzed cellulose acetate; pH values (a) 0.1, (b) 6.0, (c) 8.0, (d) 12.0.

spectrum exhibits a broad feature with an absorbance maximum at 537 nm. Equilibration times for a change in pH are comparable to those previously reported for our Congo Red optical sensor (23). Thus, this sensor exhibits a rapid, but discontinuous response over a wide range of pH values, demonstrating the feasibility of constructing multiple-component sensors that are based on the facile immobilization of direct dyes at a porous cellulosic film. Efforts to assess further the conditions that influence the response of the sensor, i.e. ionic strength of the solution and its selectivity are currently underway.

CONCLUSIONS

The combination of experimental probes employed in this study shows that permeability of a cellulose acetate film gradually increases with the duration of a base hydrolysis. The change in permeability is correlated with the hydrolytic removal of the acetate functional groups on the polymeric backbone. This results in an increase in the hydrogen bonding interactions between neighboring polymer chains, increasing the free volume of the film and opening channels and pores in its microstructure.

This study also explored the scope of the utility of direct dyes as immobilized indicators at hydrolyzed cellulose acetate films for the construction of optical sensors which exhibit a rapid response and are easily fabricated. In addition, and perhaps more importantly, the "proof of concept" design of a sensor and preliminary evaluation of characterization with a response over a broad pH range was demonstrated, illustrating opportunities provided by this approach for the fabrication of sensors with new and/or improved performance characteristics. Studies to explore the fundamental interactions that govern the formation, structure, reactivity, and stability of such sensors are currently underway.

ACKNOWLEDGMENTS

The authors express their appreciation to C. Cihal and E. Smith for their contributions to this project. MDP gratefully acknowledges the support of a Dow Corning Assistant Professorship and the Society of Analytical Chemists of Pittsburgh for a Starter Grant Award. Acknowledgment is also made to the donors of The Petroleum Research Fund, administered by the American Chemical Society, for partial support of this work. Ames Laboratory is operated for the U.S. Department of Energy by Iowa State University under contract No. W-7405-eng-82. This work was partially supported by the Office of Basic Energy Sciences, Chemical Science Division.

LITERATURE CITED

1. Wu, S. Polymer Interfaces and Adhesion, Dekker: New York, 1982.
2. Roberts, G. Adv. Phys. 1985, 34, 475.
3. Bowden, F.; Tabor, D. The Friction and Lubrication of Solids, Oxford Press: London, 1968; and references therein.

4. Baier, R.; Meyer, A.; Natiella, J.; Natiella, R.; Carter, J. J. Biomed. Mat. Res. 1984, 18, 337.
5. Gristina, A. Science, 1987, 237, 1588.
6. Lau, A.; Miller, L. J. Am. Chem. Soc. 1983, 105, 5271.
7. Guadalupe, A.; Abruna, H. Anal. Chem. 1985, 57, 142.
8. Sittampalam, G.; Wilson, G. Anal. Chem. 1983, 55, 1608.
9. Wang, J.; Hutchins, L. Anal. Chem. 1985, 57, 1536.
10. Ikeda, T.; Schmehl, R.; Denisevich, P.; Willman, K.; Murray, R. J. Am. Chem. Soc. 1982, 104, 2683.
11. Ohnuki, Y.; Matsuda, H.; Ohsaka, T.; Oyama, N. J. Electroanal. Chem. 1984, 158, 55.
12. Murray, R. in Electroanalytical Chemistry; Bard, A.J., Ed.; Dekker; New York, 1984; Vol. 13.
13.. Peterson, J.I.; Goldstein, S. R.; Fitzgerald, R. V.; Buckhold, D. K. Anal. Chem. 1980, 52, 864.
14. Kirkbright. G. F.; Narayanaswamy, R.; Welti, N. A. Analyst 1984, 109, 15.
15. Kirkbright, G. F.; Narayanaswamy, R.; Welti, N. A. Analyst 1984, 109, 1025.
16. Offenbacher, H.; Wolfbeis, O. S., Zurlinger, E. Sensors and Actuators 1986, 9, 73.
17. Saari, L. A.; Seitz, W. R. Anal. Chem. 1982, 54, 82.
18. Munkholm, C.; Walt, D. R.; Milanovich, F. P.; Klainer, S. M. Anal. Chem. 1986, 58, 1427.
19. Zhujun, Z.; Seitz, W. R. Anal. Chim. Acta 1984, 160, 47.
20. Jordan, D. M.; Walt, D. R.; Milanovich F., Anal. Chem. 1987, 59, 437.
21.. Kawabata, Y.; Tsuchida, K.; Imasaka, T.; Ishibashi, N. Anal. Sci. 1987, 3, 7.
22. Jones, T. P.; Porter, M. D. Anal. Chem. 1988, 60, 404.
23. Jones, T. P.; Porter. M. D. manuscript in preparation.
24. Chau L. K.; Porter, M. D. manuscript in preparation.
25. Saari, L. A.; Seitz, W. R. Analyst 1984, 109, 655.
26. Saari, L. A.; Seitz. W. R. Anal. Chem. 1983, 55, 667.
27. Zhujun, Z.; Seitz, W. R. Anal. Chim. Acta 1985, 171, 251.
28. Narayanswamy, R.; Russell, D. A.; Sevilla, III, F. Talanta 1988, 35, 83.
29. Smith, R.M.; Martell, A.E. Critical Stability Constants, Vol. 2, Plenum: New York, 1975.
30. Chau, L.; Pruski, M.; Porter, M. D. Anal. Chim. Acta, 1989, 217, 31.
31. Gerhardt, G.; Oke, A.; Nagy, G.; Mughaddum, B.; Adams, R. Brain Res. 1984, 290, 390.
32. Nagy, G.; Gerhardt, G.; Oke, A.; Rice, M.; Adams, R.; Szentirmay, M.; Martin, C. J. Electroanal. Chem. 1985, 188, 85.
33. Fieser, L.; Williamson, K. Organic Experiments, 5th Ed.; D. C. Heath: Lexington, MA, 1983.
34. Colour Index, The Society of Dyers and Colourists and The American Association of Textile Chemists and Colorists; Charley and Pickersgill, Ltd. Leeds, England, 2nd Ed., 1956.
35. Rattee, I.; Breuer, M. The Physical Chemistry of Dye Adsorption; Academic Press: New York, 1974.
36. Boss, R. TAPPI, 1959, 42, 185A.

37. Cellulose Chemistry and Technology; Arthur, J. Ed.; ACS Symposium Series 48; American Chemical Society: Washington, D. C., 1977.
38. Porter, M.; Bright, T.; Allara, D.; Kuwana, T. Anal. Chem. 1986, 58, 2461.
39. Li, T.; Weaver, M. J. Am. Chem. Soc. 1984, 106, 6107.
40. Albery, W. J. Electrode Kinetics, Clarendon: Oxford, 1975.
41. Cotton, C.; Smith, K.; Merrill, E.; Farrell, P. J. Biomed. Mater. Res. 1971, 5, 459.
42. Loeb, S.; Sourirajan, S. Adv. Chem. Ser. 1963, 38, 117.
43. Rubinson, K.; Baker, P. Proc. R. Soc. London, Ser. B, 1979, 205, 323.
44. Pauling, L. The Nature of the Chemical Bond, Cornell University Press: Ithaca, New York, 1960.
45. Soriaga, M.; Hubbard, A. J. Am. Chem. Soc. 1982, 104, 2735.
46. Tipson, R.; Parker, F. in The Carbohydrate, Pigman, W., Horton, D., Eds., Academic Press: New York, 1980.
47. Toprak, C.; Agar, J.; Falk, M. J. Chem. Soc. Faraday Trans. I, 1979 75, 803.
48. Nelson, M.; O'Conner, R. J. Appl. Polym. Sci. 1964, 8, 1325.
49. Ward, K.; Seib, P. in The Carbohydrates, Pigman, W.; Horton, D., Eds.; Academic Press: New York, 1970, Vol IIA.
50. Atalla, R. in Preservation of Paper and Textile of Historic and Artistic Value, Williams, J., Ed.; American Chemical Society Symposium Series, Washington, D.C. 1981.

RECEIVED March 17, 1989

Chapter 20

Fiber-Optic-Based Biocatalytic Biosensors

Mark A. Arnold

Department of Chemistry, University of Iowa, Iowa City, IA 52242

A survey of our recent efforts to develop fiber-optic biocatalytic biosensors is presented. Examples of fiber-optic biosensors based on absorbance, fluorescence, and bioluminescence phenomena are described. Absorbance-based biosensors are demonstrated with a sensor for p-nitrophenylphosphate. Alkaline phosphatase is immobilized at the tip of a bifurcated fiber-optic bundle and the biocatalytic generation of the chromophore p-nitrophenoxide is monitored. Fluorescence-based biosensors for lactate and pyruvate are used to illustrate the principles of biosensors based on the fluorometric detection of produced or consumed reduced nicotinamide adenine dinucleotide (NADH). The concept of biosensors based on a bioluminescence process is demonstrated with sensors for NADH and glutamate. In addition, a novel internal enzyme biosensor arrangement is described where the active enzyme is separated from the sample by a perm-selective membrane. An internal enzyme biosensor for ethanol is presented and the relative merits of this novel biosensor design are discussed. Finally, biosensors based on the combination of a deaminating enzyme with a fiber-optic ammonia gas-sensing probe are demonstrated with a biosensor for glutamate. Here, glutamate oxidase is the immobilized enzyme and the production of ammonia is measured fluorometrically.

Two common definitions of the term biosensor seem to exist in the scientific literature (1). In the first case, a biosensor is an analytical device that can be used to determine the concentration of a specific analyte in a "bio"-sample. Here, the emphasis is on the type of sample. Under this definition, a sodium selective glass membrane electrode must be considered a biosensor when used to determine the activity of sodium ions in a blood sample. This same

0097–6156/89/0403–0303$06.00/0

electrode, together with the same instrumentation and even the same operator, is no longer considered a biosensor when used to determine the sodium ion activity in a non-biological sample, such as in a can of soup. There is a clear inconsistency with this definition.

In the second case, a biosensor is an analytical device that incorporates a "bio"-material as a functioning part of the sensor. The best example of a biosensor of this type is the glucose electrode (2). Regardless of the type of sample (i.e., a blood sample or an aliquot from a high fructose corn syrup production line) the glucose electrode is a biosensor. Here the emphasis is on the response mechanism and required components of the sensor. This second definition will be used throughout this paper.

Biosensors are traditionally divided into two main classes. One class includes biosensors that use a biological receptor, and the other uses a biocatalyst. Examples of bio-receptors include antibodies, binding proteins, and lectins. A critical evaluation of bio-receptor-based biosensors has recently been published (1).

Various types of biocatalytic materials, such as isolated enzymes, bacterial cells, and intact mammalian and plant tissue sections, are available for the preparation of biocatalytic-based biosensors (3-7). An enzyme, or a group of enzymes, provides the required biocatalytic activity. For the bacterial cell and tissue based systems, the required enzyme is housed in these biocatalytic materials which can help stabilize the enzyme and prolong the biocatalytic activity.

Biocatalytic biosensors are kinetic devices in which the production or consumption of a detectable species is monitored. An appropriate transducer is used to monitor the reaction, and the biocatalyst is immobilized at the sensing surface of this transducer. Figure 1 schematically shows the various kinetic processes that occur at the sensing tip when a reaction product is monitored. The enzyme substrate diffuses from the sample solution to the biocatalytic layer where the detected product is generated. At some point, a steady-state concentration of product is established when the rate of product generation is counter-balanced by the rate at which the product diffuses away from the sensor surface. A steady-state signal from the transducer results and the magnitude of this signal is related to the concentration of the enzyme substrate in the sample solution.

Our research efforts center around the development of fiber-optic biocatalytic biosensors. These biosensors are based on the measurement of an optically detectable species which is either generated or consumed at the distal tip of an optical fiber sensing device. The fiber-optic device guides both the incident radiation from the source optics to the sensor tip and the resulting radiation from the sensor tip to the detector optics. The actual transduction element is the opto-electronic detector (photomultiplier tube or photodiode) and the fiber-optic device simply serves as a conduit through which light is transported to and from the sensor tip. By using the proper optical arrangement, the biocatalytic reaction can be monitored through an absorbance, fluorescence or bioluminescence process. The remainder of this chapter is a survey of our progress in the development of fiber-optic biosensors.

The first absorbance-based fiber-optic biosensor was that for p-nitrophenylphosphate (8). This biosensor uses the enzyme alkaline phosphatase which catalyzes the following reaction:

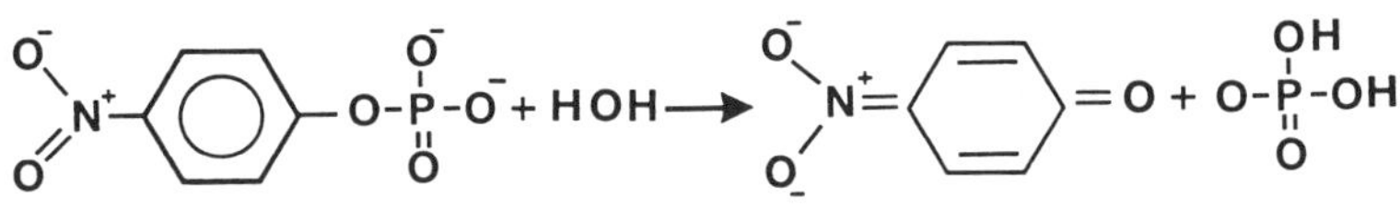

This reaction produces p-nitrophenoxide which strongly absorbs 404 nm radiation. The sensor tip is constructed with alkaline phosphatase covalently immobilized on a nylon membrane. This membrane is positioned at the common end of a bifurcated fiber-optic bundle. One arm of this bundle is connected to the source optics and the other is connected to the detector optics. Incident radiation is transported from a 100 watt tungsten-halogen lamp source to the sensor tip. A fraction of this incident radiation is back scattered off the nylon mesh and a fraction of this back scattered radiation is collected by the fiber-optic bundle and directed to a 404.7 nm interference filter and then to a photomultiplier tube detector.

As p-nitrophenylphosphate diffuses from the bulk solution to the biocatalytic layer, p-nitrophenoxide is formed at the surface of the fiber-optic bundle. A portion of the incident radiation is absorbed by this chromophore and a decrease in the amount of light that reaches the detector is recorded. A steady-state concentration of p-nitrophenoxide is established which corresponds to a steady-state absorbance value. A linear response curve is obtained when the resulting absorbance is plotted with respect to the concentration of p-nitrophenylphosphate in the bulk solution. Figure 2 shows an example of such a response curve.

In addition to the development of absorbance-based fiber-optic biosensors, we have recently demonstrated the feasibility of fluorescence-based biosensors (9, 10). Our initial work in this area has focused on the development of fiber-optic biosensors based on the fluorometric detection of reduced nicotinamide adenine dinucleotide (NADH). Here, a dehydrogenase enzyme supplies the biocatalytic activity, and either the generation or consumption of NADH is monitored.

An optical fiber device is used to supply excitation radiation from an appropriate source to the sensor tip. Figure 3 is a schematic diagram of the sensor device. A central optical fiber made of quartz is used to direct light to the sensor tip from the source optics. The source optics are configured to select the excitation radiation (350 nm) with a simple interference filter. An outer ring of plastic fibers is used to collect a fraction of the emitted radiation from the sensor tip and to guide this radiation to the detection optics. Once again, an interference filter is used to select out the emitted light (450 nm).

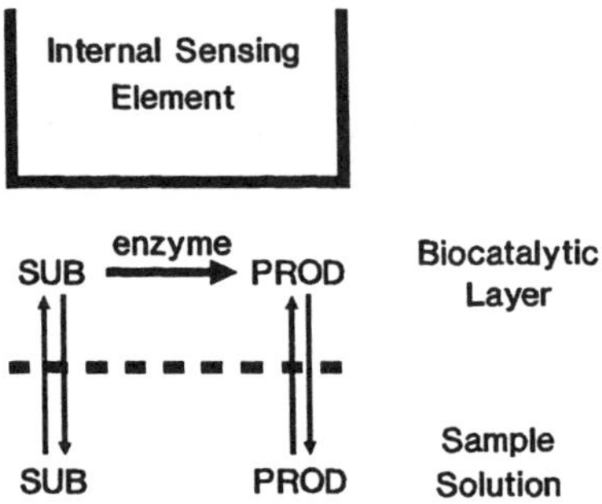

Figure 1. General schematic of a biocatalytic biosensor.

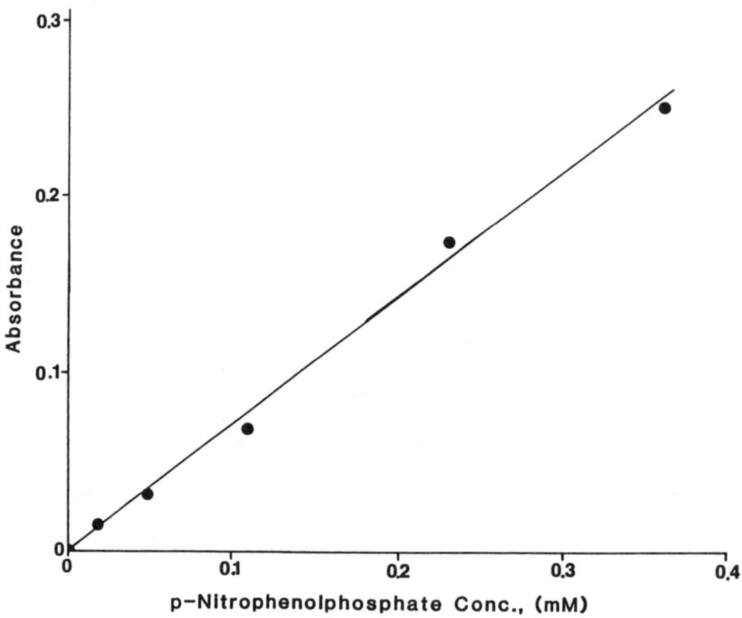

Figure 2. Response curve for the p-nitrophenylphosphate fiber-optic biosensor.

Our first NADH-based biosensor uses the enzyme lactate dehydrogenase as the biocatalyst. This enzyme catalyzed the following reaction:

$$\text{Lactate} + \text{NAD}^+ \;\;====== \;\; \text{Pyruvate} + \text{NADH}$$

The beauty of this reaction from a sensor development point of view is that the thermodynamically favored direction for the reaction can be adjusted simply by setting the solution pH. At high pH, pyruvate is the favored product, and at low pH, lactate is favored. The biosensor can be configured for lactate measurements by adjusting the solution pH to 8.6 and adding NAD^+ to the solution. As lactate enters the biocatalytic layer, the NAD^+ is converted to NADH and an increase in the fluorescence intensity is measured. For a pyruvate biosensor, the solution pH is adjusted to 7.4 and NADH is added to the solution. Pyruvate from the sample enters the biocatalytic layer and a decrease in NADH is measured as the reaction converts the NADH to NAD^+. A decrease in fluorescence intensity is measured.

Figure 4 shows a response curve for the first NADH-based biosensor configured for the measurement of lactate. A steady-state fluorescence signal is measured as a steady-state concentration of NADH is established at the sensor tip. As expected, the magnitude of this signal increases with an increase in the lactate concentration. A detection limit (S/N = 3) of 2 μM has been measured for this lactate biosensor.

Figure 5 shows the response curve for this same biosensor in the pyruvate sensing configuration. Again, a steady-state signal is obtained and, as expected, a decrease in fluorescence intensity is measured with an increase in the pyruvate concentration. The detection limit for this sensor is dependent on the amount of NADH originally present in the sample. High NADH levels generate a high background signal against which the decrease in fluorescence must be detected. For low detection limits, the initial amount of NADH must be low. A detection limit (S/N = 3) of 1 μM is possible with an initial NADH concentration of 0.05 mM.

Response times for the lactate dehydrogenase biosensor in either the lactate or pyruvate sensing mode range from 6 to 12 minutes. Faster response is obtained with higher concentrations. In addition, the lifetime of this sensor is from 3 to 7 days depending on the extent of enzyme loading and the storage conditions.

We have also found that biosensors based on the bioluminescence detection of NADH are possible (11, 12). The following reaction scheme can be used:

$$\text{NADH} + \text{FMN} + \text{H}^+ \;\; ----> \;\; \text{NAD}^+ + \text{FMNH}_2$$

$$\text{FMNH}_2 + \text{RCHO} + \text{O}_2 \;\; -----> \;\; \text{FMN} + \text{RCOOH} + \text{H}_2\text{O} + \text{LIGHT}$$

where the first reaction is catalyzed by the enzyme NAD:FMN oxidoreductase and the second is catalyzed by bacterial luciferase. FMN is flavin mononucleotide, RCHO is a long chain aldehyde (decyl aldehyde) and RCOOH is the corresponding carboxylic acid. The formation of light from NADH is detected and the magnitude of the

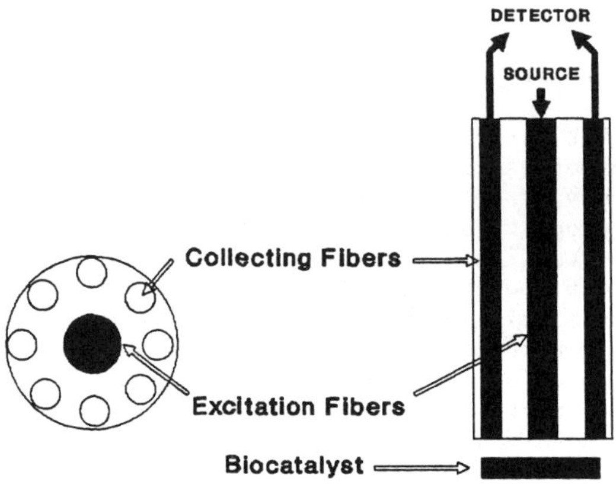

Figure 3. Schematic diagram of the fiber-optic probe for the NADH-based biosensors.

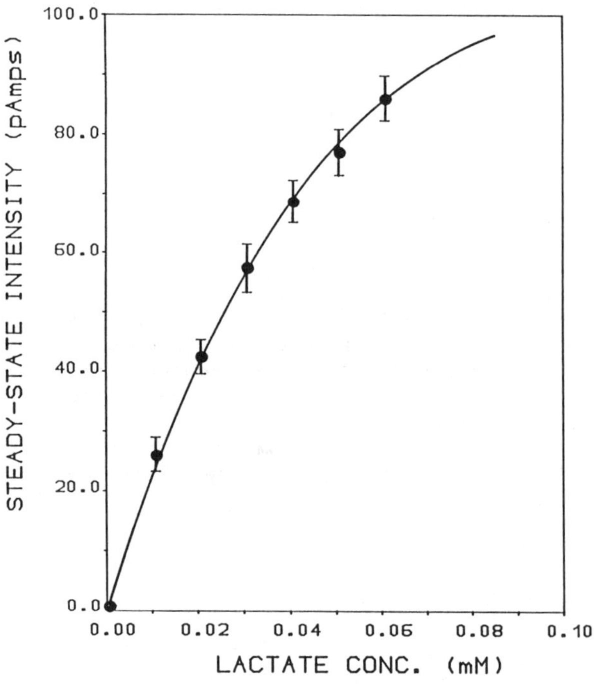

Figure 4. Response curve for the lactate from the lactate dehydrogenase-based biosensor.

resulting light intensity can be related to the concentration of NADH in the sample solution.

Figure 6 shows a schematic of an NADH biosensor based on the bacterial luciferase reaction. A micro-pipet tip is used as the outer body of the sensor. A dialysis membrane with a molecular weight cutoff of 10,000 daltons is held at the distal tip of the sensor with an o-ring. The NAD:FMN oxidoreductase and bacterial luciferase enzymes are held at the sensor tip by this dialysis membrane. The other reagents needed for the biocatalytic reactions must be added to the sample solution and enter the biocatalytic layer by passing through the dialysis membrane. NADH in the sample also crosses this membrane and the formation of light is detected by an optical fiber. Either a single fiber or a bundle of fibers is used to collect a fraction of the generated light. This collected light is guided by the fiber to a PMT detector. There is no need for a radiation source or a wavelength selection device which simplifies the optical arrangement for this biosensor.

Figure 7 shows a response curve for this NADH biosensor where the relative intensity of light produced is related to the concentration of NADH in the standard solutions. This response curve shows that the sensor's response to NADH is non-linear and that the sensor responds in the micro-molar concentration range.

A wide variety of biosensors can be developed by coupling selective dehydrogenase enzymes with the NAD:FMN oxidoreductase and bacterial luciferase system. An example is a glutamate biosensor based on the following reaction scheme:

$$\text{Glutamate} + \text{NAD}^+ \longrightarrow \alpha\text{-Ketoglutarate} + \text{NH}_4^+ + \text{NADH}$$

$$\text{NADH} + \text{FMN} + \text{H}^+ \longrightarrow \text{NAD}^+ + \text{FMNH}_2$$

$$\text{FMNH}_2 + \text{RCHO} + \text{O}_2 \longrightarrow \text{FMN} + \text{RCOOH} + \text{H}_2\text{O} + \text{LIGHT}$$

where the first reaction is catalyzed by glutamate dehydrogenase. The thermodynamically favored direction for the glutamate dehydrogenase reaction is the formation of glutamate and NAD^+. By coupling this reaction with the oxidoreductase reaction, the dehydrogenase reaction is driven in the desired direction by removing NADH. In addition, NAD^+ is recycled which also helps drive this reaction.

Figure 8 shows a preliminary response curve for a glutamate biosensor. For this sensor, glutamate dehydrogenase has been added to the biocatalytic layer in a sensor like that shown in Figure 6. As with the NADH response, a non-linear response curve is obtained. An increase in the light intensity is measured with an increase in the glutamate concentration. For both the NADH and glutamate biosensors, steady-state signals are obtained and the response times are between 1 and 2 minutes.

Each of the biosensors presented above requires that the biocatalyst be in contact with the sample solution during sensor operation. By directly exposing the enzyme to the sample, problems of endogenous activity modulators and pH incompatibility must be considered. Often times, the sample must be altered to protect the enzyme from sample components that can activate or inhibit the

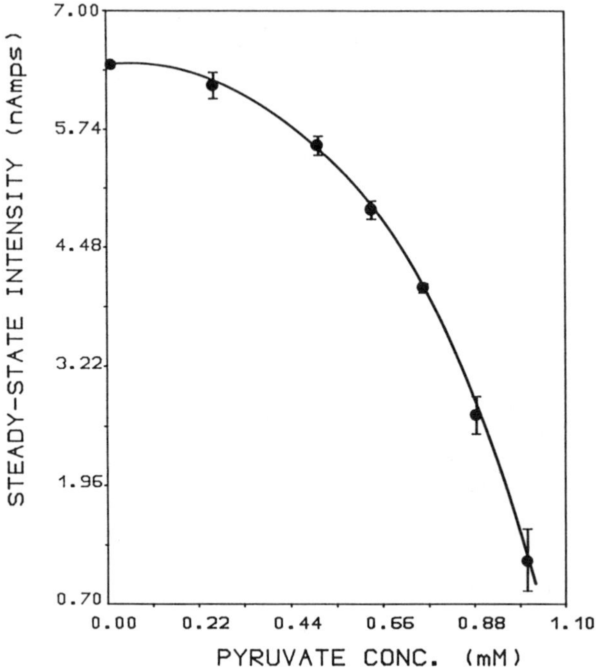

Figure 5. Response curve for pyruvate from the lactate dehydrogenase-based biosensor.

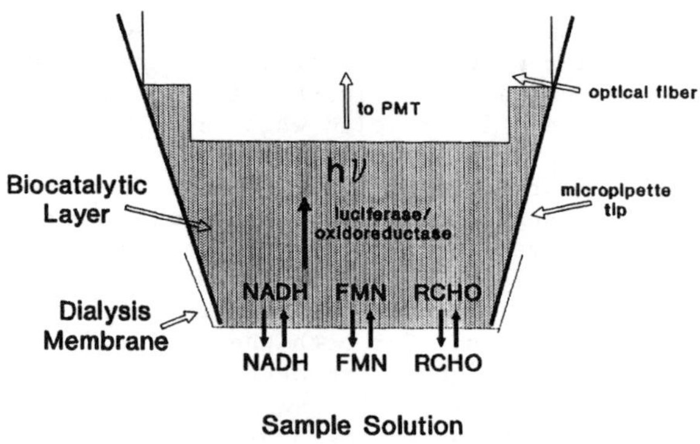

Figure 6. Schematic representation of NADH biosensors based on bacterial luciferase.

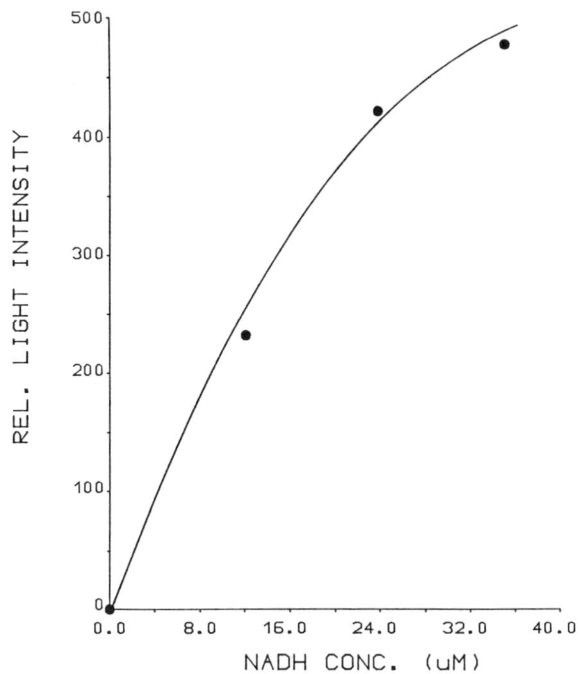

Figure 7. Response curve for NADH from the bioluminescence biosensor.

biocatalyzed reaction. The pH of the solution is also frequently adjusted to maximize the biocatalytic activity. The need to alter the sample composition limits the utility of many biosensors to samples in which such adjustments and alterations are possible.

An internal enzyme biosensor configuration has been introduced in which the enzyme is not directly in contact with the sample. Instead, the enzyme is separated from the sample by a perm-selective membrane that allows the analyte of interest to enter an internal solution. This internal solution contains all the reagents required for the analytical reaction, including the enzyme. Figure 9 is a schematic diagram of an internal enzyme biosensor with a fiber-optic detection scheme. As the sample enters the enzyme-containing internal solution, the analytical reaction takes place. The rate of this reaction is monitored and related to the concentration of the analyte in the sample solution.

The first example of a fiber-optic internal enzyme biosensor has been reported for the determination of ethanol (13). A microporous Teflon membrane is used as the perm-selective membrane for this ethanol sensor. Alcohol dehydrogenase is the enzyme and the following reactions take place in the sensor:

$$CH_3\text{-}CH_2\text{-}OH + NAD^+ \text{ ----> } CH_3\text{-}CHO + NADH$$

$$CH_3\text{-}CHO + H_2N\text{-}NH\text{-}CO\text{-}NH_2 \text{ ----> } CH_3\text{-}CH{=}N\text{-}NH\text{-}CO\text{-}NH_2$$

where the first reaction is catalyzed by the alcohol dehydrogenase. Semicarbazide is added to drive the first reaction by removing the acetaldehyde. The rate of NADH production is measured with a fluorescence measurement through a pair of optical fibers. One fiber is connected to an excitation source and the other is connected to an emission detector.

Ethanol in the sample crosses the microporous membrane and enters the internal solution where NADH is produced. Conditions are maintained so that the rate of ethanol diffusion into the internal solution is rate limiting so that the rate of NADH production is first order with respect to the amount of ethanol in the sample solution. Figure 10 shows an ethanol calibration curve where the rate of NADH production is linearly related to the ethanol concentration.

The beauty of the internal enzyme biosensor concept is that the enzyme can be used in an optimized solution environment without altering the sample solution. Moreover, response times for this type of sensor are short because a reaction rate is measured and no time is needed to establish a steady-state condition. Thus, internal enzyme biosensors seem ideally suited for _in vivo_ sensing where sample treatment is not practical and rapid response is critical. When a gas-permeable membrane is used as the perm-selective membrane, this sensor concept is limited to volatile analytes.

The last fiber-optic biosensor to be presented is based on the immobilization of a deaminating enzyme at the sensing tip of a fiber-optic ammonia probe. Figure 11 shows the various membrane phases involved in the response of the ammonia probe. The key phase is the indicator solution which contains a pH indicator dye and ammonium chloride. The indicator solution is separated from the sample by a

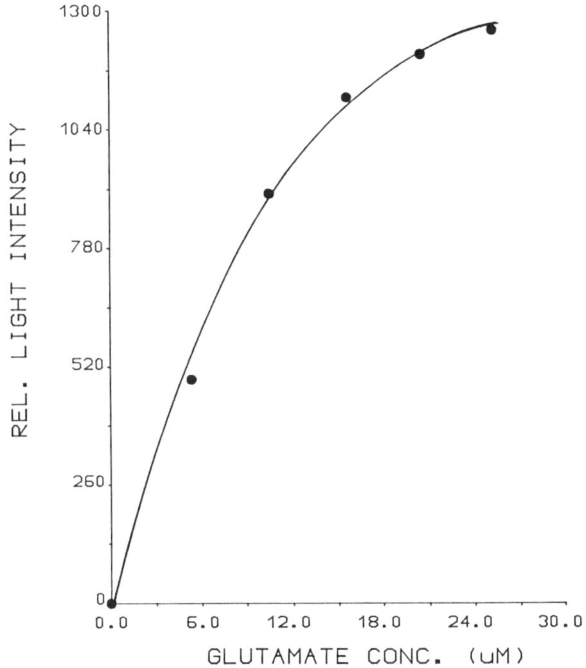

Figure 8. Response curve for glutamate from the bioluminescence biosensor.

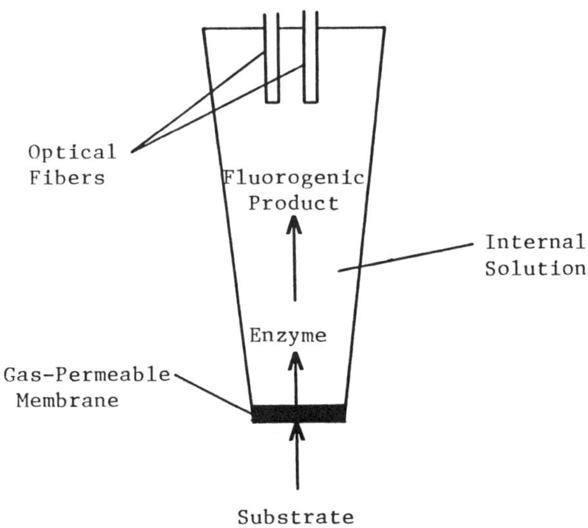

Figure 9. Schematic diagram for the internal enzyme fiber-optic biosensor.

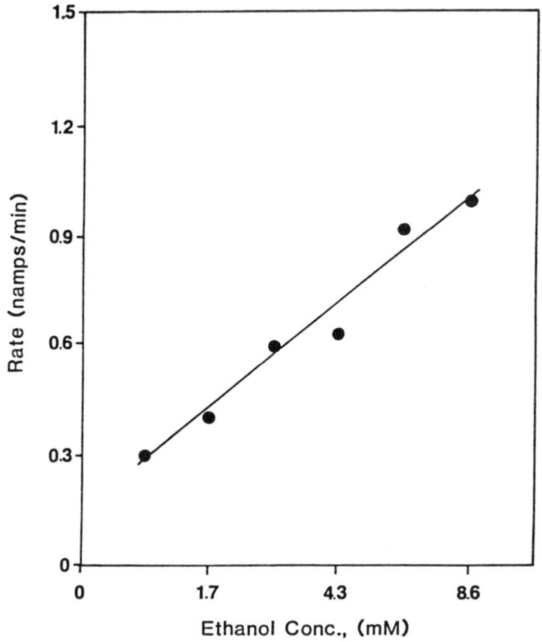

Figure 10. Ethanol calibration curve from the internal enzyme biosensor with alcohol dehydrogenase. (Reproduced with permission from ref. 13. Copyright 1988 Pergamon.)

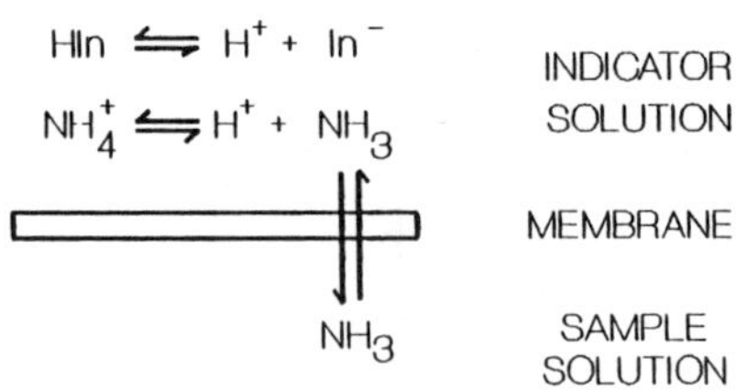

Figure 11. Membrane phases involved in the response of the fiber-optic ammonia gas-sensing probe.

gas-permeable membrane for which microporous Teflon is generally used. Ammonia crosses the Teflon membrane until the partial pressure of ammonia is equal on both sides. Changes in the ammonia concentration in the indicator solution alter the concentration of the non-protonated form of the indicator dye through the chemical equilibria shown in Figure 11. An increase in the concentration of ammonia in the sample solution corresponds to an increase in the amount of the non-protonated form of the indicator. Essentially, the sensor response is based on the titration of the indicator (a weak acid) by the ammonia (a weak base).

The amount of the non-protonated form of the indicator can be measured through either an absorbance (14, 15) or fluorescence (16) detection scheme. In both schemes, optical fibers are used to transport radiation from the source optics to the indicator solution and from the indicator solution to the detection optics.

Because the steady-state response of the fiber-optic ammonia probe is based on simple acid-base chemistry, the development of a function that describes the probe's steady-state response is straight forward (15, 16). The following equation gives the response function for the case when a fluorescent indicator dye is used:

$$I_F = \frac{\phi\ \varepsilon\ b\ K_{In}\ C_{In}\ [NH_3]}{(K_{NH3}\ C_{NH3} - K_{NH3}\ [NH_3] + K_{In}\ [NH_3])}$$

where I_F is the measured fluorescence intensity, ϕ and ε are the quantum efficiency and molar absorptivity of the non-protonated form of the indicator dye, respectively, b is the effective optical path length at the probe tip, K_{In} and K_{NH3} are the acid dissociation constants for the indicator and ammonium ions, respectively, C_{In} and C_{NH3} are the concentrations of indicator and ammonium ions in the indicator solution, respectively, and $[NH_3]$ is the concentration of ammonia in the sample solution.

Biosensors based on this fiber-optic ammonia probe use a deaminating enzyme to generate ammonia at the sensor tip. The biocatalyst is immobilized as a thin layer on the outer side of the Teflon membrane. The enzyme substrate enters the biocatalytic layer where ammonia is produced. A steady-state concentration of ammonia is established and a corresponding steady-state signal from the ammonia probe is attained. The magnitude of this signal is related to the enzyme substrate through a calibration curve.

A glutamate biosensor is being developed where glutamate oxidase is immobilized at the tip of the fiber-optic ammonia probe. This enzyme selectively catalyzes the oxidative deamination of glutamate as follows:

$$\text{Glutamate} + O_2 \longrightarrow \alpha\text{-Ketoglutarate} + H_2O_2 + NH_3$$

Figure 12 shows a typical calibration curve for glutamate with this biosensor. As expected, an increase in glutamate corresponds to more ammonia generated which is detected as more of the non-protonated form of the indicator and an increase in fluorescence intensity. Response in the 1 to 10 μM range at pH 7.8 is possible.

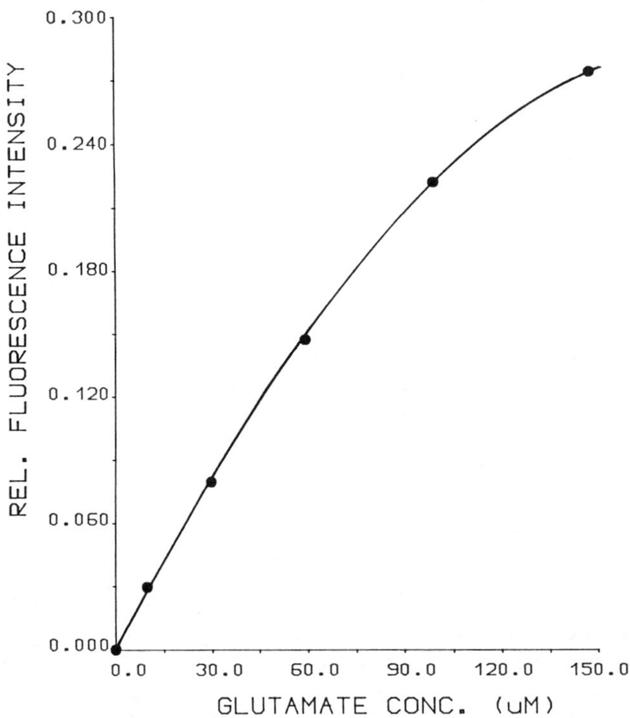

Figure 12. Glutamate response curve from the glutamate biosensor based on glutamate oxidase and the fiber-optic ammonia gas-sensing probe.

The collection of fiber-optic biosensors presented in this chapter represents a new set of analytical devices. Our challenge is to develop these biosensors so that they can be used as effective tools to investigate the fundamental chemistry behind complex biological systems.

Acknowledgments

I wish to thank the following individuals without whom this work would not have been possible: Julie Wangsa, Timothy Rhines, Maureen Stuever, and Bonnie Walters. The financial support from the National Science Foundation (BNS-8716768) and the National Institutes of Health (GM 335487) is acknowledged.

Literature Cited

1. Arnold, M. A.; Meyerhoff, M. E. CRC Critical Reviews in Anal. Chem. 1988, 20, 149.
2. Clark, L. C., Jr.; Lyons, C. Ann. NY Acad. Sci. 1962, 102, 29.
3. Rechnitz, G. A. Chem. & Eng. News 1988, 66, 24.
4. Turner, A. P. F.; Karube, I.; Wilson, G. S. Biosensors: Fundamentals and Applications; Oxford University Press: New York, 1987.
5. Arnold, M. A. Amer. Lab. 1982, 15, 34.
6. Rechnitz, G. A. Science 1981, 214, 287.
7. Carr, P. W.; Bowers, L. D. Immobilized Enzymes in Analytical and Clinica Chemistry; John Wiley & Sons: New York, 1980.
8. Arnold, M. A. Anal. Chem. 1985, 57, 565.
9. Arnold, M. A. GBF Monographs, 1987, 10, 223.
10. Wangsa, J.; Arnold, M. A. Anal. Chem. 1988, 60, 1080.
11. Arnold, M. A. Proceedings of the SPIE 1988, 906 128.
12. Stuever, M. A.; Arnold, M. A., University of Iowa, unpublished data.
13. Walters, B. S.; Nielsen, T. J.; Arnold, M. A. Talanta 1988, 35, 151.
14. Arnold, M. A.; Ostler, T. J. Anal. Chem. 1986, 58, 1137.
15. Rhines, T. D.; Arnold, M. A. Anal. Chem. 1988, 60, 76.
16. Rhines, T. D.; Arnold, M. A., University of Iowa, unpublished data.

RECEIVED March 9, 1989

Chapter 21

Design Considerations for Antibody-Based Fiber-Optic Chemical Sensors

Michael J. Sepaniak[1], Bruce J. Tromberg[1], Jean-Pierre Alarie[1], James R. Bowyer[1], Arthur M. Hoyt[2], and Tuan Vo-Dinh[3]

[1]Department of Chemistry, University of Tennessee, Knoxville, TN 37996–1600
[2]Department of Chemistry, University of Central Arkansas, Conway, AR 72032
[3]Health and Safety Research Division, Oak Ridge National Laboratory, Oak Ridge, TN 37831–6101

Fiber optic chemical sensors that employ immunochemical reagent phases are described. Relying on the specificity of antibody-antigen immune complex formation and the sensitivity of fluoroimmunoassay techniques, these sensors are capable of remotely detecting very low concentrations of biomolecules with a high degree of analytical selectivity. The type of analyte to be measured and the assay procedure that is employed influence the optimum design of the sensor. Optical and sensing termini configurations that we have successfully employed in the design of sensors based on competitive-binding and direct assay protocols are described and important experimental considerations are briefly discussed. In addition, we demonstrate how these sensors can be configured to permit relatively rapid in situ regeneration, thereby permitting pseudo-continuous operation.

Remote, in situ measurements of chemical concentration can be accomplished using fiber optic sensors. High sensitivity is possible if the measurements are based on fluorimetry. Versatility and analytical selectivity is enhanced if the fluorescence signal is the result of the interaction of the analyte with a specific reagent phase that is immobilized at the sensing terminus of the fiber optic. Commonly the interaction involves the formation of a fluorescent analyte-reagent adduct or the quenching or enhancement of the fluorescence of the reagent phase by the analyte. Fiber optic chemical sensors (FOCS) have traditionally been employed to measure small molecules or ions such as H^+(1-4), metal ions(5-8), O_2(9-12), and CO_2(13,14).

The FOCS measurement of large and macro-molecules has been accomplished recently using biologically significant "affinity" reagent phases. Affinity reagent phase/analyte combinations that have been used in this fashion include lectin/carbohydrate(15,16),

0097–6156/89/0403–0318$06.00/0

enzyme/substrate(17-22), antibody/antigen(23-26), and antibody/hapten(27-31). The specificity of immune complex formation and the sensitivity of fluorimetry have rendered fluoroimmunoassay (FIA) techniques prominent in the area of bioanalysis. This represents the motivation in developing antibody-based FOCSs. The analytical methodologies and characteristics associated with a FIA depend on the particular type of assay (direct, competitive-binding, sandwich, etc.) that is performed and the nature of the measured antigen/hapten. Concomitantly, the successful in situ measuring of antigens/haptens using antibody-based FOCSs requires strict attention to the design of the fiber optic device, with optimum design features varying considerably with the application. The configurations of the optical components (light source, fiberoptic, etc.) and the sensing terminus are the critical design features.

In this paper we present a brief discussion of FIA and fiber optic sensing principles and discuss our development of sensors based on competitive-binding and direct (of natural fluorophors) immunoassay procedures. We will also demonstrate how these sensors can be configured to permit relatively rapid in situ regeneration, thereby permitting benchtop assay procedures, including rinsing to remove interferents, to be performed repetitively without removing the sensor from the sample.

Experimental

Fluoroimmunoassay Principles. Most antibodies belong to the immunoglobulin G (IgG) class of blood plasma proteins(32). IgG is composed of two identical "light" and two identical "heavy" protein chains. Each light and heavy chain is joined by a disulfide bond. The two heavy chains are attached to one another via a disulfide linkage in the "hinge" region of the molecule. The entire structure is Y-shaped and has a molecular weight of about 160,000. The common end of IgG, sometimes refered to as the F_c portion of the molecule, has a fixed structure (i.e., does not vary between different IgG molecules). Conversely, the two separate heavy-light chain combinations in IgG, sometimes refered to as the Fab' portions of the molecule, contain variable amino acid sequences that have a tertiary structure which permits the molecule to bind strongly to a specific region, "epitope", of the foreign substance, "antigen", for which it was generated. Affinity constants (K_as) for antibody (Ab) - antigen (Ag) binding range from 10^6 to 10^{12}. An antigen must have a molecular weight of roughly 5000 in order to produce an immune response. Smaller molecules, "haptens", must be conjugated to a large protein in order to generate antibodies. Antibodies can be generated for a wide variety of antigens/haptens and in many cases are commercially available(33).

FIA methodology depends on the type of assay to be performed. FIAs are broadly classified according to whether a separation step is needed in order to distinguish antibody-bound antigens from free antigens. Heterogeneous assays require a separation step while homogeneous assays involve the measurement of a fluorescence property that changes upon formation of the immune complex and hence do not require a separation step (34). In this work antibody-based FOCSs are described that utilize heterogeneous FIA procedures. In most modern heterogeneous FIAs the separation step is accomplished by immobilizing the antibody on a solid support then rinsing to remove unbound antigen and potential interferents.

Competitive-binding, direct (of natural fluorophors), and sandwich measurement protocols are most commonly used in FIA. In competitive-binding assays fluorophor-labeled antigen competes with unlabeled antigen from the sample for a limited amount of antibody. Following the separation step the fluorescence of the antibody-bound fraction is measured and related to sample antigen concentration. An inverse relationship with a short dynamic range (about one decade in concentration) is observed. Fluorescein isothiocyanate (FITC) is the most common label and has excitation and emission maxima at 490 nm and 520 nm, respectively. Direct assays are performed by incubating sample antigen with excess antibody. Since the antigen is naturally fluorescent a label is not needed. The measured signal is directly proportional to antigen concentration and the amount of antibody employed. Dynamic range is considerably longer than for competitive-binding assays. Sandwich assays exhibit sensitivities and dynamic ranges that are similar to those observed in direct assays, but can be used to measure nonfluorescent antigens. Two types of antibody, with specificities for different antigen epitopes, are employed. The first is generally immobilized on a solid support and the second is labeled with a fluorophor. Following rinsing to remove excess labeled antibody and interferents the fluorescence of the "sandwich" immune complex is measured.

Antibody-Based FOCS Instrumentation. The rapid growth of fiber optic sensing has paralleled the commercial availability of low attenuation optical fibers. The optical fibers employed in this work are comprised of a quartz core, with a diameter of 0.2 - 0.6 mm and refractive index n_1, surrounded by a soft plastic cladding with refractive index n_2. Light transmission is based on total internal reflection (34) and depicted in Figure 1. Light rays that impinge on the core/cladding interface at an angle equal to or greater than the critical angle (θ_c), as determined by Snell's Law, can be transmitted along the fiber by total internal reflection. Accordingly, an important fiber characteristic, the half acceptance angle (β), is given by equation 1

$$\text{Sin } \beta = (n_1^2 - n_2^2)^{1/2}/n_o \qquad (1)$$

where n_o is the refractive index of the medium in which the end of the fiber resides.

Low attenuation over given spectral regions and a large β are desirable fiber characteristics. In principle, the fiber provides an optical link between the spectroscopic instrumentation in the laboratory and the remotely-located sample.

Fibers with large values of β permit the coupling of large amounts of excitation radiation from the source at the incident end of the fiber and the efficient collection of fluorescence signal at the sensing end of the fiber(34).

In this work excitation radiation and fluorescence signals are transmitted by a single fiber. The optical arrangement employed is shown in Figure 2. The radiation from either an argon ion laser (approximately 20 mW at 488 nm) or a helium-cadmium laser (approximately 10 mW at 325 nm) is passed cleanly through a 2 mm hole bored in the center of a 25 mm diameter mirror (C in the figure) and focused onto the incident end of the fiber. The advantage of using a relatively low power laser is that its highly collimated output can be very efficiently coupled onto the fiber, thereby producing larger fluorescence signals than with conventional sources(34). The focusing lens is chosen such that its diameter matches that of the mirror and its f/number matches that of the fiber. A portion of the fluorescence emission that is generated at the sensing end of the fiber is collected by the fiber and transmitted back to the laboratory instrumentation. This emission exits the incident end of the fiber at angles that are equal or less than β. Subsequently, the collected emission is collimated by the lens, reflected by the mirror (except for a few percent that is lost through the hole), and focused onto the entrance slit of an emission monochromator. Signals are processed with a photo-meter.

The configuration of the sensing end of the fiber optic is the most critical element in the design of an antibody-based FOCS. It often determines detectability, dynamic range, response rate, the effects of sample interferents, the size of the sensor, and the useful lifetime of the sensor. The two most important design features are the method used to immobilize the reagent phase (e.g., direct chemical attachment or entrapment behind a membrane) and the shape, size, and amount of reagent phase. The three different sensing termini designs shown in Figure 3 were used for the work presented herein. These sensing termini will be discussed further in applications section of this report.

Applications and Discussion

Sensing Based on Competitive-Binding FIA. The sensing terminus shown in Figure 3A was used to measure anti-rabbit IgG using competitive-binding assay procedures(26). Rabbit IgG was covalently bonded to the distal face of a 0.6 mm core diameter fiber using the activating reagent 3-glycidoxypropyltrimethoxysilane (GOPS). This activating reagent provides a link between the fiber's surface silanol groups and amine groups in the immobilized protein. This is a non-specific form of immobilization that can denature the protein and/or hinder access to binding sites on the protein. In this competitive-binding assay, involving a polyvalent antigen/antibody combination, these problems did not seem to occur. However, we subsequently conducted a comparative evaluation of various activating reagents that resulted in higher antibody affinity for reagents that react specifically with portions of the antibody that are not involved in antigen

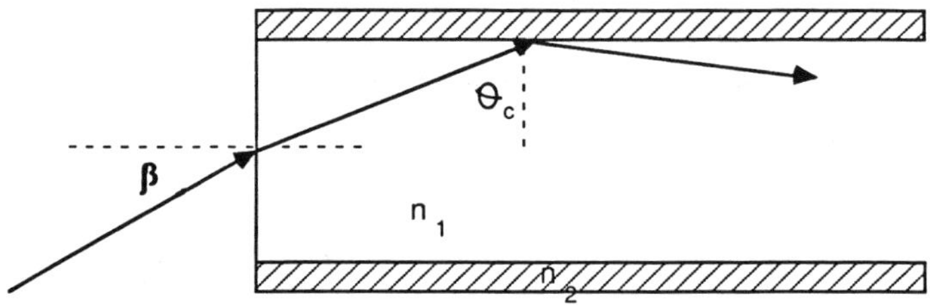

Figure 1: Depiction of total internal reflection in an optical fiber with core and cladding refractive indices of n_1 and n_2, respectively.

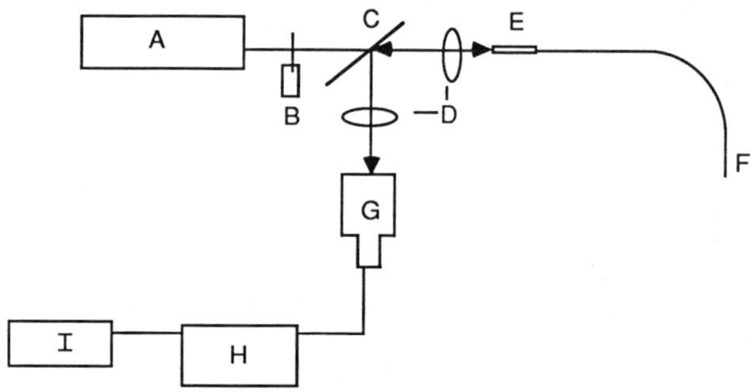

Figure 2: Optical configuration for antibody-based FOCS: (A) light source (laser in this work), (B) shutter, (C) beam splitter, (D) focusing lenses, (E) incident end of fiber in a positioner, (F) sensing end of fiber, (G) monochromator/PMT, (H) photometer, and (I) recorder.

binding(35). The small amount of protein that can be immobilized by using GOPS to link directly to the fiber's sensing face (about 3 ng in this experiment) is an advantage in competitive-binding assays as sensitivity is inversely proportional to the amount of immobilized protein(34). The protocol for these measurements involves incubating prepared fibers (a separate one for each measurement) with solutions containing varying amounts of anti-IgG and 3 x 10^{-8} M FITC-labeled anti-IgG. Using the argon ion laser for fluorescence excitation we are able to detect 25 femtomoles of anti-IgG in a 10 μL sample following a 20 minute incubation.

Ideally, a fiber optic sensor should be capable of performing remote, in situ, continuous measurements. Most of the antibody-based sensors that have been reported fall short of exhibiting these three characteristics and, hence, might be more appropriately termed probes rather than sensors. This is true of the competitive-binding sensor described above. Simultaneous incubation with sample and labeled antigen cannot be easily performed in situ. Furthermore, efforts to regenerate the sensor were not successful. Affinity chromatography columns have been successfully regenerated with chaotropic rinses or using pH changes. However, such approaches to regeneration are less successful in sensing applications since any change in affinity constant can result in an equivalent change in measured signals.

In order to enable in situ competitive-binding FIA measurements we abandoned the "simultaneous" incubation procedure described above in favor of a "pre-incubation" procedure. The results are demonstrated in Figure 4. Measurements involve incubating the fiber optic sensor with samples (this can be done in situ) containing various concentrations of anti-IgG for a period of time that does not result in complete saturation of the fiber-bound IgG (2 minutes in this case). The fiber is removed from the sample and placed in 0.5 mg/mL FITC-labeled anti-IgG prior to measuring the fluorescence signal. As expected a negative sloping calibration plot with a very short dynamic range is observed (see Figure 4). The absolute limit of detection for 1 mL samples was 100 picomoles.

Sensing Based on Direct FIA. Antibody-based FOCS were developed for certain naturally fluorescent polynucleararomatic hydrocarbons (PNAs). The sensing terminus design shown in Figure 3A was used initially(28). However, the sensitivity for the measurement of benzo(a)pyrene (B(a)P) was rather poor, presumably due to the small amount of immobilized antibody and possible denaturing upon bonding to the fiber. Subsequently, measurements of a tetraol metabolite of B(a)P, r-7,t-8,9,c-10-tetrahydroxy-7,8,9,10-tetrahydrobenzo-(a)pyrene (BPT), were performed using the terminus design shown in Figure 3B(29). This sensing terminus is constructed by forming a 40 nL liquid phase antibody chamber at the end of a 0.2 mm core diameter fiber. The chamber is terminated with a 10,000 M.W. cutoff cellulose membrane. Relative to direct covalent attachment to the fiber this design permits the immobilization of a greater amount of antibody. Furthermore,

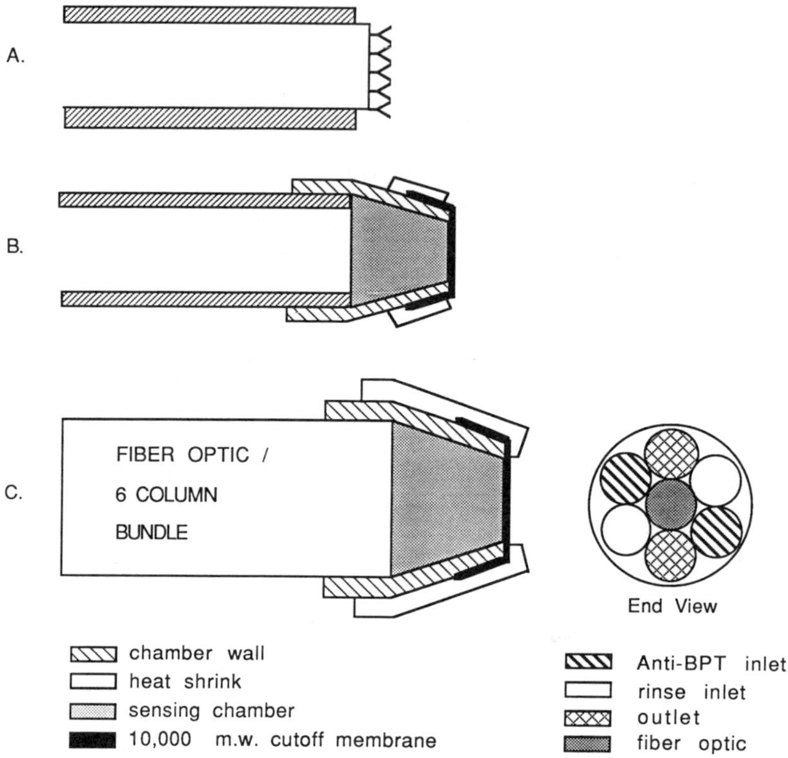

Figure 3: Sensing termini for antibody-based FOCS: (A) direct covalent attachment of immunochemical, (B) membrane entrapment of immunochemical, and (C) a regenerable immunosensor.

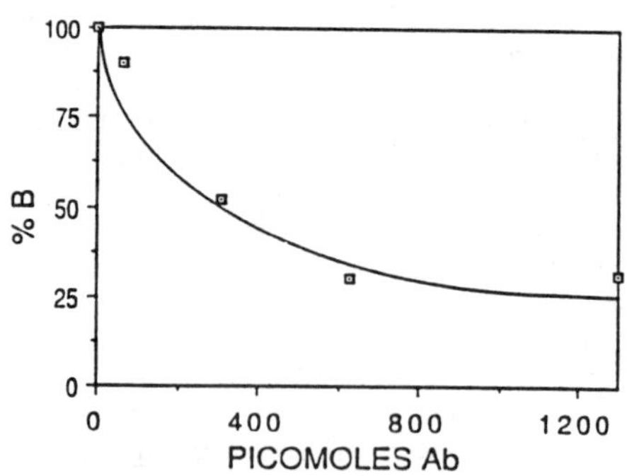

Figure 4: Antibody-based FOCS calibration plot for a pre-incubation, competitive-binding measurement of anti-IgG: Percent bound (%B) labeled anti-IgG vs amount of unlabeled anti-IgG (Ab) for 1.0 mL stirred solution, 2-minute incubation. (Reproduced from ref. 26. Copyright 1987 American Chemical Society.)

there is essentially no denaturing or hindering of access to antigen binding sites.

In this work the chamber is filled with a 2 x 10^{-6} M solution of anti-BPT. When placed in sample solutions, BPT diffuses into the chamber were it forms an immune complex that is trapped by virtue of its size. In this manner the anti-BPT serves to concentrate the BPT; fluorescence signals generated by using the helium-cadmium laser for excitation are observed to increase linearly over a period greater than 2 hours. In a period as short as 15 minutes signal levels are significantly greater than for bare fiber measurements without antibody. A linear calibration plot over two decades of BPT concentration is shown in Figure 5. In addition to excellent sensitivity (the absolute limit of detection for 60 minute incubations in 10 μL samples is about 1 femtomole), using very short incubations, the sensor is capable of short duration continuous operation. Excellent response selectivity to BPT in the presence of numerous spectrally interfering PNAs was also demonstrated(29). However, this required that the sensor be removed from the sample and dialyzed for several minutes in solvent to remove unbound impurities.

There are several limitations or problems associated with the direct assay sensor described above. First, its applications are limited due to the fact that most assays are performed on non-fluorescent antigens (using competitive-binding or sandwich assay procedures). Second, continuous, in situ operation is only possible over a short time span and the selectivity attribute described above requires that the sensor be removed from the sample. Third, the only mode of sampling is via diffusion which can be prohibitively slow for large molecules. Finally, the membrane that is needed to retain the large antibody molecule can exhibit memory and concentration polarization effects. Most of these limitations or problems can be potentially reduced or eliminated by using the regenerable antibody-based FOCS described below.

Regenerable Antibody-Based FOCS. By combining FOCS technology with a capillary reagent delivery system we have constructed microscale sensors that are capable of performing a variety of heterogeneous FIA procedures repetitively, remotely, and in situ. These procedures include adding solid or liquid phase antibody, adding secondary reagents (e.g., the labeled "second" antibody when performing sandwich assays), and rinsing to remove unbound impurities. In addition to delivering reagents, the sensor has the capability of sampling analyte through a membrane via either diffusion or aspiration. The later mode of sampling could be very beneficial in the eventual use of the sensor for the measurement of large molecules.

The configuration of a regenerable sensor that is designed for the repetitive measurement of BPT using a direct assay procedure is shown in Figure 3C. The sensor is constructed by surrounding a 0.2 mm core diameter fiber with six 0.2 mm i.d. fused silica capillary columns (these columns are routinely used in capillary chromatography). The diameter of the fiber/6-column-bundle is about 1 mm. The chamber at the sensing end of the

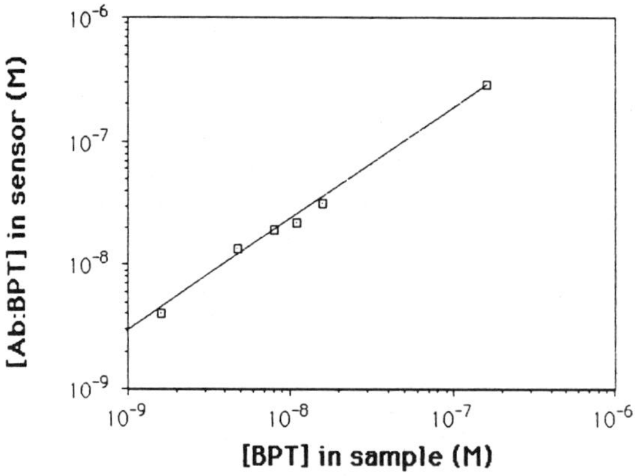

Figure 5: Antibody-Based FOCS calibration plot for a direct measurement of BPT.

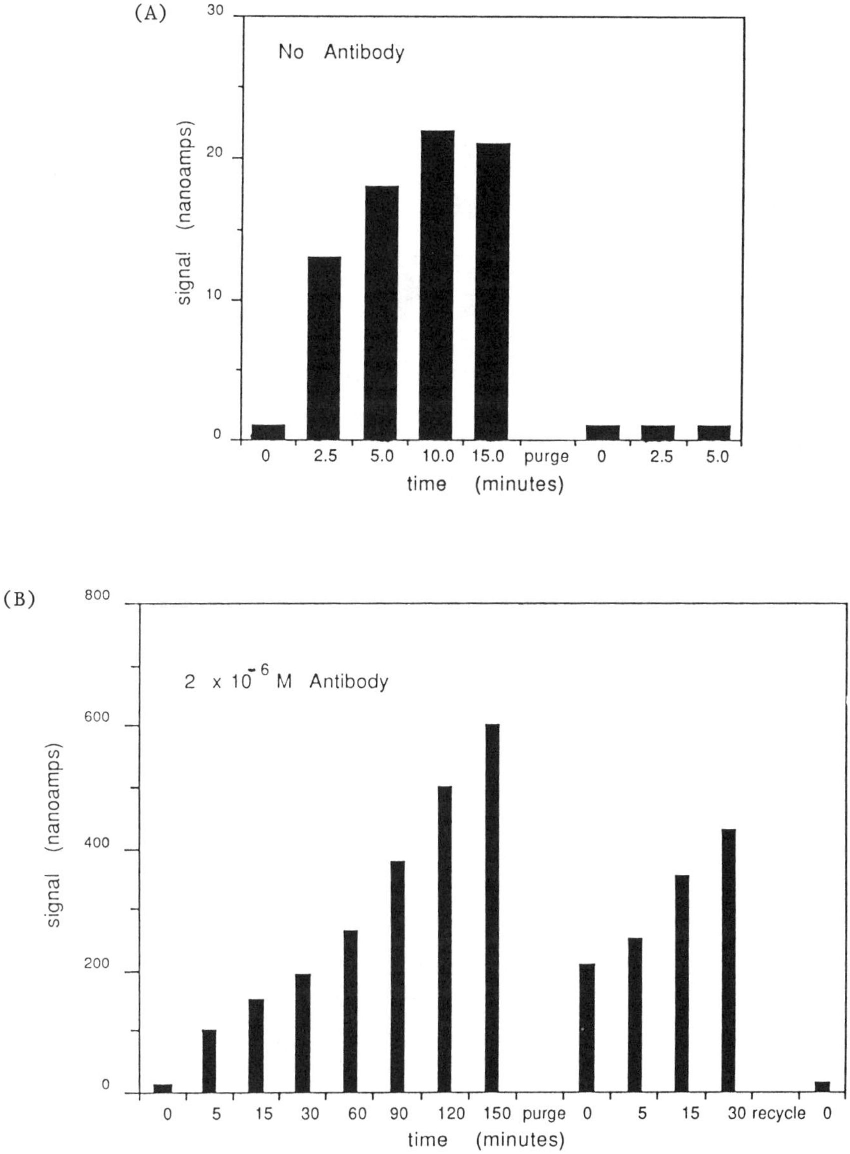

Figure 6: Direct measurement of BPT using a regenerable immunosensor: (A) without entrapped anti-BPT and (B) with entrapped anti-BPT (2 x 10^{-6} M).

bundle is comprised of a chamber wall, which extends less than 0.5 mm beyond the fiber, that is capped with a membrane (a 10,000 M.W. cutoff cellulose membrane is used for the measurement of BPT). The volume of the chamber is less than 1 μL.

We are in the process of developing and testing protocols for performing sandwich assays using solid phase antibody (immunobeads) and direct assays using liquid phase antibody. That work will be reported later. However, the results of preliminary experiments demonstrating the function of the sensor in the direct measurement of BPT are shown in Figure 6. In this experiment two capillary columns are filled with 2 x 10^{-6} M anti-BPT, two capillary columns are filled with rinse solution, and the remaining capillary columns are used as outlets. The delivery of these solutions is accomplished using syringe pumps. The capillary columns are terminated at the pump end with a leur syringe fitting that contains a shut-off valve. In the first experiment the chamber is filled with rinse solution (distilled water) and the sensor is placed in a 1.0 x 10^{-7} M BPT solution. As can be seen in Figure 6A, the signal over background increases rapidly and comes to equilibrium after about 10 minutes. With the anti-BPT and outlet capillary columns sealed the chamber is purged through the membrane using the rinse solution capillary columns. The signal rapidly drops to baseline and when removed from the sample and placed in a rinse solution does not recover after the purge (i.e. the unbound BPT was effectively removed from the chamber by the purge).

In the next experiment the chamber is filled with antibody solution and the sensor is placed in a 1.0 x 10^{-7} M BPT solution. As can be seen in Figure 6B, the signal rapidly increases to values much higher than for the no antibody measurement. This response is similar to that observed with the direct assay sensor described in the previous section. However, this regenerable sensor is capable of in situ purging (to remove potential interferents) and regeneration. In this experiment the signal drops dramatically during the purge step due to the movement of the immune complex to the membrane where it is not effectively "seen" by the fiber(34). Nevertheless, the "purge" signal is much greater than the equilibrium signal for the no antibody case. Unlike the no antibody case, the signal recovered after the purge (due to diffusion of the immune complex from the membrane) indicating that the immune complex had been retained. The sensor was then recycled (to the original baseline) by flushing the chamber with rinse solution with the outlet columns open, illustrating that the sensor can be regenerated. The time required for filling, purging, and recycling is only several minutes.

Acknowledgments

This research was supported by the National Institutes of Health under Contract GM 34730 with the University of Tennessee, Knoxville, and the Office of Health and Environmental Research, U.S. Department of Energy, under Contract DE-AC05-840R21400 with Martin Marietta Energy Systems, Inc.

Literature Cited

1. Peterson, J. I.; Goldstein, S. R.; Fitzgerald, R. V.; Buckhold, D. K. Anal. Chem., **52**, 864 (1980).
2. Saari, L. A.; Seitz, W. R. Anal. Chem., **54**, 821 (1982).
3. Munkholm, C.; Walt, D. R.; Milanovich, F. P.; Klainer, S. M. Anal. Chem., **58**, 1427 (1986).
4. Jordan, D. M.; Walt, D. R.; Milanovich, F. P. Anal. Chem., **59**, 437 (1987).
5. Saari, L. A.; Seitz, W. R. Anal. Chem., **55**, 667 (1983).
6. Zhujun, Z.; Mullin, J. L.; Seitz, W. R. Anal. Chimica Acta, **184**, 251 (1986).
7. Wyatt, W. A.; Bright, F. V.; Hieftje, G. M. Anal. Chem., **59**, 2272 (1987).
8. Wolfbeis, O. S.; Schaffar, B. P. H. Anal. Chimica Acta, **198**, 1 (1987).
9. Peterson, J. I.; Fitzgerald, R. V.; Buckhold, D. K. Anal. Chem., **56**, 62 (1984).
10. Urbano, E.; Offenbacher, H.; Wolfbeis, O. W. Anal. Chem., **56**, 427 (1984).
11. Zhujun, Z.; Seitz, W. R. Anal. Chem., **58**, 220 (1984).
12. Wolfbeis, O. S.; Posch, H. E.; Kroneis, H. W. Anal. Chem., **57**, 2556 (1985).
13. Hirschfeld, T.; Miller, F.; Thomas, S.; Miller, H.; Milanovich, F.; Gaver, R. W. IEEE J. of Lightwave Technol., **5**, 1027 (1987).
14. Zhujun, Z.; Seitz, W. R. Anal. Chimica Acta, **160**, 305 (1984).
15. Schultz, J. S.; Mansouri, S.; Goldstein, I. J. Diabetes Care, **5**, 245 (1982).
16. Meadows, D.; Schultz, J. S. Talanta, **35**, 145 (1988).
17. Wolfbeis, O. S. Anal. Chem., **58**, 2876 (1986).
18. Wangsa, J.; Arnold, M. A. Anal. Chem., **60**, 1080 (1988).
19. Kulp, T. J.; Camins, I.; Angel, S. M.; Munkholm, C.; Walt, D. R. Anal. Chem., **59**, 2849 (1987).
20. Opitz, N.; Lübbers, D. W. Talanta, **35**, 123 (1988).
21. Arnold, M. A. Anal. Chem., **57**, 565 (1985).
22. Walters, B. S.; Nielsen, T. J.; Arnold, M. A. Talanta, **35**, 151 (1988).
23. Sutherland, R.; Dähne, C.; Place, J. F.; Ringrose, A. S. Clin. Chem., **30**, 1533 (1984).
24. Petrea, R. D.; Sepaniak, M. J.; Vo-Dinh, T. Talanta, **35**, 139 (1988).
25. Andrade, J. D.; Vanwagenen, R. A.; Gregonis, D. E.; Newby, K.; Lin, J.-N. IEEE Transactions on Electron Devices, **32**, 1175 (1985).
26. Tromberg, B. J.; Sepaniak, M. J.; Vo-Dinh, T.; Griffin, G. D. Anal. Chem., **59**, 1226 (1987).
27. Anderson, F. P.; Miller, W. G. Clin. Chem., **34/7**, 1417 (1988).
28. Vo-Dinh, T.; Tromberg, B. J.; Griffin, G. D.; Ambrose, K. R.; Sepaniak, M. J.; Gardenhire, E. M. Appl. Spectrosc., **41**, 735 (1987).

29. Tromberg, B. J.; Sepaniak, M. J.; Alarie, J. P; Vo-Dinh, T.; Santella, R. M. Anal. Chem., **60**, 1401 (1988).
30. Vo-Dinh, T.; Tromberg, B. J.; Sepaniak, M. J.; Griffin, G. D.; Ambrose, K. R.; Santella, R. M. in "Optical Fibers in Medicine III", A. Katzir, Ed.; Proceedings of SPIE 910-18: Los Angeles, 1988.
31. Tromberg, B. J.; Sepaniak, M. J.; Vo-Dinh, T. in "Optical Fibers in Medicine III", A. Katzir, Ed.; Proceedings of SPIE 906-06: Los Angeles, 1988.
32. Nisonoff, A.; Hopper, J. E.; Spring, S. B. The Antibody Molecule; Academic Press: New York, 1975.
33. Linscott's Directory of Immunological and Biological Reagents, 40 Glen Drive, Mill Valley, CA 94941.
34. Sepaniak, M. J.; Tromberg, B. J.; Vo-Dinh, T. Prog. Anal. Spectros., 11, 481 (1988).
35. Alarie, J. P.; Sepaniak, M. J.; Vo-Dinh, T. Anal. Chem. submitted for publication.

RECEIVED March 20, 1989

Chapter 22

Acetylcholine Receptor-Mediated Optical Transduction from Lipid Membranes

U. J. Krull, R. S. Brown, K. Dyne, B. D. Hougham, and E. T. Vandenberg

Chemical Sensors Group, Department of Chemistry, Erindale College, University of Toronto, Mississauga, Ontario L5L 1C6, Canada

Chemical transduction in a neural synapse by acetylcholine receptor (AChR) occurs by modulation of the conductivity of an ion channel. A physical model of this chemoreceptive process can be established by reconstituting natural AChR into artificial bilayer lipid membranes, and subsequently observing electrochemical activity induced by an agonist. The receptor protein or the membrane lipid may be labeled with 4-nitro-2,1,3-benzoxadiazole (NBD), providing a fluorescent tag which can be used to optically monitor receptor binding with analytes. This work reports the use of fluorescence properties of the specific and generic labeled systems to provide quantitative analytical measurements of carbamylcholine at micromolar concentrations, and α-bungarotoxin at nanomolar concentrations. The signals from both systems are similar, fitting the same binding constant relationship, and providing enhancements in fluorescence intensity of up to 200 percent.

The basis for a generic transduction mechanism for biosensor development is exemplified by natural chemoreception, where only the selectivity of the sensing surface need be altered to accommodate detection of a wide variety of analytes(1,2). Chemical communication in living organisms is mediated by selective binding events associated with lipid membranes. Lipid membranes provide an environment which is in part responsible for maintenance of the conformational properties of active binding proteins, and the membranes also provide the ion and solute selective permeability necessary for establishing chemical concentration gradients for transduction based on ion conduction and secondary messenger phenomena(3). Significant experimental evidence has been collected which supports the proposal that lipid membranes are not passive elements in chemical transduction events, and that protein-mediated selective binding processes can

0097–6156/89/0403–0331$06.00/0

lead to both structural and electrostatic perturbations of the supportive lipid matrix(4).

Natural chemoreception for processes such as olfaction operates in an analytical sensing mode by means of fast electrochemical events. While the rapid response and extreme sensitivity of the electrochemical ion translocation mechanisms associated with gated processes provide some "ideal" analytical characteristics, any single sensing structure is limited by background noise and a lack of information which can define a unique selective binding event. Non-selective binding events and spontaneous ion current transients are processed in natural organisms by means of complicated sequences of multiple sensor arrays and by signal processing of frequency and amplitude data in a manner analogous to a combination of hardware and software applications of chemometric approaches(5,6). This technology represents a horizon of research in the area of sensor development, and presently limits the analytical potential of biosensors based on lipid membrane electrochemistry.

Lipid membrane perturbations by selective binding events or non-selective interactions have been studied previously by fluorescence spectroscopy, where fluorophores associated with the lipid matrix or the selective agent could provide indications of changes of structure and environmental conditions(7,8). The use of fluorescence as an indicator of membrane perturbations offers the opportunity for multidimensional analysis based on concurrent observations of intensity, wavelength, polarization and lifetime. The combination of information associated with each of these properties can provide sufficient data for unique identification of selective binding processes and quantitative evaluation of non-selective interferences(9). A reduction in analytical sensitivity is intrinsic to transduction by fluorescence response in contrast to the single "channel" events which can be observed electrochemically. However this is offset by: the general sensitivity of the fluorescence experiment; the potential of selective binding events to cause concerted phase transitions leading to large changes in optical properties of membranes; reduction of spontaneous noise in the system; the superfluous sensitivity of the electrochemical experiment in terms of the ambient concentration of analytes of common clinical, industrial and environmental interest.

Lipid membrane perturbations by products of enzyme reactions or antibody-antigen-complement aggregation interactions have been previously reported, and represent processes which modify membrane structure by the introduction of new chemical species to the membrane environment(10). A more subtle perturbation and therefore a more rigorous test of the ability of lipid matrices to act as generic transducers can be derived from molecular receptors that function to gate transmembrane ion conductivity. Selective interactions resulting in ion gating are common for neurotransmitter receptors, and apparently involve the opening of conductive "pores" or "channels" in the interior of large protein assemblies when binding of agonist occurs(11). Little perturbation of the lipid environment is expected during gating events from the seemingly minor structural and electrostatic changes associated with "channel" opening, if the electrochemical conditions are established such that no net transport of ions occurs. Large scale conformational changes of such proteins have not been observed, however such molecular receptors can have specific physical associations with lipids, and may be involved in aggregative events that are related to selective binding of ligands(3).

Acetylcholine receptor (AChR) is the best characterized neurotransmitter receptor with respect to both structure and function. The protein is a dimer in a natural environment, where each monomer consists of five discrete protein subunits which together arrange to form a cylinder capable of ion conduction through lipid membranes(12). It operates as a classical ion gating system, being stimulated by the natural agonist acetylcholine and by many other similar compounds(13).

Fluorescence has been used to investigate receptor function through the use of fluorescent analogues of acetylcholine(14), by monitoring the intrinsic fluorescence of the protein(15) and by first attaching a fluorescent molecule (fluorophore) to the receptor and monitoring the subsequent fluorescence(16). The general aim of these investigations has been the study of binding kinetics or equilibria between the AChR and various agonists and antagonists. Fluorescence was used since conventional methods for studying receptor function, such as radio-labeled ligand and ion influx studies, are unable to indicate response in seconds or milliseconds, such as is desirable in a stopped-flow experiment.

Recent investigations of the AChR, some involving fluorescence studies, have focused on the interactions of the protein with the surrounding lipid matrix and the effects of the lipid components on the activity of the protein(17,18). Apparently, some lipid components are tightly associated with the surface of the protein that is exposed to the hydrocarbon region of a supportive bilayer lipid membrane (BLM)(19). This indicates that effects of selective binding processes along the length of the 11 nm cylinder may be transmitted to the lipid matrix, even though the binding site for the agonist is located 5 nm beyond the surface of the supportive lipid matrix.

Fluorescence experiments not using fluorescent agonist or antagonist molecules have exclusively used fluorescence of the protein itself, whether from labeled or intrinsic moieties. Changes in fluorescence have been attributed to complexation induced conformational changes in the protein which affect the microenvironment of the fluorophore, though the exact nature of these changes is not known. Studies of the role of the supporting lipid are generally based on lipids modified to act as quenchers of the AChR fluorescence, where changes in fluorescence are attributed to conformational changes which alter the accessibility of the protein molecule to the quenching lipid components. Few studies have discussed either the analytical potential of the signal or the labeling of the lipid component to obtain the signal, each of which is significant to the development of a generic sensing system.

This report deals with our investigation of the analytical response by acetylcholine receptor to agonist and antagonist species. Transduction of selective binding events is monitored by the fluorescence response derived from perturbations of the environment of a fluorophore which was either directly labeled onto the AChR system, or was previously labeled only to lipid components. The direct labeling used the probe 4-[N-(iodoacetoxy)ethyl-N-methyl]amino-7-nitrobenz-2-oxa-1,3-diazole (IANBD). The fluorescent lipid component used was N-(-7-nitrobenz-2-oxa-1,3-diazole-4-yl)-N-phosphatidyl ethanolamine (NBD-PE). In both cases, the fluorophore and receptor protein were located in bilayer lipid membranes of a vesicular structure.

Materials and Methods

AChR was obtained from Torpedo Californica and purified according to standard methods(20). Reconstitution was into soybean lecithin (Avanti Polar Lipids, Birmingham, AL, USA) in 10 mM sodium phosphate at pH 7.4, which was then stored as aliquots in liquid nitrogen. Activity of the stock receptor solution was determined by incubation with ^{125}I α-bungarotoxin (α-btx)(ICN, Irvine, CA, USA) to be 1.3 μM in toxin sites(21). Protein concentration was determined to be 0.3 $mg.ml^{-1}$ as described elsewhere(22), yielding a specific activity of 4.3 $nmoles.mg^{-1}$ in toxin sites. All chemicals were reagent grade and used without further purification. Water was obtained from a Milli-Q filtration system (Millipore, Mississauga, ON, CANADA) with a specific resistivity of 18 $Mohm.cm^{-1}$.

IANBD labeling of the AChR was done according to previous methods(23). IANBD (Molecular Probes, Eugene, OR) was added in excess as a solid and the solution was stirred on ice and in the dark for two hours. Controls of soybean lecithin and soybean lecithin/bovine serum albumin (BSA) were prepared identically to the AChR sample. Dry lecithin was suspended as vesicles by evaporation from chloroform followed by addition of PBS and sonication for 2 hours with a probe-tip sonicator.

A solution of 1 $mg.ml^{-1}$ NBD-PE (Avanti) in HEPES buffer (10 mM HEPES, 35 mM sodium nitrate, pH 7.4) was prepared in vesicle form. 60 ul of this was added to 100 ul of stock AChR solution and sonicated for different periods of time, yielding a final system of vesicles approximately 2 mol% NBD-PE in lipid component, assuming a mean molecular weight of 750 $g.mol^{-1}$ for the soybean lecithin.

A second receptor solution of 2 mol% NBD-PE was prepared by addition of the NBD-PE during the reconstitution stage of the AChR purification. The appropriate amount of NBD-PE was added to affinity-purified receptor in 2% sodium cholate detergent before dialysis to yield vesicular structures. The sample was centrifuged at 9500 rpm to remove residual NBD-PE solid.

"Poisoning" of the receptor was accomplished by pre-incubation with a minimum three-fold excess of α-BTX for at least 2 hrs. over ice.

Fluorescence spectra were obtained using an in-house assembled fluorescence spectrometer consisting of: an atmospheric pressure nitrogen laser (Model LN 103, PRA, London, ON, CANADA) for excitation at 337.1 nm; a Bentham M300 monochromator with SMD

3B stepper motor controller (Optikon, Waterloo, ON, CANADA); a Hammamatsu R928 photomultiplier tube (PMT) (Hammamatsu, Bridgewater, N.J.) in a linearity-optimized housing (PRA) operated at -1250 V.; and a SR 250 Gated integrator/Boxcar averager (SRS, Stanford, CA) operated by an IBM PC through a SR 245 interface. A short length of optical cable placed in the laser output carried a small portion of the excitation pulse directly to a second photomultiplier tube for triggering of the gated integrator.

Signal collection was by integrating the PMT output for 30 nsec triggered at the beginning of emission. Ten pulses were averaged for each point.

Results and Discussion.

IANBD Labeled AChR System. The IANBD labeling of the AChR vesicle system was first done in an attempt to reproduce the fluorescence enhancement experiment of Dunn and Raftery(23). One system was obtained by labeling of crude receptor preparation, and produced the spectrum seen in Figure 1, curve A. The "crude" preparation contained protein-rich membranes, isolated from a bulk mixture by centrifugation. The non-specific nature of the labeling procedure was likely to have resulted in a mixture of reactions with nucleophiles on AChR, other proteins and lipid membrane components, as well as with the surrounding water.

Enhancement was measured as the increase in peak area observed on addition of the agonist carbamylcholine (carb), which is commonly used to generate an ion-gated response from AChR. The enhancement versus concentration of agonist was modeled by:

$$F([carb]) = \frac{F(max) \times [carb]}{K_d + [carb]} \qquad [1]$$

where F([carb]) is the measured enhancement, F(max) is the limit of enhancement, [carb] is the concentration of the agonist and K_d is the dissociation constant of the complex. Peak areas were obtained from the spectra either by integration of the entire peak (500nm to 650nm), by integration of the shorter wavelength component (500nm to 550nm) or by calculation of the ratio of the areas of the shorter and longer wavelength components (500nm to 550nm / 550nm to 650nm). Typical enhancement of these values plotted versus concentration of agonist is shown in Figure 2. The fitted curve is presented with the observed data, where K_d and F(max) were calculated using a least squares iteration. Reproducibility between two trials using the same receptor preparation is illustrated in Table I. Observation of the spectra before and after enhancement showed that the intensity increase (F(max)) was greatest on the shorter wavelength portion of the spectrum, corresponding to emission from components in a relatively non-polar environment. The existence of those fluorescent components which were not affected by the binding event provided a convenient method for internal compensation for other variables such as intensity of the source, sensitivity of

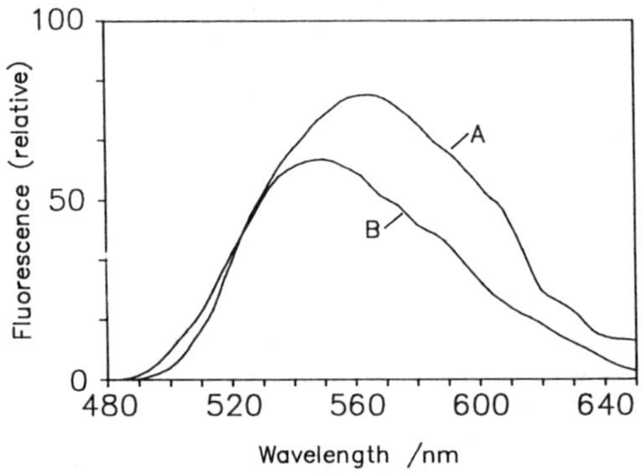

Figure 1. Fluorescence spectrum of IANBD labeled AChR. A - unpurified receptor preparation, showing component on long-wavelength side. B - purified AChR preparation. NBD-PE/AChR system gives identical spectrum to B.

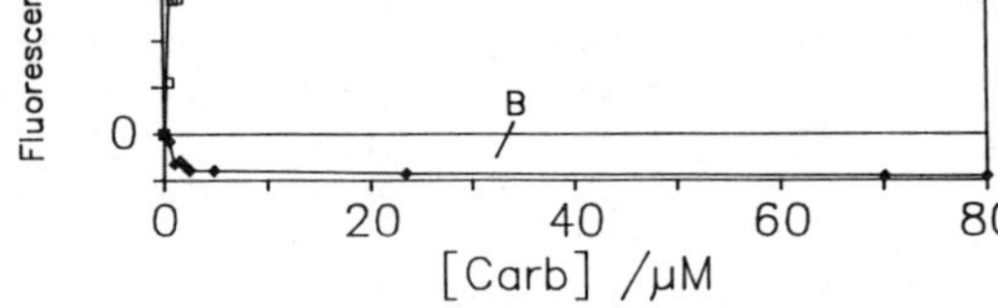

Figure 2. Fluorescence enhancement of purified IANBD labeled AChR with addition of carbamylcholine. A - enhancement of peak area from 500 nm to 650 nm. Fitted to K_d=1.62μM, F(max)=218. B - control experiment, AChR pre-incubation with α-btx.

the detector and variations in signal associated with the inhomogeneity of vesicular preparations. The enhancement of the ratio values yielded much greater reproducibility between trials.

Table I. Comparison between two trials of K_d and F(max) parameters measured in the fluorescence enhancement titration of IANBD labeled "crude" AChR preparation with carbamylcholine between two trials

Parameter		Trial 1	Trial 2
Peak area 500 nm-550 nm	K_d F(max)	5.15 52.5	13.5 79.6
Peak area 500 nm-650 nm	K_d F(max)	4.98 12.5	23.5 44.5
Ratio areas 500 nm-550 nm/550 nm-650 nm	K_d F(max)	8.15 37.7	12.9 35.1

IANBD labeling of purified, reconstituted receptor preparation produced the spectrum in Figure 1, curve B. This narrower spectrum corresponded to the shorter wavelength portion of the IANBD-labeled 'crude preparation', suggesting that the longer wavelength components present in the unpurified system were removed in the purification process. There was uniform enhancement of the entire spectrum of the purified system on agonist addition (see Figure 2), and there was no change observed in the ratio measurement used above. While this system was better purified and therefore better characterized in terms of composition, poor reproducibility in percent enhancement between trials indicated that it may be desirable for fluorescent species which are not perturbed by the complexation event to be incorporated to facilitate internal compensation of the signal.

Selective receptor mediation of the enhancement process was demonstrated by pre-incubation of the receptor system with α-bungarotoxin (α-btx). This is a large protein which is a known selective antagonist for AChR. It is presumed to bind irreversibly with the receptor on or near the agonist binding site, prohibiting agonist binding. The expected result, a negation of the agonist response, was observed with the IANBD-AChR system (Figure 2, curve B).

In addition, it was observed that the fluorescence signal was enhanced by the initial addition of α-btx at very low concentratrations (40% enhancement at 10 nM, data not shown). This indicated an antagonist-associated conformational change which may or may not be similar to the change observed for agonist addition, but which may also permit antagonist detection.

The enhancements observed in our experiments resembled those of Dunn et al. with some exceptions. The previous researchers worked in the high-µM to mM agonist concentration region, and attributed the binding phenomenon to a 'low affinity' site on the receptor with a dissociation constant (K_d) of approximately 1 mM. We observed the same magnitude enhancement while working at concentrations in the low µM regime, indicating high affinity binding with K_d approximately 1 µM. The tendency in the literature is to explain such discrepancies by invoking multiple receptor states, binding sites and complexation pathways to account for changing affinities and binding rate constants. Using this reasoning, it is possible that the receptor in our system was in a high-affinity, desensitized state(16), which also precludes any direct association of the fluorescence response with pharmacological 'channel opening' behaviour. This was supported by the inability to open ion channels for a Na^+ efflux assay with the same AChR preparation and agonist. It has also been noted that measured binding constants vary with lipid composition in reconstituted systems. Differences in the lipid preparations may therefore explain some of the discrepancy between results.

Incorporation of NBD-PE. The desirability of a generic strategy for transduction of receptor complexation and recent investigations of the role of membrane lipid components in the function of AChR led us to propose that a fluorescent lipid component could participate in the fluorescence enhancement mechanism. It seemed reasonable to suspect that some of the membrane lipids could have been labeled in the IANBD experiment, especially considering the presence of nucleophilic phosphate groups in the headgroup region, though the relative amount of lipid to protein labeling was not determined.

To investigate the response of a system containing only fluorescent lipid components, preparations were made by vesicle fusion between unlabeled receptor systems and vesicles of the lipid fluorophore NBD-PE. The resultant emission spectrum was identical to the purified system labeled with IANBD. Fluorescence enhancement was observed on agonist addition, as shown in Figure 3. Dissociation constants were similar to those measured in the IANBD labeling case and a reasonable fit was obtained using equation 1. Enhancement was also seen on addition of low levels of the α-btx.

Selectivity was demonstrated in this system by the elimination of the agonist response after pre-incubation with α-BTX. To further confirm receptor mediation of the enhancement, controls were performed using pure lipid vesicles, vesicles with other protein (bovine serum albumin) and vesicles containing degraded (by extensive sonication) AChR. All of these controls yielded no fluorescence enhancement.

Significant difficulties were encountered in this system with regards to reproducibility and stability of the signal with time. It was suspected that some of the variability was due to incomplete exchange between the NBD-PE vesicles and the AChR vesicles prior to beginning the agonist addition, resulting in fluorescence changes due to subsequent lipid exchange. Sonication of the sample in a sonic bath for 20 minutes eliminated the gradual signal increase, presumably by homogenization of the lipid system, but caused extensive degradation of the receptor (e.g. from 11 nM to 530 pM activity in α-btx binding sites).

Some inconsistency was encountered in the pre-incubation with α-btx. While the control result indicated in Figure 4, curve B was repeatable, on occasion enhancement occurred after the α-btx treatment. This enhancement was likely due to further lipid exchange processes in the inhomogeneous vesicle system.

To circumvent some of the difficulties of the vesicle fusion technique, flourescence incorporation was recently accomplished by addition of fluorescent lipid to stock lipid before reconstitution of AChR from detergent phase into vesicles. This yielded a more stable signal and a consistent control with α-btx pre-incubation. Thus far, the enhancement observed with this sample has been less than with previous systems, and it has yielded a more poorly described curve in terms of equation 1 (see Figure 4).

While the mechanism of the fluorescence response is yet to be identified, in all cases it provides some insight into the nature of membrane-receptor interactions which may play a part in receptor function. The NBD-PE probe has been used to study phase domain structure in membrane systems by classical spectrofluorometric measurements such as enhancement/self-quenching(24) and by the relatively new technique of fluorescence microscopic imaging of lipid monolayers at an air-water interface(4). Both techniques are possible as a result of the natural tendency of the NBD-PE molecules to partition between phases in a mixed phase system such that they concentrate in the less ordered domains. In the case of microscopy, this allows structures in monolayers to be visualized as bright and dark regions, provided such structures are of the order of um in dimension.

When in locally concentrated regions, the NBD moieties will "self-quench", causing a reduction in the overall measured intensity. Any changes in the relative areas of phase structure types will thus result in a change in the overall intensity. Such changes may be proposed for the system under study as the NBD-PE probe has been shown to indicate perturbation of lipid phase structure by incorporated proteins(25), and recently in our laboratory by the presence of AChR (unpublished results).

The macromolecular nature of the observed structural perturbations suggests that the protein-membrane interactions may not arise from simple steric interactions, but may propagate over some distance across the membrane surface. This is consistant with electrostatic models of protein-lipid interaction, and may also be attributed to phenomena involving aggregation of protein molecules. A macromolecular perturbation would also seem to be necessary to explain the enhancement observed in this work, as it is difficult to envision a localized perturbation capable of altering the emission of a membrane fluorophore by, in some instances, over 200 percent.

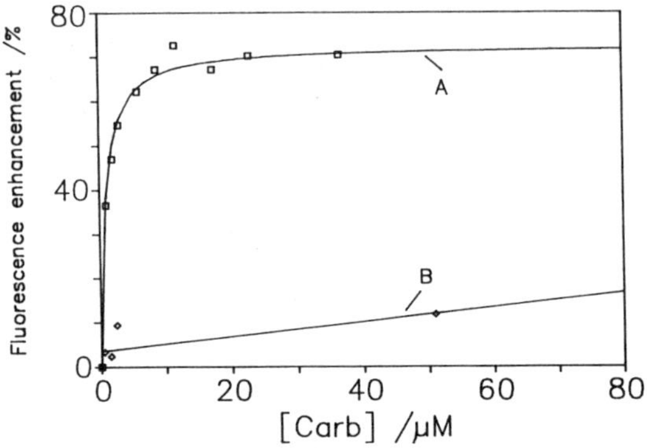

Figure 3. Fluorescence enhancement of purified AChR mixed with vesicles of NBD-PE and sonicated with addition of carbamylcholine. A-enhancement of peak area from 500 nm to 650 nm. Fitted to K_d=.888μM, F(max)=72.5. B - control experiment, AChR pre-incubation with α-btx.

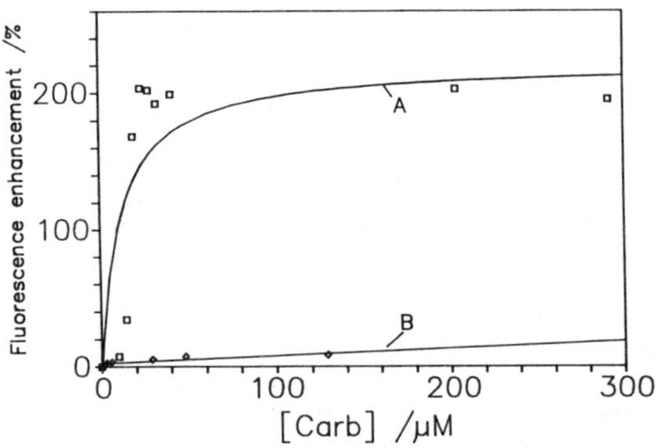

Figure 4. Fluorescence enhancement of purified AChR reconstituted into lipid containing 2 mol% NBD-PE with addition of carbamylcholine. A - enhancement of peak area from 500 nm to 650 nm. Fitted to K_d=11.8μM, F(max)=221. B - control experiment, AChR pre-incubation with α-btx.

Summary

A generic scheme for transduction of membrane protein binding events has been demonstrated based on the ability of fluorescent membrane components to respond to complexation of AChR. While reproducibility between trials is poor, quantitation should be possible based on the ability to fit the response to the expected binding parameters, and may involve internal calibration using the multivariate nature of the fluorescence signal. Further investigations of fluorescence lifetime or polarization may provide additional means of signal generation and calibration without a need for the additional fluorescent components used in a wavelength-based calibration.

A simplified approach to quantitation used a linear approximation of the first few data points in Figure 2. The linear fit (see Figure 5) yielded a sensitivity of 43%·μM^{-1} and a limit of detection of 300 nM over a dynamic range of 3 µM with a correlation coefficient of 0.978.

Future Work

Development of a sensor for neurologically active compounds using the vesicle system characterized here necessitates stabilization of a functional receptor system onto a solid substrate. Possible methods for this stabilization are direct covalent attachment via the protein or the lipid components, fusion of vesicular structures with the substrate(26), or incorporation of the protein into a monolayer at an air/water interface and transfer of the monolayer onto the substrate by Langmuir-Blodgett deposition techniques(27).

The surface deposition must provide a stable system which will not quickly desorb or degrade under a variety of working conditions, e.g. in-vivo. It must also provide an environment in which the receptor will be able to function, both in terms of binding to the analyte and transduction of the binding event into a change in fluorescence. These two conditions are often directly opposed, as many of the models of signal transduction require mobility of the matrix containing the protein molecule whereas the most stable deposited structures are immobile. Preventing loss of biological activity during stabilization may favour certain approaches such as: in the covalent attachment case, attachment of only a portion of the lipid component may result in sufficient residual mobility of unbound lipid and protein for signal generation to occur; in the monolayer deposition case, deposition onto modified surfaces or multilayer assemblies(28) may allow for sufficient mobility in the layers containing the protein for transduction of the binding event. In all cases, the orientation of the protein is critical, which may necessitate some directional control of the protein during any surface stabilization procedure.

Since the signal generated in this work does not depend on the complete biological function of the protein, (i.e. ion channels need not be formed) a further consideration is that fragments of the receptor or artificial analogues of the receptor

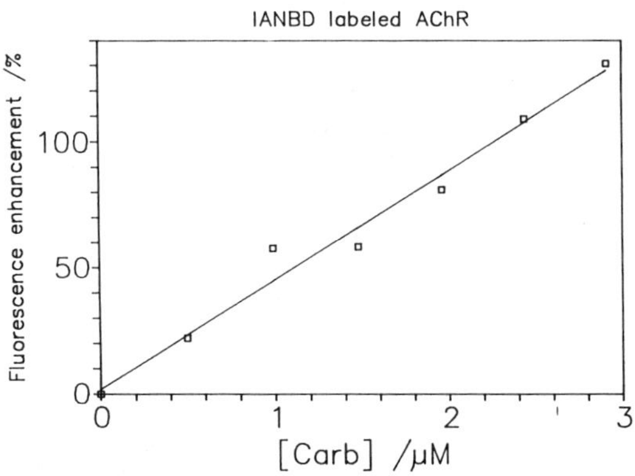

Figure 5. Linear approximation of fluorescence enhancement for first 3 μM addition of carbamylcholine from Figure 2. Linear regression gives slope = 43.4±2.9 and constant 1.9±7.4 with correlation coefficient 0.978.

may be capable of transducing similar binding processes into a fluorescence signal. This may circumvent some of the difficulties of working with the actual AChR, and illustrates the advantage of the generic approach, where the membrane carries the fluorescent component, in being able to function with a variety of receptors.

Pending stabilization of a receptor system onto a solid substrate, sensor configuration is likely to copy that previously established for intrinsic mode fiber-optic sensors(29). The specific configuration would probably involve modification of the surface of the fibre, as in fluoroimmuno optical sensors(30), and using the "evanescent" model to interact optically with excitation radiation being propagated inside the fibre. The resultant fluorescence would then be returned by the same fiber to a detector.

Acknowledgments

The authors would like to acknowledge the Defense Research Establishment of Ottawa, the Natural Science and Engineering Research Council (Canada), the Ontario Ministry of the Environment, and Imperial Oil Canada for sponsorship of this work.

References

1. U.J. Krull and M. Thompson, IEEE Trans. Electron. Dev., 1985,32,1180.
2. M. Thompson and W.H. Dorn, in Chemical sensors, T.E. Edmonds, Ed., Blackie and Sons, Glasgow, 1987.
3. J.L. Tedesco, U.J. Krull and M. Thompson, "Molecular Receptors and Biosensors", Biosensors, in press.
4. W.M. Heckl, Laterale Organisation von Lipidmonoschichten bei Einbau von amphiphilen Fremdstoffen und Proteinen, Ph.D. Thesis, Department of Physics, Technical University of Munich, 1988 (and subsequent publications).
5. P.P.C. Graziadei, in The Ultrastructure of Sensory Organs, I. Friedman, Ed., Elsevier, New York, 1984, Ch. 4.
6. M. Thompson, W.H. Dorn, U.J. Krull, J.S. Tauskela, E.T. Vandenberg and H.E. Wong, Anal. Chim. Acta, 1986, 180, 251.
7. J. Yguerabide and M.C. Foster, in Molecular Biology, Biochemistry and Biophysics: Membrane Spectroscopy, E. Grell, Ed., Springer-Verlag, New York, Vol. 31, p. 199.
8. Spectroscopic Membrane Probes, L.M. Loew, Ed., CRC Press, Boca Raton, 1988.
9. U.J. Krull, R.S. Brown, R.F. DeBono and B.D. Hougham, Talanta, 1988, 35, 129.
10. M. Thompson and U.J. Krull, Anal. Chim. Acta, 1983, 147, 1.
11. B. Hille, Ionic Channels of Excitable Membranes, Sinauer Associates, Sunderland, Massachusetts, 1984.
12. J.-P. Changeux, A. Devillers-Thiery and P. Chemouilli, Science, 1984, 225, 1335.
13. F.B. Abramson, R.B. Barlow, M.G. Mustafa and R.P. Stephenson, Br. J. Pharmac., 1969, 37, 207.

14. H.-W. Meyers, R. Jurss, H.R. Brenner, G. Fels, H. Prinz, H. Watzke and A. Maelicke, Eur. J. Biochem., 1983, 137, 399.
15. T. Endo, M. Nakanishi, S. Furukawa, F.J. Joubert, N. Tamiya and K. Hayashi, Biochem., 1986, 25, 395.
16. S.M.J. Dunn, S.G. Blanchard and M.A. Raftery, Biochem., 1980, 19, 5645.
17. T.M. Fong and M.G. McNamee, Biochem., 1986, 25, 830.
18. J.R. Zabrecky and M.A. Raftery, J. Receptor Res., 1985, 5, 397.
19. J.M. Gonzales-Ros, M.C. Farach and M. Martinez-Carrion, Biochem., 1983, 22, 3807.
20. J. Lindstrom, R. Anholt, B. Einarson, A. Engel, M. Osame and M. Montal, J. Biolog. Chem., 1980, 255, 8340.
21. R.A. Kohanski, J.P. Andrews, P. Wins, M.E. Eldefrawi and G.P. Hess, Anal. Biochem., 1977, 80, 531.
22. M.A.K. Markwell, S.M. Haas, N.E. Tolbert and L.L. Bieber, Meth. Enzym., 1981, 72, 296.
23. S.M.J. Dunn and M.A. Raftery, Proc. Natl. Acad. Sci. USA., 1982, 79, 6757.
24. A.E. McGrath, C.G. Morgan and G.K. Radda, Biochim. Biophys. Acta, 1976, 426, 173.
25. W.M. Heckl, M. Losche, H. Scheer and H. Mohwald, Biochim. Biophys. Acta, 1985, 810, 73.
26. T.H. Watts, H.E. Gaub and H.M. McConnell, Nature, 1986, 320, 179.
27. H. Kuhn, Thin Sol. Films, 1983, 99, 1.
28. V. von Tscharner and H.M. McConnell, Biophys. J., 1981, 36, 421.
29. U.J. Krull and R.S. Brown, in _Laser Remote Chemical Analysis_, R.M. Measures, Ed., Wiley-Interscience, New York, 1988, p. 505.
30. K. Newby, W.M. Reichert, J.D. Andrade and R.E. Brenner, Appl. Opt., 1984, 23, 1812.

RECEIVED March 21, 1989

Chapter 23

New Developments and Applications of Fiber-Optic Sensors

S. M. Angel, M. N. Ridley, K. Langry, T. J. Kulp, and M. L. Myrick

Environmental Sciences Division, Lawrence Livermore National Laboratory, Livermore, CA 94550

Optrodes that provide trace-level detection of trichloroethylene and chloroform with high accuracy have been developed. High accuracy is obtained by providing an internal intensity reference. Also, a diode laser has been used to measure surface-enhanced Raman spectra (SERS) of low levels of fluorescent molecules. Near-infrared SERS is shown to be a useful technique for remote spectroscopic measurements of aqueous contaminants. In addition, a temperature optrode has been used to obtain real-time temperature profiles in a geothermal well.

Fiber-optic sensors (optrodes) are becoming increasingly popular because of their potential to measure samples remotely and in situ. Their use has become sufficiently widespread that several literature reviews have recently appeared (1-3). Fiber-optic sensors are small, immune to electrical and radio frequency noise, and permit measurements to be made using remotely located instrumentation that minimally disturbs the sampling region. Fiber-optic sensors are basically of two types: those that rely on a change in the physical characteristics of the optical fiber itself, and those that use the optical fiber simply as a light pipe to connect a spectrometer to a remotely located optical sensor.

Several areas of fiber-optic sensor research are being pursued by the Environmental Sciences Division at Lawrence Livermore National Laboratory (LLNL). These include the development of new chemically specific optrodes (4, 5) and the development of new remote spectroscopic techniques (6). Also, we are investigating some applications of these devices both inside and outside the laboratory. New laboratory applications that are being investigated include monitoring anaerobic microbial degradation (manuscript in preparation) and measuring transport properties of groundwater contaminants in porous soil systems (7). Field applications that have been recently demonstrated include measuring organochlorides in contaminated aquifers (8), measuring chlorophyll

0097–6156/89/0403–0345$06.00/0

fluorescence from marine algae in situ using a ship-towed fiber-optic probe (Cowles, T.J.; Moum, J.N.; Desiderio, R.; Angel, S.M. Appl. Optics, in press), and measuring temperature profiles in a geothermal well (Angel, S.M.; Garvis, D.G.; Sharma, S.K.; Seki, A. Appl. Spectrosc., in press). This report describes recent progress in some of these areas of research.

New Chemically Specific Optrodes

Dual-Wavelength Absorption Optrode for Trichloroethylene and Chloroform

Recently, a new type of trichloroethylene (TCE) and chloroform ($CHCl_3$) specific optrode was developed for the purpose of measuring groundwater contaminants. This device shows dramatically improved performance over previously reported fluorescence-based sensors (8). The new optrode measurement is based on the absorption of light by a colorimetric indicator that is held in a small-diameter capillary tube (see Figure 1). Two 320-µm-core optical fibers are sealed in a 2-mm-diameter glass capillary tube that holds about 10 µL of the optrode reagent. The open end of the capillary tube is sealed with a white Teflon membrane. The analyte vapor diffuses rapidly through the Teflon membrane and reacts with the optrode reagent, producing a highly colored product that absorbs at 530 nm. One optical fiber delivers a very-low-intensity probe light to the optrode. The light passes through the reagent and is attenuated at 530 nm by the colored product that develops. The light then reflects from the Teflon membrane and passes through the reagent a second time, providing further attenuation (and higher sensitivity). The attenuated light is then collected by the second optical fiber and is delivered to a filter fluorimeter for analysis. An internal intensity reference is provided by measuring light at a wavelength that is transmitted through the optrode without attenuation (610 nm). The ratio of the two signals (transmission–T) is proportional to the analyte concentration and is independent of the light intensity. The absorbance (A), -log(T), is linearly proportional to the analyte concentration for low A values, assuming that the product accumulates in the optrode linearly with time.

The optrode reagents are different for TCE and $CHCl_3$. For $CHCl_3$, the reagent is a mixture of 80% pyridine (by volume), and 20% tetrabutylammonium hydroxide–TBAH (40% aqueous solution). For TCE, the reagent is 60% by volume of the above mixture and 40% tetraethyleneglycol dimethylether (TTEG). In the presence of strong base (TBAH), both analytes form transient species that react with the pyridine to form colored products. The amount of color that develops is proportional to the vapor-phase concentration of the analyte. Both reagents are very stable and can be used for several weeks without loss of response.

The fiber-optic spectrometer used for this work is shown in Figure 2. The illumination block consists of a 5-watt quartz-halogen lamp whose output is chopped at 30 Hz and imaged onto the excitation optical fiber. The UV wavelengths are removed from the excitation light with a long-wavelength pass

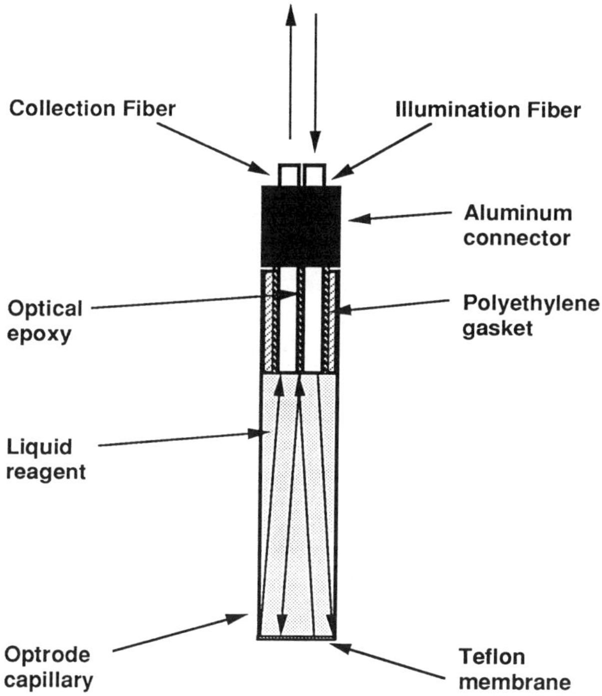

Figure 1: Dual-wavelength absorption optrode for trichloroethylene (TCE) and chloroform ($CHCl_3$).

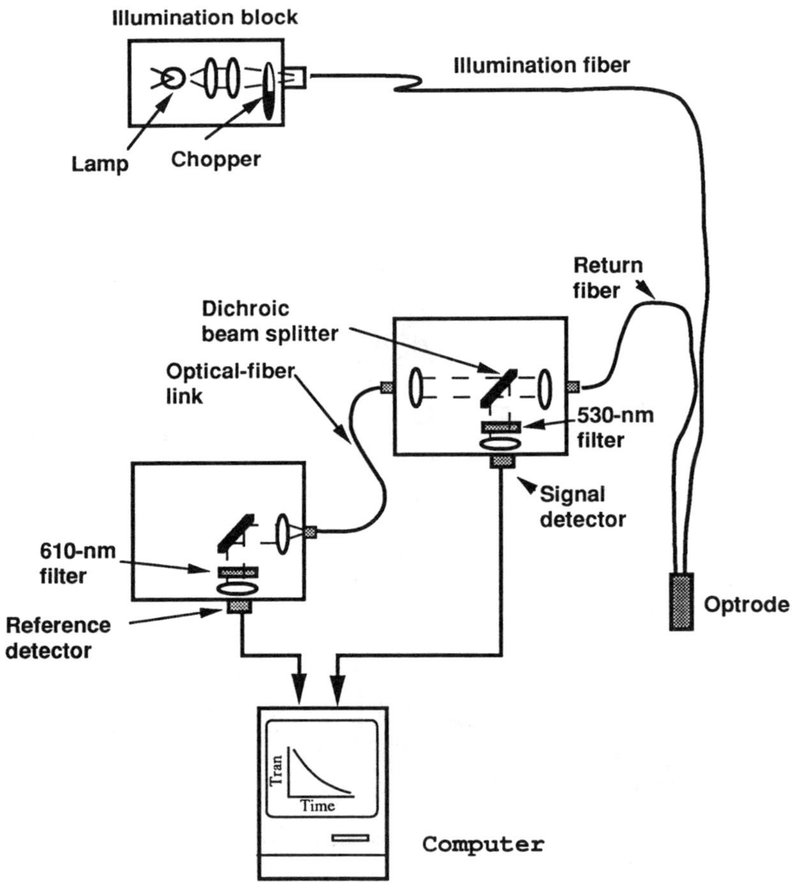

Figure 2: Instrumental setup used to measure the dual-wavelength absorption optrode.

filter. The light that returns from the optrode is analyzed at 530 nm and 610 nm using two single-channel fiber-optic fluorimeters (Douglas Instruments Model II Fiber-Optic Fluorimeter) whose detector electronics are synchronized to the lamp-modulation frequency. A dichroic beam splitter in the first instrument separates the incoming optrode signal into a red-wavelength component (wavelengths longer than 550 nm are transmitted) and into a green-wavelength component (wavelengths shorter than 550 nm are reflected). The reflected component is then passed through a 530-nm band pass filter (~30-nm band width) and is detected by a silicon photodiode. The component that is transmitted through the dichroic beamsplitter is imaged onto a short (2-m) 200-μm core optical fiber and routed to the input of the second instrument. This light is then passed through a 610-nm band pass filter (~25-nm band width) and is detected by a second silicon photodiode. The optrode response is taken as the ratio of the 530-nm and 610-nm signals. The normalized ratio is equivalent to transmission (T). The decrease of T is monitored as a function of time. The initial value (time=0) of T is normalized to 1.

In the laboratory experiments, all measurements were made using a 160-ft-long optical fiber cable. About 100 mL of analyte was used in a 250-mL Erlenmeyer flask. The optrode was held about 1/2 inch above the analyte solution by a Teflon stopper that sealed the flask during the measurement. The contents of the stoppered flask were stirred 5 min prior to and during the optrode measurement. It was found that changes in the stir rate and changes in the height of the optrode above the solution significantly affected the optrode response. Therefore, care was taken to reproduce stirring conditions and optrode position.

Figure 3 shows the response (T) of the $CHCl_3$ optrode plotted versus exposure time for aqueous concentrations of $CHCl_3$ that range from 10 ppb to 5 ppm. (All concentrations referred to in the text are the original aqueous-phase concentrations.) The response at 10 ppb is much larger than the noise level, indicating that lower $CHCl_3$ concentrations can be measured. The vapor-phase $CHCl_3$ concentrations for these solutions were about 10% of the original solution concentrations. The amount of $CHCl_3$ in the vapor phase was independently measured using ^{14}C-labeled $CHCl_3$, and also by GC. The vapor-phase concentrations differed slightly from the expected Henry's law equilibrium values.

A plot of -log(T) versus $CHCl_3$ concentration produces a linear calibration curve (see Figure 4) for concentrations below 1 ppm. However, at higher concentrations, the response is not linear because the reference wavelength is attenuated by the intensely colored product that develops in the optrode. Attenuation of the reference wavelength leads to artificially high values of T and low values of -log(T).

The accuracy of the $CHCl_3$ optrode was determined by making 5 separate measurements of a 500-ppb $CHCl_3$ solution using 5 different optrodes. The sample standard deviation of the optrode responses was taken after 15 min. The average response after 15 min was 89.8 ±6.2%.

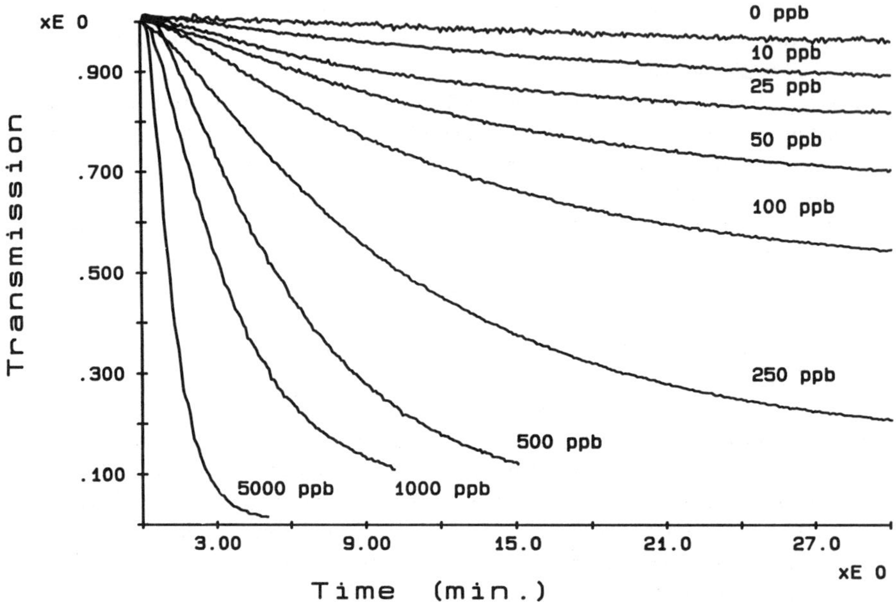

Figure 3. Response of the $CHCl_3$ optrode plotted versus exposure time. The concentrations refer to the original aqueous-phase concentrations.

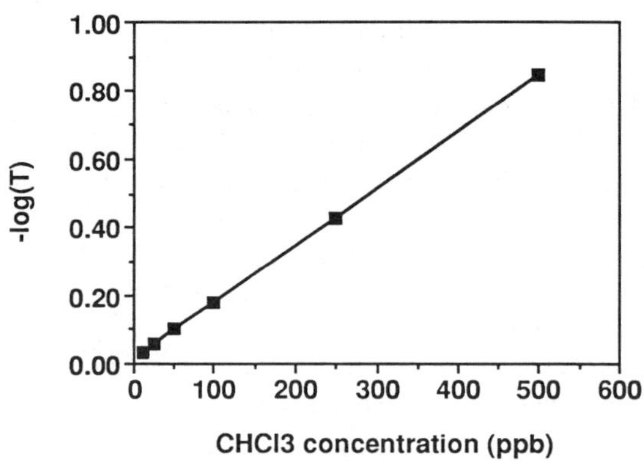

Figure 4. Calibration curve for the $CHCl_3$ optrode.

Figure 5 shows the response (T) of the TCE optrode for aqueous concentrations of TCE that range from 25 ppb to 1000 ppb plotted versus exposure time. The response to the 25-ppb TCE solution is well above the noise level. The vapor-phase concentrations of TCE were about 10% of the original solution concentrations. This was confirmed by ^{14}C-labeling and also by GC analysis of vapor samples. The accuracy of the TCE optrode was determined as indicated above using a 500-ppb TCE solution, except measurements were taken after 30 min. The average response after 30 min was 62.4 ±8.3%.

The response curves for $CHCl_3$ (Figure 3) and TCE (Figure 5) are qualitatively different. The $CHCl_3$ optrode responds very rapidly and the response decreases at an exponential rate (at least for concentrations below 1 ppm). The TCE optrode response is slower and it is nonexponential. We have found that the TCE product that forms in the optrode is not stable and degrades over a period of several minutes. The addition of TTEG to the optrode reagent slows down the rate of product degradation, improving the response to TCE by about a factor of two (while decreasing the response to $CHCl_3$ by the same amount). There is some evidence that part of the degradation is photochemical because the response to TCE also improves by using lower lamp powers.

A plot of -log(T) versus TCE concentration is shown in Figure 6. It is not as linear as the $CHCl_3$ calibration curve because of degradation of the TCE product. However, the calibration curve is reproducible as indicated above. The most linear calibration curves are obtained by using the lowest possible lamp power. However, even at low lamp powers the calibration is still nonlinear above 1 ppm because of attenuation at the reference wavelength by the intensely colored product that develops in the optrode.

Future Organochloride Optrode Research

The performance of the colorimetric organochloride optrode is sufficient for many applications. We are currently evaluating its use in monitoring microbial degradation of selected environmental contaminants. For example, some microbes can degrade carbon tetrachloride to $CHCl_3$ under anaerobic conditions and are being selected for remediation of contaminated aquifers. We are using the $CHCl_3$ optrode to monitor the rate of microbial degradation in laboratory microcosms with the hopes of using this technique to monitor microbial degradation remotely in situ in aquifers in the future.

In late 1989, we will begin evaluating the use of the TCE optrode to measure TCE vapors in contaminated soils at LLNL. Initially, the optrode will be used in combination with other techniques. However, eventually we would like to use the optrode for rapid screening of subsurface contaminants. For this purpose, we are developing a portable instrument that uses light-emitting diodes as the light source. Not only will this make the instrument smaller but it should also improve the precision of our measurements because of the stability of light-emitting diode sources (9).

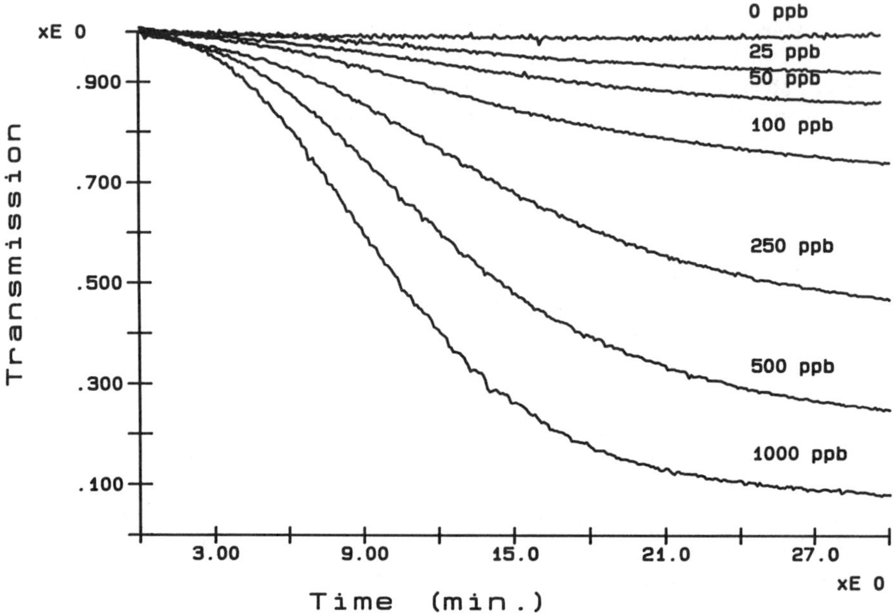

Figure 5. Response of the TCE optrode plotted versus exposure time. The concentrations refer to the original aqueous-phase concentrations.

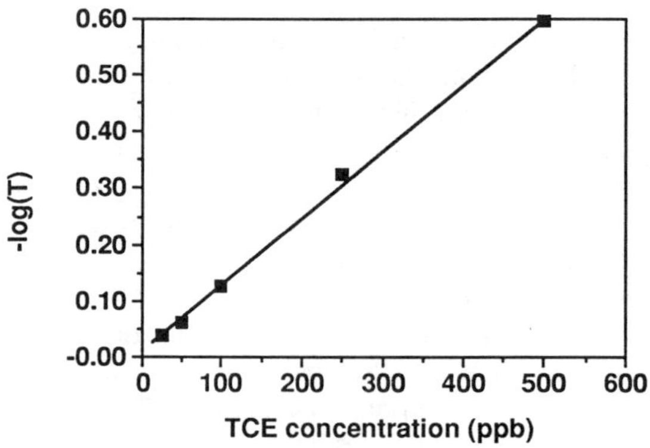

Figure 6. Calibration curve for the TCE optrode.

Remote Spectroscopy

Near-Infrared Surface-Enhanced Raman Spectroscopy

Surface-enhanced Raman spectroscopy (SERS) has been a focus for much study since it was first reported by Fleischmann and coworkers in 1976 (10). Thus far, a considerable amount of this work has been aimed at understanding the physical origin of the phenomenon (10-13), but recently there have been many investigations of potential analytical uses for SERS (14-18). Although it has not yet come into general use, SERS is an attractive analytical technique that can detect certain types of molecules with a high level of sensitivity. Furthermore, because SERS is a vibrational spectroscopy, a SER spectrum contains a lot of molecular information.

With the introduction of Fourier-Transform (FT) Raman instruments (19, 20), near-infrared (NIR) Raman spectroscopy has become an excellent technique for eliminating sample fluorescence and photochemistry in Raman measurements. Recently, the range of NIR Raman techniques was extended to include NIR SERS (6, 21). Most SERS studies to date have been performed using visible excitation sources such as Ar-ion lasers; the demonstration of NIR SERS offers the possibility of using solid-state Nd:YAG and diode lasers.

Potential applications of NIR SERS are numerous. A principal interest in this laboratory has been remote spectroscopy using fiber optics. NIR SERS offers many advantages for this application. In addition to eliminating fluorescence problems from the optical fiber and sampling region, NIR excitation should significantly reduce the intensity of fiber-optic Raman bands due to the λ^{-4} dependency of the Raman signal intensity. Also, the absorption losses of many optical fibers are very low for NIR radiation. Furthermore, diode laser excitation offers the potential for developing portable instruments.

The purpose of our initial SER research was to determine if SER spectra could be obtained using NIR-wavelength excitation and to quantitate the magnitude of the effect relative to visible-wavelength excitation. For this purpose, an FT Raman spectrometer was used with a Nd:YAG (1.064-µm) excitation laser. In initial experiments with pyridine, it was found that Cu and Au electrodes both give very large SER enhancements that are comparable in magnitude to reported enhancements using visible-wavelength excitation (6). Sub-ppm concentrations of pyridine were measured on Cu electrodes. In later work, it was found that Cu and Au colloids also gave very large NIR SER enhancements using 1.064-µm excitation (Angel, S.M.; Katz, L.F.; Archibald, D.D.; Honigs, D.E. Appl. Spectrosc. in press).

FT SER spectra have been obtained of many different types of compounds including some environmental contaminants. Initially, for environmental applications, very simple monosubstituted aromatic compounds are being studied. One of the simplest of these is 3-chloropyridine (CP). This compound gives very intense NIR-SER spectra on Cu colloids and on Cu electrodes (see Figure 7). Plot A of Figure 7 shows the SER spectrum of 1 mM CP on a Cu electrode and Plot B shows the SER spectrum of 0.1 mM CP in a Cu colloid solution. Both methods give very large SER enhancements. The small peaks shown in Plot B of Figure 7 are probably overtone and combination bands. The CP SER enhancement in Cu colloid solution is about 10^6 and the sensitivity is sufficient to measure sub-ppm concentrations.

FT SER spectra have also been measured for mixtures of CP and picoline with surprising results (Angel, S.M.; Archibald, D.D. Appl. Spectrosc. in press). Plot A of Figure 8 shows the SER spectrum of 1 mM CP on a Cu electrode at a potential of -0.6 V versus the saturated calomel electrode (SCE). The most intense SER spectra were obtained for this compound at this potential. The SER spectrum of 1 mM picoline (Plot C of Figure 8) also shows the greatest intensity at -0.6 V versus SCE. SER spectra obtained for a mixture containing 0.5 mM of each of these compounds is potential dependent (Plot B of Figure 8). Furthermore, the optimal potential for observing both compounds is about -1.0 V. SER spectra measured at potentials near 0 V show mainly picoline bands. At more negative potentials, CP bands appear. These types of measurements are important because contaminated ground water usually contains mixtures of related compounds. Therefore, it is important to determine what factors effect the SER spectra of mixtures.

NIR SERS Using a Diode Laser

Semiconductor lasers are ideal excitation sources, and they have already been demonstrated to be remarkably versatile and useful in a number of spectroscopic applications, as indicated by Ishibashi and coworkers (22, 23). These solid-state devices are inexpensive, small, easy to use, long-lived, and require little maintenance. The problem with semiconductor lasers is that powerful single-mode lasers are only available at NIR wavelengths, although this is likely to change in the near future. Until this work, however, no report had been made of the use of diode lasers for SERS. Recently, we demonstrated NIR SERS with a diode-laser excitation source and investigated the characteristics of the technique (Angel, S.M.; Myrick, M.L. unpublished data).

The 785-nm line of a GaAlAs diode laser was used for this study (D.O. Industries, Model GALA-078-16-8). A very strong broad-band emission from the diode laser was removed using an 830-nm band-pass interference filter that was tilted at a large angle to allow maximum transmission of the 785-nm laser line. This filter attenuated the output power of the diode laser from 10 mW to 4.3 mW. The spectrometer used for these studies consisted of an f/4 double monochromator (SPEX Model 1680B) with 1200 grooves/mm holographic gratings and an f/1 collection lens. The detection system was a GaAs photomultiplier (RCA Model 31034) operated at -1800 V with a photon counting system (E.G.&G Model 1121A amplifier discriminator with Model 1112 counting system). All SER spectra were recorded with a two-second integration time and a nominal spectral resolution for a 1000 cm^{-1} Raman band of 10 cm^{-1}.

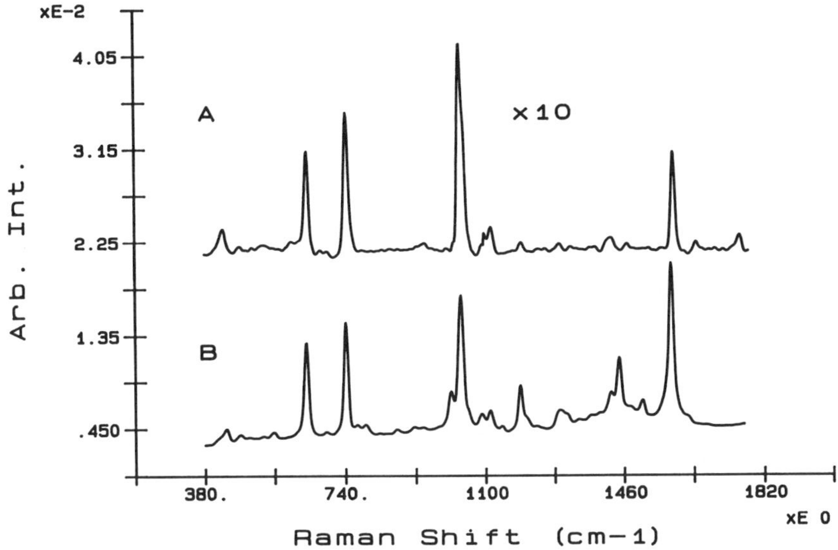

Figure 7. NIR SER spectra of 3-chloropyridine (CP) using 1.064-µm excitation from a Nd:YAG laser: (A) 1 mM CP on a Cu electrode at a potential of -0.6 V versus SCE; (B) 0.1 M CP in a Cu colloid solution.

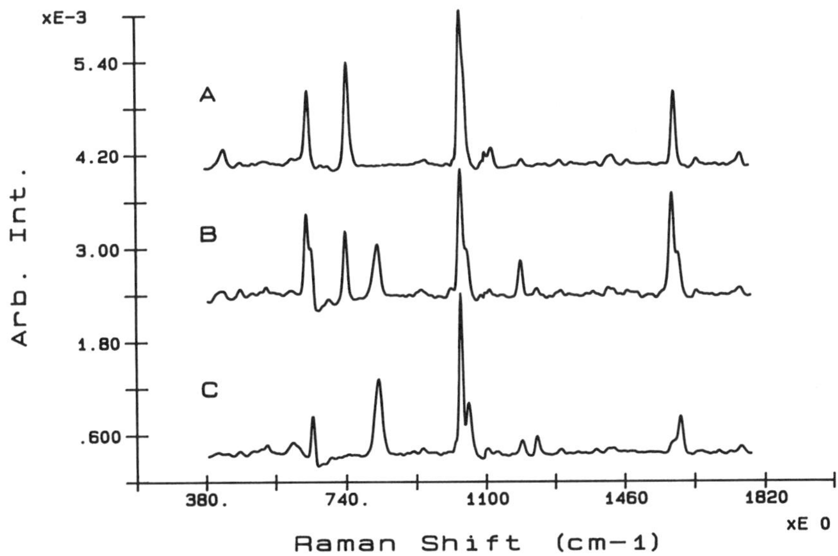

Figure 8. NIR SER spectra of 3-chloropyridine (CP) and picoline on copper electrodes using 1.064-µm excitation from a Nd:YAG laser: (A) 1 mM CP at a potential of -0.6 V versus SCE; (B) a mixture containing 0.5 mM each of CP and picoline at a potential of -0.999 V; (C) 1 mM picoline at a potential of -0.6 V.

Even with the low output power of the diode laser, SER spectra were easily obtained of a variety of different compounds on Ag and Cu electrodes. For 0.1 M pyridine on an Ag electrode, typical signal levels were about 10,000 cps. Figure 9 shows the SER spectrum of 0.1 M pyridine on Cu (Plot A) and Ag (Plot B) electrodes. In each case, the SER enhancement is about 10^5 to 10^6. However, it is somewhat lower on the Cu electrodes than on the Ag electrodes.

The 785-nm line of the diode laser used in these experiments is transmitted very efficiently by optical fibers. Moreover, this wavelength is ideal for measuring SER spectra of highly luminescent compounds that have visible absorption bands. This wavelength is far removed from most electronic absorption bands and thus produces little or no fluorescence. Furthermore, the spectral range of a typical Raman spectrum using 785-nm excitation is within the range of photomultiplier tubes; thus, diode lasers can be used with conventional spectrometers.

The elimination of sample luminescence using a diode laser was demonstrated by measuring SER spectra of tris(2,2'-bipyridine)ruthenium(II), [RB_3], in bulk solution. RB_3 is a highly luminescent compound that has received considerable interest because of its unique excited state properties (24-26). However, because of its intense visible luminescence, it is difficult to obtain luminescence-free Raman spectra even with SER enhancement. Further, visible-wavelength Raman spectra of this compound are usually complicated by resonance Raman contributions to the spectrum.

Figure 10 shows the diode-laser excited SER spectra of 6 mM RB_3 solutions on an Ag electrode (Curve A) and on a Cu electrode (Curve B). Clearly, even at this concentration, the diode-laser-excited SER spectra can be obtained with good signal to noise ratios. What is remarkable about these spectra is that they were obtained in bulk solution with no evidence of luminescence or attenuation of the laser beam, even though there was a 1-cm thickness of solution on the excitation and collection sides of the electrode. SER spectra could not be obtained using 514.5-nm excitation under these same conditions because of intense RB_3 luminescence and attenuation of the laser line. These spectra agree well with data previously published for this complex (27). By comparison, spectra obtained by excitation at 632 nm with a more intense HeNe laser, are complicated by background luminescence despite the fact that the peak absorbance of the complex is at 455 nm in room temperature aqueous solution and decreases rapidly at lower energy.

An additional point to be made concerning the spectrum on Ag (Figure 10A) concerns the relative enhancements obtained at the two wavelengths for RB_3. Using the same electrode and solution, we found that the absolute magnitudes (in counts per second) of 785-nm-excited spectra are slightly larger than 632-nm-excited spectra. Relative enhancements are difficult to obtain quantitatively due to differences in detector response, optical alignment, and excitation intensity. However, the relative enhancements obtained for the NIR-excited spectra seem qualitatively much larger than for the visible-excited spectra. In addition, the detector response falls rapidly in the NIR region, and the excitation power

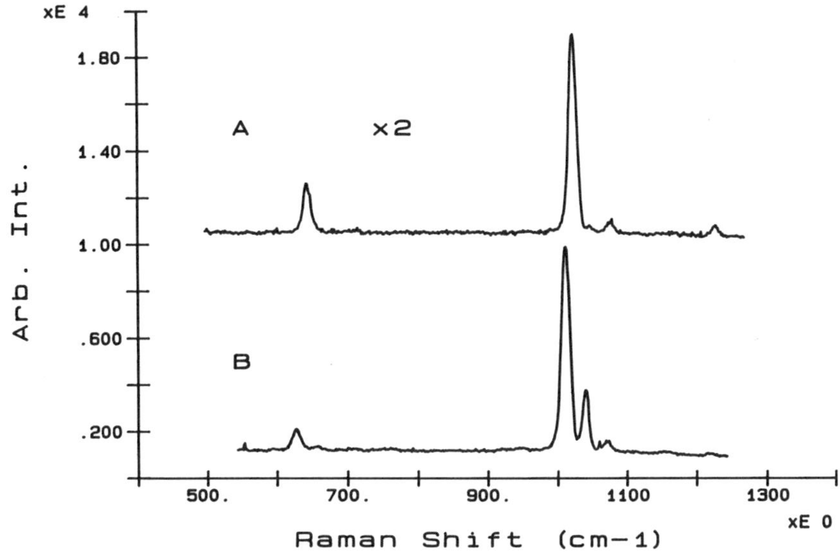

Figure 9. NIR SER spectra of 0.1 M pyridine using the 785-nm line of a diode laser: (A) Cu electrode at -0.6 V versus SCE; (B) Ag electrode at -0.7 V.

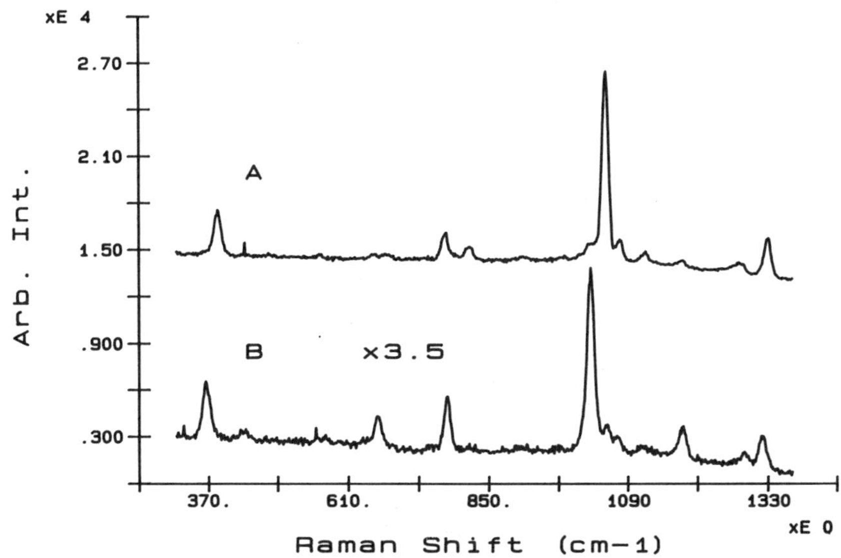

Figure 10. NIR SER spectra of 6 mM tris(2,2'-bipyridine)ruthenium(II) using the 785-nm line of a diode laser: (A) Ag electrode at -0.4 V versus SCE; (B) Cu electrode at -0.4 V.

available from the NIR diode was only 51% that of the 632-nm line. To obtain comparable intensities for the RB_3 spectra, the enhancement obtained on an Ag electrode using 785-nm excitation must be approximately an order of magnitude larger than the enhancement obtained on an Ag electrode using 632-nm excitation. Enhancements for this compound on the Cu electrodes appeared smaller than on Ag electrodes, though of the same general magnitude.

Future Remote SERS Research

The measurement of SERS over optical fibers is straightforward when short fibers (about 10 m) are used. However, the fiber Raman background for longer optical fibers can be larger than the SER signal itself. The reason for this is that a considerable amount of Raman signal is generated in the fiber and it is reflected by the SERS substrate back into the collection fibers. The amount of fiber Raman background is dependent on the length of the fiber. We are investigating methods of eliminating fiber Raman using miniature optics in a fiber-optic SERS probe. We have found that the Raman background can be virtually eliminated by the proper use of filters between the excitation and collection fibers and the SERS substrate.

The application of remote NIR SERS to environmental measurements looks promising. However, much research is needed to make this a useful technique for routine analytical analysis.

Applications: Geothermal Well Measurements

Recently, a program was started at LLNL to develop optical-fiber sensors for use in logging and monitoring geothermal wells. Initial research focused on developing a temperature optrode that could be used to obtain real-time information at high temperatures in very deep geothermal wells. This optrode has been completed and thoroughly tested in the laboratory at temperatures up to 320°C and pressures of 10,000 psi.

The geothermal temperature optrode is based on the temperature-dependent phosphorescence of the Eu(III) ion (5). Briefly, Eu(III) and Er(III) are doped into lattice sites within an inert ceramic-like carrier matrix ($CaZrTi_2O_7$). The two dopants phosphoresce when excited with the 488-nm line of an argon ion laser. In the case of Eu(III), the intensity of the phosphorescent bands at 610 nm and 595 nm increases as the temperature increases, while, in the case of Er(III), the intensity of the phosphorescent bands at 548 nm and 560 nm decreases as the temperature increases (Figure 11). The ratio of the Eu(III)- and Er(III)- band intensities is plotted as a function of temperature to produce a calibration curve for the optrode (Figure 12). The precision of this optrode is about ±0.2°C in the range 100 to 300°C, using a short (1- to 2-m) optical fiber and a laboratory spectrometer.

The well-logging tests were made in a geothermal exploratory well (GTW-2, USGS No. 2686-02) located on the island of Hawaii. Well GTW-2 was drilled in

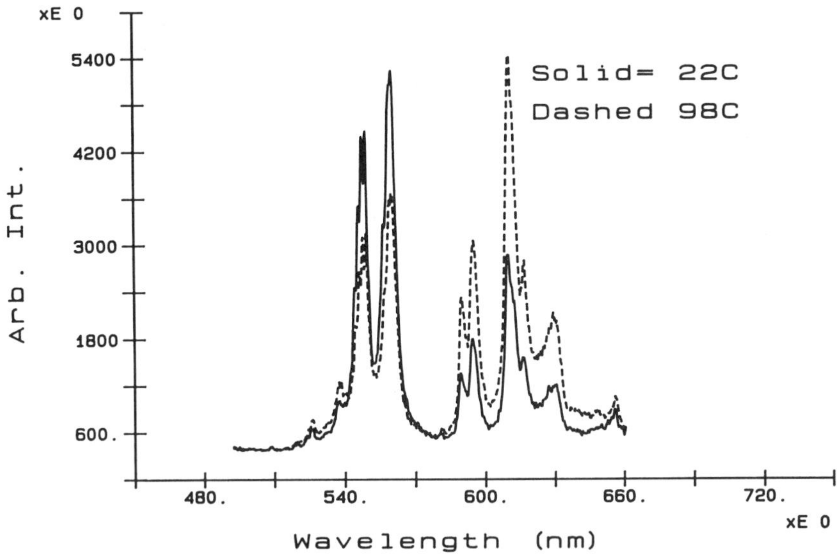

Figure 11. Temperature optrode spectral response using 488-nm excitation at two different temperatures. The 548 and 560 nm Er(III) band intensities decrease as the temperature increases. The 595 and 610 nm Eu(III) band intensities increase as the temperature increases.

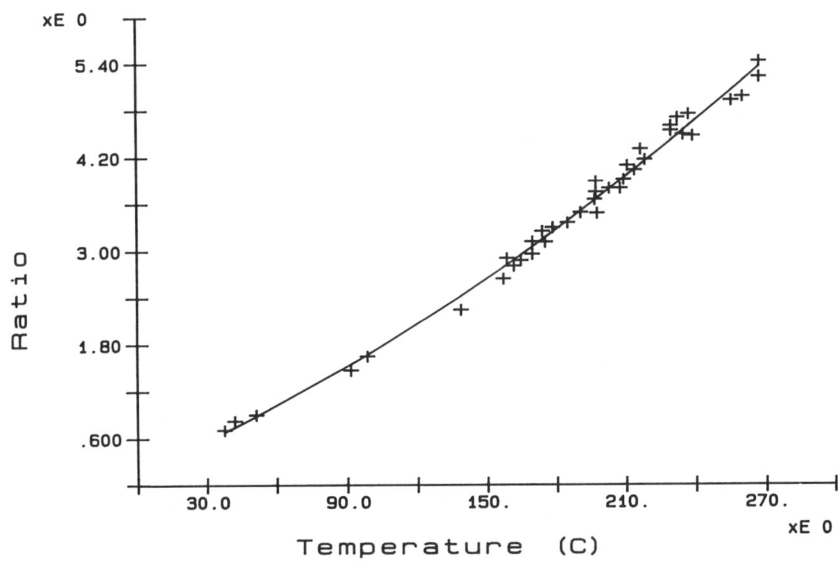

Figure 12. Response of the temperature optrode as a function of temperature using 488-nm excitation. The response is plotted as the ratio of the 610 nm and 560 nm band intensities.

1961 to a depth of about 169.5 m. All well measurements were made over a single, 300-m length of high-temperature, polyimide-coated, fused-silica optical fiber (Polymicro, FHP 320/385/415) with a 320-µm-diameter core. The details of the optrode construction are described elsewhere (Angel, S.M.; Garvis, D.G.; Sharma, S.K.; Seki, A. Appl. Spectrosc., in press).

A laser-based fluorimeter with an intensified diode-array detector was used for all measurements. The instrument was installed in a large moving van that was parked along the roadside near the well site. The fiber-optic cable was laid out from the van to the well site and was protected by placing it inside a Teflon tube. The Teflon tube joints were connected by heat-shrink tubing reinforced by wire wrap. A Chromel-Alumel (K-type) thermocouple was used as a reference in the data collection, and also served to provide rigid support for the fiber-optic cable. This was important because the force of the steam from the well was strong enough to cause damage to the fiber-optic cable. A two-pound lead weight was attached to the end of the cable to facilitate the insertion of the fiber into the well against the force of the escaping steam. The optical fiber cable assembly was lowered into the well with a pulley equipped with a wire-line depth gauge that was used to measure the depth of the probe in the well.

The optrode was calibrated over the range of 30 to 100°C prior to the well tests using the same 300-m optical fiber that was used in the experiments. A polynomial equation was fit to the resulting data. The polynomial equation used to fit the temperature optrode response is [Response = $0.4684+(0.0069 \times T)+(1.57 \times 10^{-5} \times T^2)$]. The resulting equation was used to relate the optrode response to the temperature in the well. The accuracy of the resulting calibration curve was about ±2°C.

Figure 13 shows temperature profiles that were obtained for well GTW-2 using the optrode and the thermocouple. The optrode temperature value is plotted as open diamonds. The thermocouple readout is plotted as open squares. Measurements were made to a depth of 150 m (about 490 ft). The agreement between the optical measurements and the thermocouple measurements is within the error of the calibration data (±2°C).

Summary

Optical fibers offer tremendous potential for remote and in situ measurements and are beginning to move out of the laboratory for certain special applications. Optrodes that monitor physical parameters (e.g., temperature) and that make direct spectroscopic measurements are the most extensively developed. Optrodes that are designed to monitor specific chemical species (e.g., TCE) are not as well developed because of the complexity of developing chemically-specific chemical indicators that can be immobilized on an optical fiber. Much research is still needed in this area.

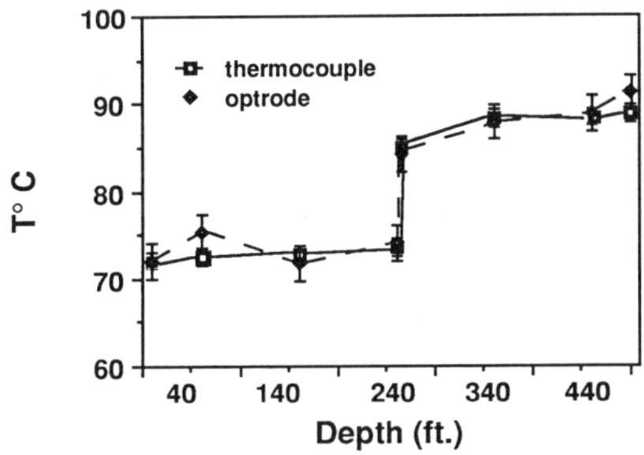

Figure 13. Temperature in Well GTW-2 plotted as a function of depth. The optrode values are indicated by open diamonds. The thermocouple values are indicated by open squares.

Acknowledgments

The work at Lawrence Livermore National Laboratory was performed under the auspices of the U.S. Department of Energy under Contract W-7405-Eng-48. The authors would like to thank John Mathur of the office of Defense Programs and Jim Epler of the Hazardous Waste Remedial Actions Program for funding the TCE and $CHCl_3$ work and we would also like to thank Gerald Goldstein and Paul Duhamel of the Office of Health and Environmental Research and Mort Mendelsohn of LLNL for supporting the SERS work. The authors would also like to express thanks to Marshall Reed of DOE and Paul Kasameyer of LLNL for their support of the geothermal work. Finally, we express thanks to Paul Daley for help in preparing this manuscript.

Literature Cited

1. Seitz, W.R. Anal. Chem. 1984, 56, 16A.
2. Wolfbeis, O.S. Trends in Analytical Chemistry 1985, 4, 184.
3. Angel, S.M. Spectroscopy 1987, 2, 38.
4. Kulp, T.J.; Camins, I.; Angel, S.M.; Munkholm, C.; Walt, D.R. Anal. Chem. 1987, 59, 2849.
5. Angel, S.M. Geothermal Resources Council, Transactions 1987, 11, 155.
6. Angel, S.M.; Katz, L.F.; Archibald, D.D.; Lin, L.T.; Honigs, D.E. Appl. Spectrosc. 1988, 42, 1327.
7. Kulp, T.J.; Bishop, D.; Angel, S.M. J.Soil Sci.Soc.Am. 1988, 52, 624.
8. Milanovich, F.P.; Garvis, D.G.; Angel, S.M.; Klainer, S.K.; Eccles, L. Anal. Inst. 1986, 15, 137.
9. Smith, B.W.; Jones, B.T.; Winefordner, J.D. Appl. Spectrosc. 1988, 42, 1469.
10. Fleischmann, M.; Hendra, P.; Mcquillan, A. Chem.Phys.Lett. 1974, 26, 163.
11. Furtak, T. In Advances in Laser Spectroscopy; B.Garetz and J. Lombardi, Eds.; Wiley: New York, 1983; Vol. 2, p. 175.
12. See Surface Enhanced Raman Scattering; R. Chang and T. Furtak, Eds.; Plenum Press: New York, 1982.
13. Jeanmarie, D.;van Duyne, R. J.Electroanal.Chem.Interfacial Electrochem. 1977, 84, 1.
14. Vo-Dinh, T.;Hiromoto, M.;Begun, G.;Moody, R. Anal.Chem. 1984, 56, 1667.
15. Enlow, P.;Buncick, M.;Warmack, R.;Vo-Dinh, T. Anal.Chem. 1986, 58, 1119.
16. Alak, A.;Vo-Dinh, T. Anal.Chem. 1987, 59, 2149.
17. Torres, E.;Winefordner, J.D. Anal.Chem. 1987, 59, 1626.
18. Carrabba, M.;Edmonds, R.;Rauh, R. Anal.Chem. 1987, 59, 2559.
19. Hirschfeld, T.;Chase, B. Appl.Spectrosc. 1986, 40, 133.
20. Chase, D.B. J.Am.Chem.Soc. 1986, 108, 7485.
21. Chase, D.;Parkinson, B. Appl.Spectrosc. 1988, 42, 1186.
22. Nakanishi, K.;Imasaka, T.;Ishibashi, N. Anal.Chem. 1985, 57, 1219.

23. Imasaka, T.;Yoshitake, A.;Ishibashi, N. Anal.Chem. 1984, 56, 1077.
24. Crosby, G.A.; Elfring Jr., W.H. J. Phys. Chem. 1976, 80, 2206.
25. Blakley, R.A.; Myrick, M.L.; DeArmond, M.K. J. Am. Chem. Soc. 1986, 108, 7843.
26. Myrick, M.L.; Blakley, R.L.; DeArmond, M.K. J. Am. Chem. Soc. 1987, 109, 2841.
27. Virdee, H.; Hester, R. J. Phys. Chem. 1984, 88, 451.

RECEIVED March 17, 1989

Chapter 24

Optical Waveguide Chemical Sensors

John F. Giuliani

Chemistry Division, U.S. Naval Research Laboratory, Washington, DC 20375–5000

Optical waveguide chemical sensors have demonstrated their practical value not only for the detection of toxic vapors and gases in air, but also their adaptability to liquid phase detection of biomolecules. The basis for the success of our sensor program has been the strong multidisciplinary research effort into identifying and understanding the physical and chemical properties of organic materials including dyes and polymers, which form the reagent waveguide sensor coatings that give these devices their required sensitivity and specificity. In addition to the exploitation of cylindrical glass optical waveguide geometries, our work has widened to include the fabrication of multi-waveguide structures on glass substrates for the purpose of obtaining chemical sensor arrays on a single support.

The last six years at NRL have seen the rapid development and diversification of optical waveguide chemical sensors for the detection of toxic gases and vapors in air, and small molecules in aqueous solutions. Our multidisciplinary approach has lead to the successful utilization of a large number of selective, chemically reversible and highly sensitive solid organic reagent film coatings. In addition to the exploitation of hollow cylindrical glass optical waveguide substrates, our work has widened to include the fabrication of integrated planar sensors. Although the development of practical optical sensors is our primary goal, both types of devices have been very useful in detecting subtle dynamic equilibrium surface interactions produced at the reagent film/waveguide glass interface.

The basic principles and a general survey of chemical sensor applications based on fiber optics, have been adequately given by Seitz (1) and Angel (2). It is quite evident from the numerous references cited, that optical sensors are rapidly taking their place as the sensor of choice for a variety of chemical detection applications.

I will confine this report to optical waveguide sensor development at our laboratory. The research effort to be presented discusses briefly some geometrical design considerations and the reasoning behind the particular geometry finally chosen. Extensive vapor and solution phase detection data taken with our sensors will be summarized. Finally some preliminary data relating to the fabrication of planar structures and multi-element sensor "arrays" will be presented.

Waveguide Geometries

One of the useful properties of optical sensors is that they can be formed into a variety of geometrical shapes depending on their desired use. For example, Figure 1 displays three different useful geometric configurations. The solid rod shape is used generally in the form of a drawn-out fiber. Early in our work we decided on the hollow-cylindrical shape based on ease of coating, mechanical strength, and the simplicity of coupling commercial LED (light sources) and photodetectors. Moreover, it has been shown by Kapany and Pontarelli, (3) that this geometry leads to high differential sensitivity for detecting small surface changes in refractive index. Also, as seen in Figure 1, the number of optical reflections per unit length for a thin-walled hollow cylinder can be several hundred for the 90 mm x 1.1 mm x 0.8 mm capillary tube. Most of our chemical sensor studies have been performed using this hollow-cylindrical geometry, in which one end is closed in a lens-like surface for focusing the multiply reflected light into a small commercial detector (4). We have, however, in the last few years explored the micro-planar geometry, both from the viewpoint of fabrication and application to sensing vapors. More will be said about this later on in this report.

Hollow-Cylindrical Waveguide Sensor

Figure 2 shows a schematic of our reagent-coated sensor incorporated in a gas chamber in which the vapor or gas can be admitted and allowed to flow over the waveguide surface. Also shown is the light source and circuitry for an LED which is coupled into one end of the cylinder, and a photodetector coupled to the other end, with its associated electronics. A full description of this device has been given as well as its application to a variety of vapor detection situations (5,6,7,8).

Application of the NRL Sensor to Selected Vapors

Table I summarizes the application of the NRL sensor for the detection of four toxic vapors (Column 1), various selective reagent coatings (Column 2), the lowest vapor concentration detected (Column 3), and the probable reaction mechanism (Column 4). The types of vapors besides ammonia, include an organophosphonate, and two sulfur compounds. The reagent coatings included a variety of dyes, (i.e., oxazine, lead-phthalocyanine (Pb-Pc), lead tetraphenylprophrin, (Pb-

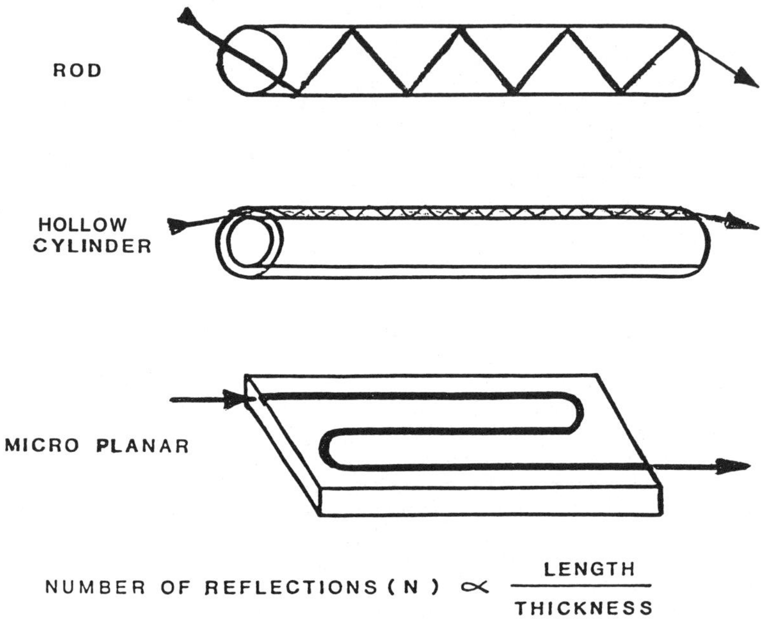

Figure 1. Various optical waveguide geometries.

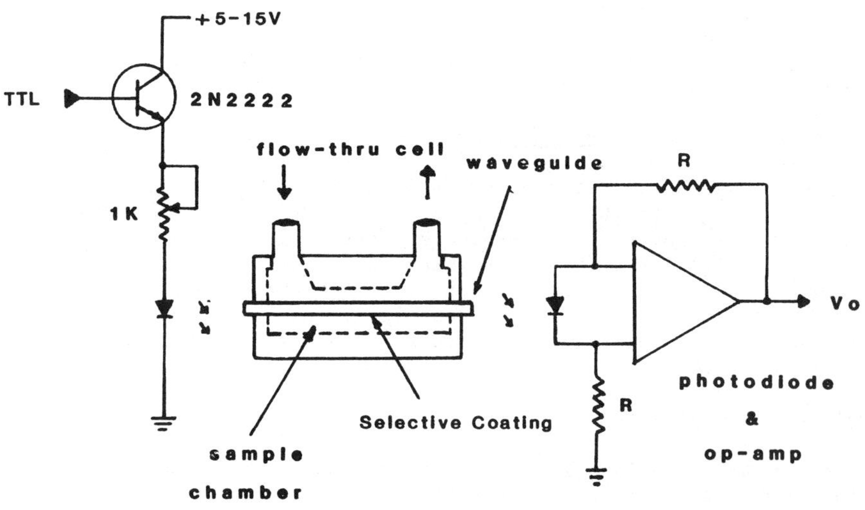

Figure 2. Schematic of the optical, electronic, and flow chamber components for the NRL optical waveguide sensor.

TTP), a Michlor's Ketone "EMKO", and an oxime dye incorporated with a catalyst. The polymer coatings tested were polyvinylpyrrolidone (PVP), poly-methylmethacrylate (PMM), polyepichlorohydrin (PEH), and polyethylene maleate (PEM). Note that with the exception of the oxazine dye, all other dyes show no color reversibility. However, all the polymers show reversibility to these vapors indicating physical adsorption predominates the vapor/film interaction.

The optical coupling of the evanescent wave at the interface is thought to be affected by solute-solvent, surface wetting, and film/glass adhesion interactions. More will be said about this effect later in connection with solubility measurements in polymer films.

Table I. Optical Waveguide Detection of Toxic Vapors

Vapor	Coating Reagent	Lowest Concentration Detected,Δ	Reaction Mechanism
Ammonia	Oxazine Dye+	8 ppm	Colorimetric Changes
Dimethyl Methyl-Phosphonate	Pb-Pc	8 ppm	
	Pb-TTP	20 ppm	
	*PVP+	5 ppm	Solubility Interactions
	*PMM+	25 ppm	
	*PEH+	10 ppm	
	*PEM+	20 ppm	
Methane Sulfonyl Chloride	"EMKO"	9 ppm	Colorimetric Changes
	Oxime/Catalyst	6 μg/liter	
Benzene Sulfonyl Chloride	Oxime/Catalyst	8 ppm	

+ Reversible Charge
Δ Dry Air
* Polymers

<u>Solution Phase Small Molecule Detection</u>

Lubbers and Optiz (9) and Andrade et al (10) have employed fiber optic sensors for the continuous measurement of chemical reactions in biological systems. High sensitivity may be achieved using these sensors to measure fluorescent-tagged antigens or antibodies in competitive binding immunoassay reactions in solution.

The NRL work involves the covalent attachment of a colorimetric redox dye indicator film to the outer surface of the hollow cylindrical waveguide, which when immersed in aqueous solution can reversibly detect various reducing species. The method of coating the cylindrical waveguide and experimental demonstration of the detection of redox agents has been described in detail (11,12). Figure 3 shows a typical dynamic reversible colorimetric response when the viologen coated waveguide surface is exposed alternately to distilled water/sodium dithionite ($Na_2S_2O_4$) solutions, where $Na_2S_2O_4$ is a strong reducing species.

Figure 4 displays the dynamic reversible colorimetric detection of the formate ion in which a successful linking of both the redox-viologen indicator dye and the formate dehydrogenase (FDH) enzyme to the waveguide surface leads to the reaction of the bound enzyme with the formate ion $HCOO^-$ in the presence of a co-enzyme NAD. The production of NADH by the enzyme-catalyzed reaction results in a reduction of the positive charge on the bound viologen which is detected by a change in color (i.e., from colorless to blue).

Our hollow cylindrical waveguide structure has also been coated with a human h-IgG antibody enzyme and used to detect a fluorescent labeled antigen Ag*/Dye in a immunoassay-type reaction.

All of the surface phase reactions detected by the hollow cylindrical optical waveguide sensor are summarized in Table II, which like Table I shows four columns depicting the detected analyte, the selective waveguide coating, the lowest concentration detected, and the reaction mechanism.

Fabrication of Integrated Optical Waveguide Sensors

We have successfully photopolymerized two commercial oligomers by means of lasers and incoherent ultra-violet light sources using the technique of photolithography. Our procedure has been described in (13, 14). The polymerized waveguide channel is a commercial

Table II. NRL Optical Waveguide Solution Phase Studies

Analyte	Coating	Concentration	Reaction
Ag*/Dye Labeled	h-Ig G	66 nMoL/L	Antigen-Antibody Complex-Fluorescence
HCO_2H (Formic Acid)	Viologen/FDH	7.5 mMol/L	Enzyme Catalyzed Charge Transfer-
HCO_2Na (Sodium Formate)	Viologen/FDH	2 nMol/L	Colorimetric
$Na_2S_2O_4$	Viologen	>0.005 mg/ml	Charge Transfer
H_2 gas/water	Viologen	1 μMol/L	Absorption Dependent-
NADH/H_2O	Viologen	0.5 g/L	Fluorescence

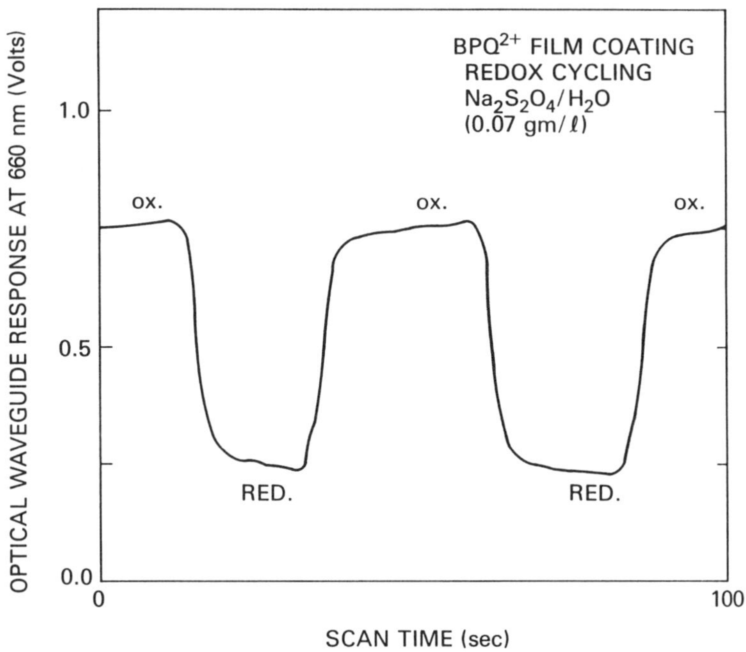

Figure 3. Typical dynamic response of the viologen-coated waveguide exposed alternately to distilled water/sodium dithionite solutions.

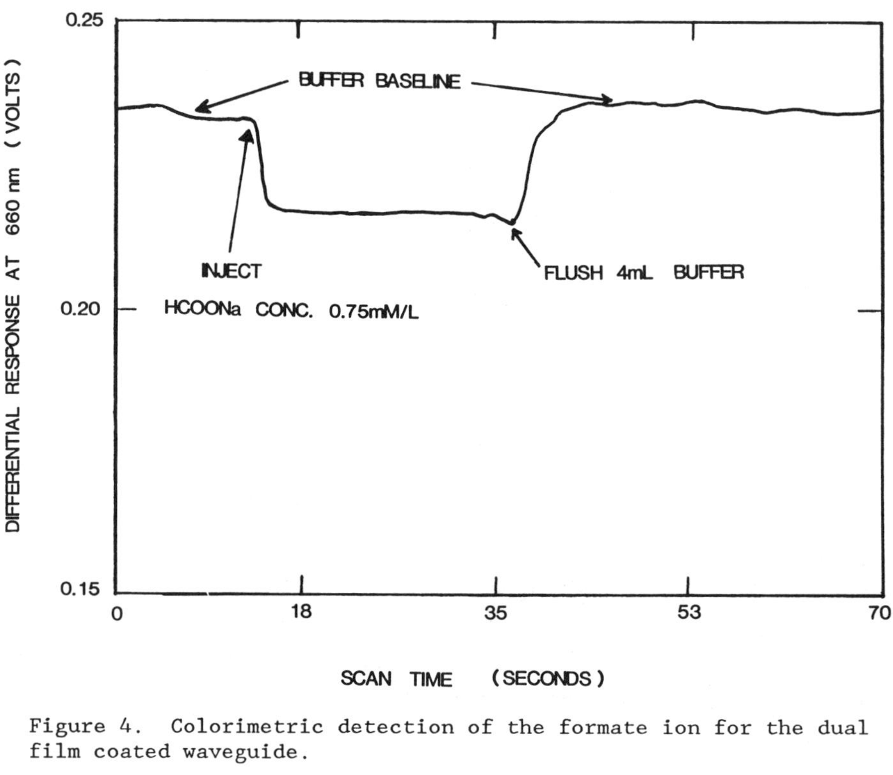

Figure 4. Colorimetric detection of the formate ion for the dual film coated waveguide.

thiolene compound which has the photo-initiator incorporated in the starting material. Fiber optic "pigtails" are attached to the initially unpolymerized material (14). Figure 5 displays the optical response of this detector to acetone vapors at concentrations of 10,000 ppm, probed by a neon incoherent light source.

Table III lists a series of saturated vapors produced at 22°C to which this sensor was exposed. As can be inferred from the table, solubility apparently plays an important role in the selective optical detection. A more detailed investigation of the solubility properties of polymers and their optical response to selected vapors will be discussed in the next section.

Table III. Optical Response of the Photopolymerized Thiolene Strip Shaped Waveguide Structure on a Glass Substrate to Various Vapors in a Nitrogen Carrier Flow System

Vapors	Optical Response
2-Propanol	No
Toluene	No
Acetone+	Yes
Water	No
Ammonium Hydroxide	No
Chloroform+	Yes
Xylene	No
Benzene	No

Source: Reprinted from ref. 17.
+Solvent for the monomer

Vapor/Polymer Solubility Interactions

As was mentioned in the last section, the solubility of the polymerized planar sensor material for a specific vapor appeared to be important. In this connection we have carried out a series of experiments in which two polymers having known solubility parameters (δ) have been deposited on the hollow cylindrical waveguide structure as thin films produced by dipping the waveguide into the polymer/solvent solution. Rapid evaporation of the solvent results in a thin film (15). The polymers investigated, were polyethylene maleate (PEM) and polyfluoropolyol (PFP) whose indices of refraction are 1.484 and 1.413, respectively. These films were exposed separately to a series of vapors having a broad range of solubility

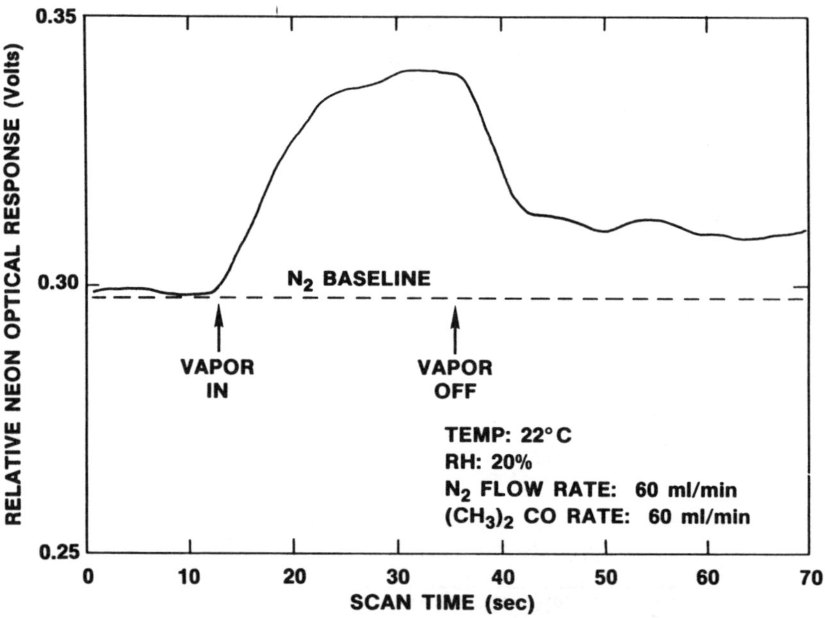

Figure 5. Relative dynamic response of the thiolene waveguide sensor to saturated acetone vapors. (Reprinted from ref. 17.)

parameters (δ) between 8.6 $(cal/cm3)1/2$ and 23.4 $(cal/cm^3)^{1/2}$. The vapor concentrations were on the order of 10,000 ppm. Figures 6 and 7 show the measured change in optical transmission as a function of the solubility parameters for the various vapors. Note that there is a significant dip in transmission for both polymer films when the solubility of the vapor matches that of the polymer within several percent. The two compounds DMMP (dimethyl methylphosphonate) and DMAC (N, N-dimethylacetamide) have solubility parameters which lie within a few percent of the two polymer films tested. The optical changes detected are believed to be caused by the softening of the bulk polymer film which in turn produces a significant change in its light-guiding properties. Our dynamic response data show that in all cases the surface phenomena is reversible when the optical signal is monitored prior to and after the vapor is removed from the dynamic equilibrium flow system.

Multi-Element Optical Waveguide Sensor Arrays

A single chemical sensor with a single coating would not be a practical chemical detector for a particular gas or vapor dispersed in a mixture of interferents. Hence, multiple sensors, and a few selective reagent coatings would be desirable, so that discrimination between classes of compounds may be obtained using pattern recognition techniques (16). Our initial array of sensors made use of the previously characterized hollow-cylindrical structure already discussed (4). Figure 8 outlines two general methodologies (i.e., single vs multiple photodetectors) coupled to eight single cylindrical waveguides. In this report we will describe experiments using a single photo-diode detector and an array of eight color emitting LED sources attached to eight separate hollow cylindrical waveguides. The initial experiments to be described, involve the detection of pH and redox reactions in aqueous solutions, using two dye coatings (i.e. alizarin yellow and the previously discussed viologen redox indicator).

A block diagram of the multi-element system is shown in Figure 9. For this configuration, a small computer can monitor and control normal system operations. When the computer is not directly monitoring the control interface control board (CCIB), it can execute system control routines. As can be seen in Figure 9, the CCIB is an interface which has direct influence over all input/output operations, while its control parameters are monitored by a 6502 based microcomputer (Commodore 64). The main functions therefore, of the computer are those of parameter control and data manipulation. The CCIB is a free-running hardwired controller complete with its own clock for running the multi-element system in a real time mode. It interrogates the channels on a variable time basis. The data channels are monitored and updated within the computer and passed to the CCIB via digital outputs. The system software senses when the CCIB is making the amplified photo-detector signal available for further processing and storage. For this system, the minimum on-times for the LED output stabilization are about 1 to 2 milliseconds. For a typical data-cycle a single LED is activated and after a suitable delay, the amplified photo-detector

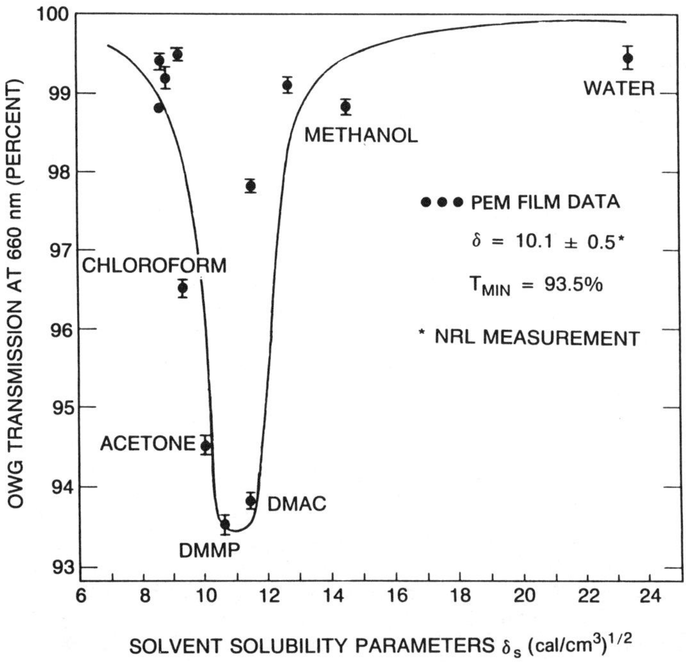

Figure 6. Optical response of PEM coated cylindrical waveguide to a series of vapors encompassing a range of solubility parameters. The names of some of the vapors tested, are shown in the figure. (Reprinted from ref. 15.)

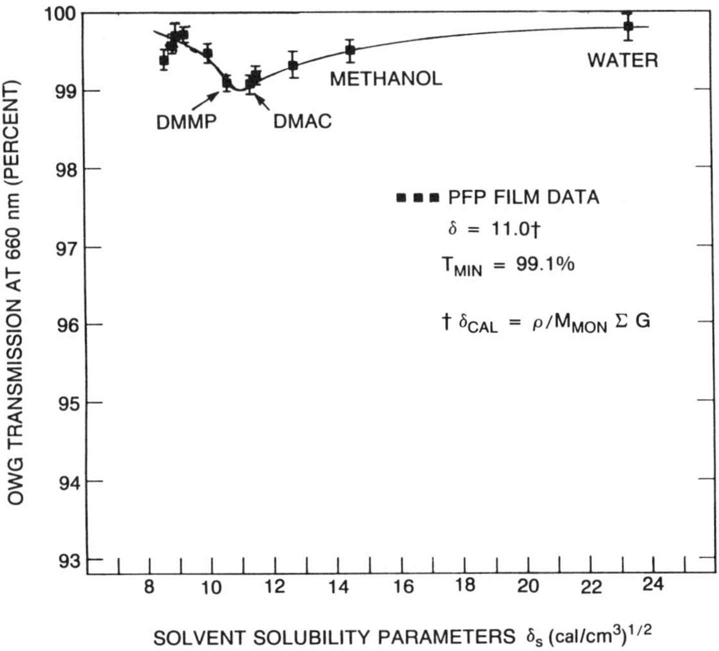

Figure 7. Optical responses of PFP coated cylindrical waveguide exposed to the same range of vapors and solubility parameter range as shown in Figure 6. (Reprinted from ref. 15.)

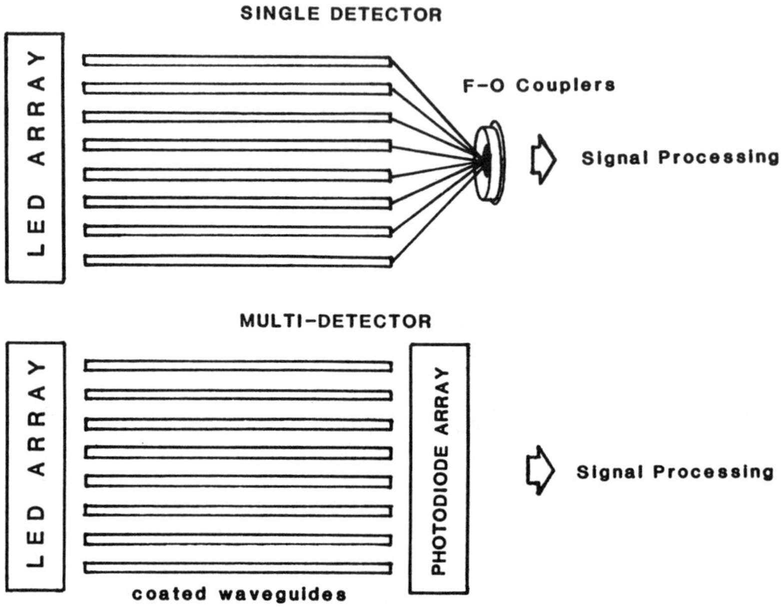

Figure 8. Multi-element array configurations.

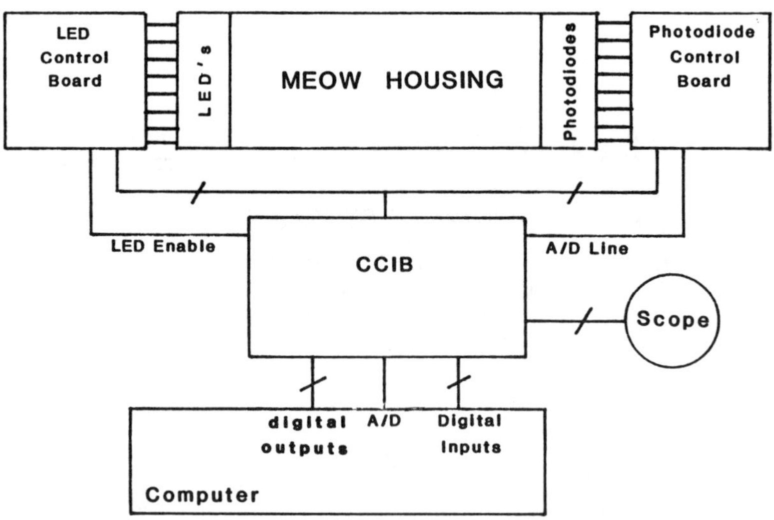

Figure 9. Block diagram of a proposed multi-element optical waveguide chemical sensing system.

output is digitized and stored in memory. Shortly thereafter, the LED is deactivated and the remaining data channels are processed in like fashion.

Some preliminary data taken with this multi-element array system are shown in Figure 10. For display purposes, the various channels of a single photo-detector were connected to an analog oscilloscope to produce a bar-graph pattern with individual bar heights being proportional to the detector output for that particular channel. Two of the eight channels (i.e., numbers 3 and 6) represent optical response data obtained from optical sensors coated with a 500Å film of a silanized-glass coupled viologen dye

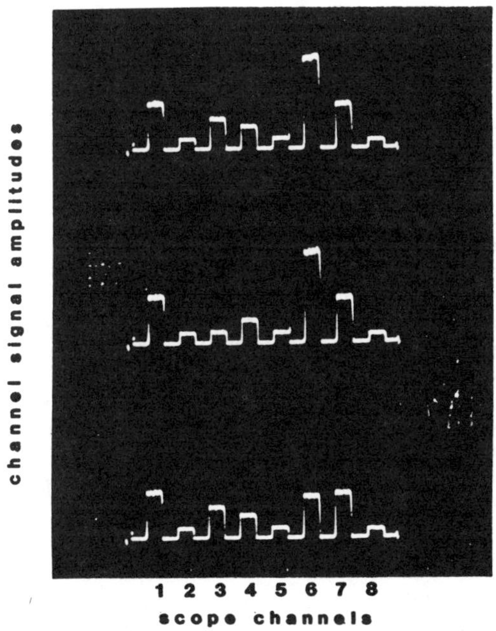

Figure 10. Oscilloscope traces displaying the colorimetric response for two of the eight-channels in the array, to neutral (top trace) reducing (middle trace) and basic (bottom trace) pH and redox solutions.

(i.e. redox indicator) and an alizarin yellow dye (i.e., pH indicator), respectively. Exposing the redox-coated sensor to a reducing solution (70 mg/L $Na_2S_2O_4$) produces a colorless to blue colorimetric change due to a charge transfer reaction. The red-emitting LED throughput hence is attenuated as is shown in the middle trace of channel 3. When the reducing solution was removed, and replaced with distilled water, the film becomes transparent again and the signal for channel 3 (i.e. lower trace) returns to the same amplitude level as before (see upper trace of channel 3 for comparison). The optical sensor displayed in channel 6 was coated with a micron-thick insoluble pH-sensing alizarin yellow dye film. This film undergoes a colorimetric change from yellow to red over a pH 10 to 12 range. This is readily seen in the upper and lower scope traces of channel 6, where the reducing $Na_2S_2O_4$ solution (pH 6) had been purged and replaced by a basic (pH 10) aqueous solution. The green LED-emitting diode used for channel 6 shows a strong attenuation due to the color change, as can be seen by comparing the upper and lower scope amplitudes for this channel. As previously noted all other channels for the six uncoated waveguide sensors show no change in amplitude for the three sets of scope traces monitored, as would be expected.

Summary and Conclusions

The optical waveguide sensors research at NRL has shown that these devices can be applied with equal facility, economy and high sensitivity to the detection of vapors, gases and small molecules in liquid media. Moreover, our preliminary results obtained in the fabrication of planar polymer structures on glass substrates, have demonstrated that not only can the polymerized organic structure function as a waveguide channel, but it can also serve as the selective chemical sensing element itself, thus eliminating the need for a separate reagent coating process. This points to the potential of batch processing small inexpensive, throw-away chemical sensors or sensor arrays.

Acknowledgments

The author wishes to thank the Office of Naval Technology for their support of this work, and to Paul P. Bey, Jr. of our Laboratory for all of the design and testing of the electronics employed in our single and multi-element array proto-type sensor systems.

Literature Cited

1. Seitz, W.R., Anal. Chem. 1984, 56, 16A-33A.
2. Angel, S.M., Spectroscopy, 1986, 2, No. 4, 38-48.
3. Kapany, N.S. and Pontarelli, D.A., Appl. Opt., 1963, 2, 425-430.
4. Giuliani, J.F. and Bey, Jr., P.P., Appl. Optics, 1988, 27, 1353-57.
5. Giuliani, J.F. Wohltjen, H., and Jarvis, N.L., Opt. Lett., 1983, 8, 54-56.

6. Ballantine, Jr., D.S., and Wohltjen, H., Anal. Chem., 1986, 58, 2883-85.
7. Giuliani, J.F. and Jarvis, N.L., J. Chem. Phys., 1985, 82, 1021-24.
8. Giuliani, J.F. J. Chem. Phys., 1985, 83, 5998-6000.
9. Lubbers, D.W. and Opitz, N., Sensors and Actuators, 1983, 4, 641-54.
10. Andrade, J.D., Vanwagenen, R.A., Gregonis, D.E., Newby, K., and Lin, J.N., IEEE Trans. on Electron. Devices, 1985, ED-32 No. 7, 1175-79.
11. Giuliani, J.F., Dominguez, D.D., Barger, W., Bey, Jr., P., Barbano, E., and R. Smardzewski, Proc. of 1985 Conf. on Chem. Def. Res., 1986, p. 621.
12. Dominguez, D.D. and Giuliani, J.F., J. of Bio. Phys., 1987, 15, 75-80.
13. Giuliani, J.F., Kim, K.H., Butler, J.E., Appl. Phys. Lett., 1986, 48, 1311-13.
14. Giuliani, J.F., J. of Optical Sensors, 1987, 2, 11-16.
15. Giuliani, J.F., J. of Poly. Sci., 1988, 26, 2197-2201.
16. Ballantine, Jr., D.S., Rose, S.L., Grate, J.W., and Wohltjen, H., Anal. Chem., 1986, 58, 3058-66.
17. Giuliani, J.F., Sensors and Actuators, 1985, 15, 25-31.

RECEIVED March 9, 1989

Chapter 25

Multidimensional Fluorescence with a Simple Fiber-Optic-Based Probe

Kevin S. Litwiler and Frank V. Bright

Department of Chemistry, State University of New York at Buffalo, Buffalo, NY 14214

Multifrequency phase and modulation measurements have been performed through a single optical fiber up to 175 meters in length. The recovered lifetimes agree very well with non-fiber-optic-based approaches and the instrument is capable of resolving complex decay kinetics. We show also the first phase-resolved fluorescence measurements through a fiber-optic probe and demonstrate that this powerful resolution technique can be used to determine the individual components in remotely located samples. In addition, it is demonstrated that phase-resolved techniques can be used to eliminate the stray light problems often encountered in fiber-optic-based sensing. Finally, phase-resolved fluorescence and lifetime measurements are used to resolve two distinct species of rhodamine 6G in a rhodamine 6G-impregnated Nafion film.

Over the last few decades, numerous research groups have exploited the inheret multidimensional nature of fluorescence processes with the goal of increasing analytical selectivity (1-2). In addition to its great selectivity, fluorescence spectroscopy has gained additional popularity because of its intrinsically-high sensitivity and relative simplicity.

Weber (3) and Ainsworth (4) are recognized as the first to employ multidimensional wavelength selectivity to resolve spectrally similar components. By incorporating video-electronic detection, the Christian (5) and Warner (6) groups have extended this approach significantly. In addition to simple spectral selectivity, McGown and co-workers (7) have employed temporal selectivity to simultaneously quantitate spectrally overlapping species in multicomponent mixtures. Most recently, Brand et al. (8) and Bright (9) have described component resolution based on differences in the rotational diffusion rates of the individual fluorescent species.

In spite of such advances, all these extremely powerful fluorescence-based methods of analysis were necessarily restricted

0097–6156/89/0403–0380$06.00/0

to the laboratory site. That is, the ability to perform these same types of information-rich measurements remote from the laboratory location was generally impossible. Fortunately, over the last decade, the telecommunications industry has developed inexpensive, small, rugged, high-light transmitting optical fibers. Because optical fibers provide a convenient means of transporting light from one location to another many, once laboratory bound, spectroscopic techniques such as ultra-violet and visible absorption, visible and near-infrared reflectance, Raman spectroscopy, and molecular fluorescence have been implemented remotely using fiber-optic probes. Angel (10) and Seitz (11) have recently reviewed fiber-optic-based sensing in chemical analysis.

With respect to fiber-optic-based fluorescence sensing, most of the past approaches have employed either excitation or emission wavelength selectivity (10,11). The other selectivity parameters: fluorescence lifetime, steady-state polarization, and rotational diffusion rates have received little if any attention. The acquisition of fluorescence lifetime information via optical fibers has been demonstrated previously (12); however, the time resolution and ability to resolve multiexponential decays of fluorescence were neither demonstrated nor possible with this earlier instrument. Recently, our own group (13,14) has developed and described the first fiber-optic-based fluorescence lifetime instrumentation capable of unequivocally determining single, double, and/or triple exponential decays of fluorescence in remotely-located samples.

In this paper, we employ a new version of our own fiber-optic-based multifrequency phase and modulation fluorometer. Specifically, we show results for the accurate determination of fluorescence lifetimes through a <u>single</u> fiber-optic probe up to 175 meters in length. In addition, we show the first phase-resolved fluorimetric measurements with a fiber-optic probe and we demonstrate the power of multidimensional measurements for the elucidation of the complex decay kinetics of rhodamine 6G-impregnated Nafion films and the simultaneous resolution of the individual spectral components in binary mixtures.

Theory

<u>Lifetimes</u>. The theory of frequency-domain lifetime determinations has been described in detail elsewhere (15-19). Briefly, a high-frequency (MHz - GHz) sinusoidally-modulated light source is used to excite the fluorescent sample. The time-dependent mathematical representation of the excitation waveform (Ex(t)) is given by:

$$Ex(t) = A_{ex}\,(1 + m_{ex}\sin(\omega t)) \qquad [1]$$

where A_{ex} is the dc intensity, m_{ex} is the depth of modulation for the excitation waveform, t is time, ω is the angular modulation frequency ($\omega = 2\pi f$), and f is the linear modulation frequency in hertz.

The resulting time-dependent fluorescence (Em(t)) is equal in frequency to the exciting sinusoid (forced response), but is demodulated by a factor M and phase shifted by an angle ϕ:

$$Em(t) = A_{ex}(1 + m_{ex}M\sin(\omega t - \phi)) \qquad [2]$$

to an extent that depends critically on the decay rate of the fluorophore (15-19). This is analogous to the frequency-dependent response of an RC-filter circuit to the input of a high-frequency sinusoidal waveform. The only difference is that the RC time constant is replaced directly by the fluorescence lifetime, τ. For a single emitting population, the phase angle and/or demodulation factor are/is used to calculate the apparent fluorescence lifetimes by phase (τ_p) and modulation (τ_M):

$$\tau_p = 1/\omega \tan\phi \qquad [3]$$

$$\tau_M = (1/\omega)\ [(1/M^2) - 1]^{1/2} \qquad [4]$$

Throughout the previous discussion we have assumed the sample is accurately described by a single exponential decay process. Unfortunately, many cases exist where the sample is not accurately described by such a simple process (e.g., fluorophore mixtures and proteins). When the sample emission is described as a multiple exponential decay, consisting of n fluorescent components, the frequency-dependent phase ($\phi_c(\omega)$) and demodulation ($M_c(\omega)$) values are given by:

$$\phi_c(\omega) = \arctan\ (N(\omega)/D(\omega)) \qquad [5]$$

$$M_c(\omega) = [N^2(\omega) + D^2(\omega)]^{1/2} \qquad [6]$$

where the subscript c denotes the calculated values and $N(\omega)$ and $D(\omega)$ are the sine and cosine transforms:

$$N(\omega) = \sum_{i=1}^{n} \frac{a_i \omega \tau_i^2}{1+\omega^2\tau_i^2} \Big/ \sum_{i=1}^{n} a_i\tau_i \qquad [7]$$

$$D(\omega) = \sum_{i=1}^{n} \frac{a_i \tau_i}{1+\omega^2\tau_i^2} \Big/ \sum_{i=1}^{n} a_i\tau_i \qquad [8]$$

where a_i and τ_i are the fractional contributions to the total emission and lifetimes of the individual components, respectively. The goodness of fit between the calculated (c subscripts) and the measured values (m subscript) is given by the chi-squared (χ^2) value:

$$\chi^2 = \sum_{\omega=1}^{k} \frac{1}{e_\phi^2(\omega)} (\phi_M(\omega) - \phi_c(\omega))^2 + \sum_{\omega=1}^{k} \frac{1}{e_M^2(\omega)} (M_n(\omega) - M_c(\omega))^2 \qquad [9]$$

where $e_\phi(\omega)$ and $e_M(\omega)$ are the associated variances for the phase angle and demodulation factor measurements, respectively, and the other terms are as described previously. With our instrument $e_\phi(\omega)$ and $e_M(\omega)$ are approximately 0.25 and 0.000025, respectively. To recover the kinetic parameters for a multiexponential decay, non-linear regression is used over parameter space to minimize the chi-squared function (Eqn. 9) (17,19). When the assumed model is in agreement with the experimental data a χ^2 value approaching unity is achieved and a random distribution of the residual errors results.

Phase-Resolution. Phase-resolved fluorescence spectroscopy (PRFS) (7) is based on the phase-modulation method described above. Briefly, the time-dependent ac portion of the fluorescence (Eqn. 2) is selectively detected using lock-in type detection (7). The result is a time-independent dc intensity, the phase-resolved intensity (PRI), that is directly proportional to the cosine difference between the detector phase angle (ϕ_D) and the fluorescent species (ϕ) (7):

$$PRI = A_{em} m_{ex} M_{em} \cos(\phi_D - \phi) \qquad [10]$$

where A_{em} is the emission spectrum for the fluorescent species and all other terms are as described previously. Clearly, the maximum PRI is achieved when ϕ_D and ϕ are equal and a minimum is achieved when $\phi_D = \phi \pm 90°$. Of course, intermediate values of PRI are achieved at intermediate detector phase angle settings.

For a multicomponent system of j independent species, the phase-resolved intensities are linearly additive:

$$PRI = \sum_{i=1}^{j} A_{em,i} m_{ex,i} M_{em,i} \cos(\phi_D - \phi_i) \qquad [11]$$

Thus, for a two component system (A and B), in which both species have different lifetimes, one can selectively eliminate the contribution form one component (A) by adjusting the detector phase so that it is 90° out of phase with A ($\phi_D = \phi_A \pm 90°$). The resulting wavelength-dependent PRI is thus due only to the second component (B); however, the signal is attenuated by a factor of sin ($\phi_A - \phi_B$). The result is that the spectrum for component B is acquired devoid of interference from component A. To collect the unperturbed spectrum for component A one simply suppresses component B and repeats the scan.

Instrumentation

Figure 1 shows a schematic of the new single fiber multifrequency phase and modulation fluorometer. All components are rigidly mounted on a 5'x10'x8" laser table (Newport Research). Single-line excitation at 351.1 or 457.9 nm is achieved using a cw argon-ion laser (Coherent, model 90-6). The laser output is directed via a beam translator (BT; Newport Research, model 670) and a pair of mirrors (M) through a polarizing beam splitter (PBS), into a Pockel's cell electro-optic modulator (PC; Lasermetrics, model 1042). The remainder of the modulated laser beam (over 90%) passes through a bandpass filter (BPF) to eliminate plasma-tube discharge, is focused by a lens, through a perforated mirror (PM) into a 200 μm core diameter UV-grade multi-mode fiber optic (General Fiber Optics, cat. #14-200). The fiber optic is mounted in a fine-adjustment

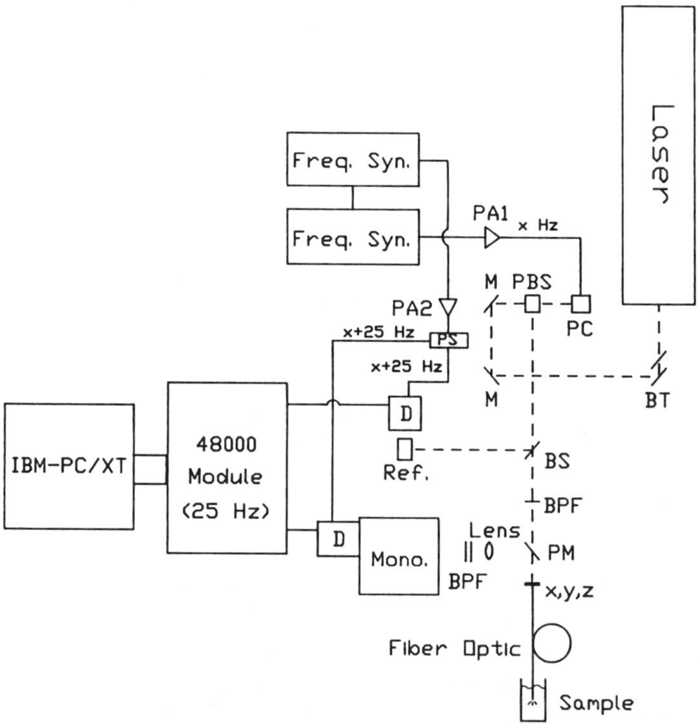

Figure 1. Schematic of the new single-fiber-based multifrequency phase and modulation fluorometer. Abbreviations represent: BT, mirrored beam translator; M, mirrors; PBS, polarizing beam splitter; PC, Pockel's cell; Freq. Syn., frequency synthesizers; PA1 and PA2, power amplifiers; BS, beam splitter; PM, perforated mirror; D, photomultiplier tube detector; BPF, band-pass filters; x,y,z, x,y,z translation stage; PS, power splitter.

x,y,z translation stage (x,y,z; Newport Research, model LP-05B) using an Amphenol connector. The optical fiber is either 10 or 175 meters in length.

The back-scattered fluorescence is collected back up the same fiber, reflected from the mirrored face of the perforated mirror (PM), and imaged by another lens onto the entrance slit of a monochromator (SLM-Aminco, Inc.) with a spectral band-pass of 8 nm. Fluorescence is detected with another photomultiplier tube (D), identical to the reference channel unit. The cross-correlation (heterodyne) frequency (16,18) is produced with a second PTS 500 frequency synthesizer amplified by a 5W power amplifier (PA2; Amplifier Research, model 5W1000). The output from this amplifier is directed simultaneously to the second dynodes of the both reference and detection photomultiplier tubes using a simple power splitter (PS; Adam Russell, model H-9). The cross-correlation frequency employed for this instrument is 25 Hz.

Control of the entire instrument is achieved with a SLM 48000 control module (SLM-Aminco, Inc.) that is in turn controlled via an IBM-PC/XT. All data regressions were performed off-line with an IBM-PC/AT computer using SLM-Aminco software or a global analysis package (19).

Experimental

Unless otherwise noted all reagents were used as received and all samples were prepared in distilled-deionized water or ethanol (Aaper Chemical Co.). Benzo[a]pyrene, fluorescein, dimethylPOPOP, rhodamine B, rhodamine 6G, Nafion, and rubrene were all from Aldrich. 2,7-dichlorofluorescein was purchased from Kodak. All samples were prepared in the appropriate solvent and sonicated for at least 15 minutes.

Rhodamine 6G-impregnated Nafion films were prepared by first casting a thin film of Nafion on microscope cover slips from the Nafion solution as purchased. After the solvent had completely evaporated, the films were dipped in aqueous Rhodamine 6G (1 m*M*) for about 10 minutes to make a "concentrated" film. For more dilute films (more pink in appearance than red) a more dilute solution (10^{-5} *M*) of rhodamine 6G was used.

For fluorescence lifetime measurements, data collection is very similar to collection with a conventional phase-modulation fluorometer. The fiber-optic probe is immersed in an appropriate reference fluorophore-containing solution (20) and the relative phase angle and demodulation factor are determined. Then the fiber optic is rinsed completely and immersed in the sample (i.e., the unknown) and the relative phase angle and demodulation are again determined for the sample. In the case when a rhodamine 6G-impregnated Nafion film was studied, the fiber was either coated with the film material or the fiber was placed in direct contact with the cast film. The sample/reference measurements are repeated in a cyclical fashion and the absolute phase angle and demodulation factor for the unknown are subsequently calculated as described elsewhere (13,14,20). Following this a new modulation frequency is selected and these measurements are repeated. In general, between

3-5 sample/reference pairs are run at each frequency and between 10-20 frequencies employed for each sample. With 15 frequencies the data collection time is approximately 45 minutes for a moderately strong fluorescent sample. Importantly, by employing reference fluorophores we also totally eliminate any color effects (13,14,20) inherent in the detectors and more importantly in this case, the intrinsic multimodal propagation within the fiber are directly and totally compensated. That is, the wavelength-dependent, dispersive nature of the multimode fiber-optic is compensated because we employ a reference fluorophore that emits over the same spectral (wavelength) region as our unknown.

For phase-resolved measurements, the fiber is simply immersed in the sample or placed in direct contact with the film. The various resolution protocols employed are identical to those described elsewhere (7,21).

Results and Discussion

Our initial concern with the new instrument is its detection abilities. In our laboratory, using approximately 100 mW of laser power (457.9 nm) launched into the fiber-optic probe we could routinely perform dynamic measurements on samples containing as little as 10 and 180 pmoles of 2,7-dichlorofluorescein with the 10- and 175-meter probes, respectively. We found further that the linear dynamic range for each fiber-optic probe extended well over 5 orders-of-magnitude. Thus, we concluded that the new instrument demonstrated sensitive detection abilities and a good linear dynamic range.

In an effort to lower the detection limits still further, we pursued the use of a gradient-index (GRIN) lens mounted at the distal end of the probe. Unfortunately, our initial attempts failed to yield better results. In fact, the detection limits were higher if anything. We have traced this poorer performance of the GRIN-lens-based system to a significant fluorescence from the attaching epoxy (Figure 2). From phase-resolved measurements we determined that this emission is from at least two components (results not shown).

We next investigated the new instrument for the determination of fluorescence lifetimes. Figures 3 and 4 show multifrequency phase and modulation plots for 1 μ*M* 2,7-dichlorofluorescein using the 10- and 175-meter probes, respectively. Fluorescein in 0.1 *M* NaOH is used as the reference fluorophore for these measurements and a lifetime of 4.00 ns used. In both cases, the recovered lifetimes agree quite well with the lifetime determined without the fiber-optic probe (3.71 ns). However, the precision of the 175-meter results are clearly much poorer owing to the lower total light flux through the optical system. Of course, this imprecision could be offset simply by averaging for a longer period of time.

To demonstrate the temporal resolving power of our instrument Figure 5 shows multifrequency phase and modulation plots for a binary mixture of fluorescein and rubrene. A single exponential fit to the data (not shown) yields a χ^2 of 49 and residual errors that deviate in a systematic pattern. However, when the data are fit to a double exponential decay model, the recovered lifetimes of 3.4 and

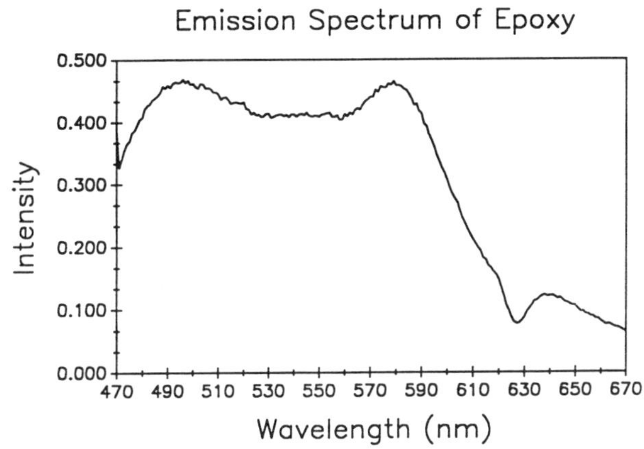

Figure 2. Steady-state emission spectrum for the epoxy used to attach the GRIN lens to the distal end of the fiber-optic probe. The pronounced dip at 630 nm is due to the Wood's anomaly in the monochromator. Laser excitation at 351.1 nm.

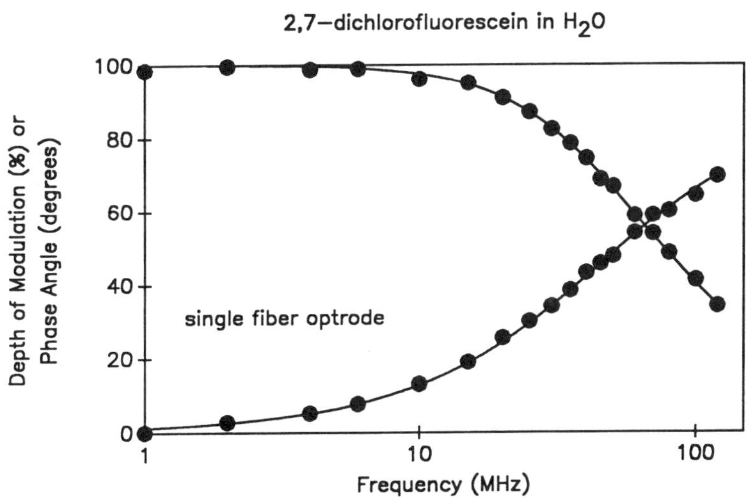

Figure 3. Multifrequency phase and modulation plot for 1 μM 2,7-dichlorofluorescein determined with the 10-meter fiber-optic probe. Laser excitation at 457.9 nm and emission monitored at 520 nm (8 nm bandpass). The recovered lifetime is 3.61 ns (χ^2 = 1.03).

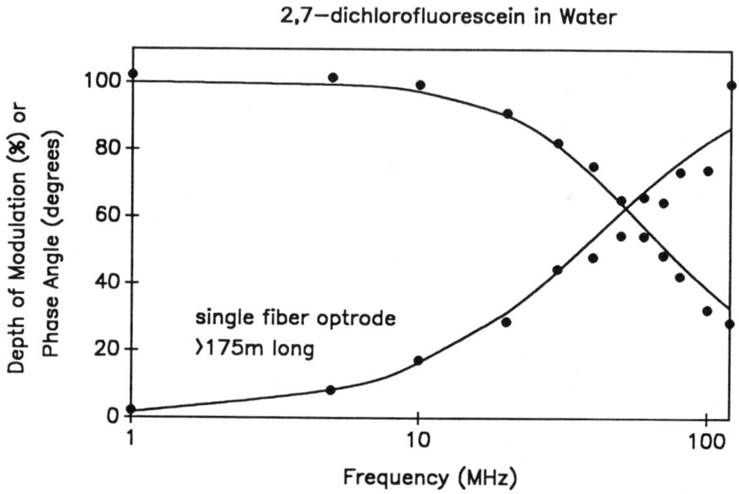

Figure 4. Multifrequency phase and modulation plot for 1 μM 2,7-dichlorofluorescein determined with the 175-meter fiber-optic probe. Laser excitation at 457.9 nm and emission monitored at 520 nm (8 nm bandpass). The recovered lifetime is 3.80 ns (χ^2 = 1.51).

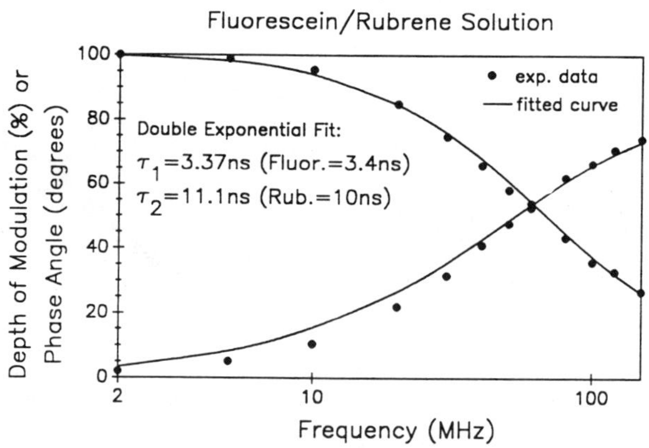

Figure 5. Multifrequency phase and modulation plot for a mixture of fluorescein and rubrene determined with the 10-meter fiber-optic probe. Laser excitation at 457.9 nm and emission monitored at 550 nm (8 nm bandpass). The recovered lifetimes of 3.4 and 11.1 ns (χ^2 = 1.26) agree quite well with the actual values of 4.0 and 10.0 ns.

11.1 ns correspond quite well with the actual values of 4.0 and 10.0 ns for fluorescein and rubrene, respectively. Moreover, a χ^2 value of 1.26 results and the residuals (not shown) oscillate randomly about zero.

One of the problems that often plagues fiber-optic-based measurements is the presence of stray or scattered light. For example, in a clinical environment one has the room lights to contend with and in remote analysis one has stray light from the environment. Phase-resolved fluorescence has been used previously to minimize scattered light affects so we felt that it could be used to great advantage in fiber-optic-based measurements also. Figure 6A shows the steady-state emission spectra for a rhodamine 6G-impregnated Nafion film (see further discussion below) sample and the associated "blank" collected using our instrument with the laboratory fully illuminated. Any obvious fluorescence from the sample is completely obliterated by the overwhelming background from the room lights. Fortunately, when we employ phase-sensitive detection (Figure 6B), null the room light contribution, and again scan we can directly recover the unperturbed fluorescence spectrum for the sample. Thus, phase-resolution can be used to eliminate the contribution of stray light to fiber-optic-based fluorescence measurements.

Phase-resolved fluorimetric measurement had also been employed previously for the simultaneous quantification of the individual components in complex mixtures (7). However, as with most spectroscopic techniques it has not been used for remote analysis. By coupling our multifrequency fluorometer to a fiber-optic probe we are capable of performing phase-resolved fluorimetric measurements on remotely located samples also. Figure 7 shows un-normalized steady-state and phase-resolved emission spectra for a binary mixture of rhodamine 6G (3.7 ns) and rhodamine B (1.8 ns). The steady-state spectrum (curve a) is clearly composed of contributions from both rhodamine 6G (curve c) and B (curve b) and it is not possible to resolve these two components using wavelength selectivity alone. Again, by employing phase-resolution we are easily able to spectrally resolve the contributions from the two components directly (Figure 7). An additional example of multicomponent phase-resolution is shown for fluorescein and rubrene mixtures (Figure 8). By choice of detector phase angle (Figure 8) we can completely (selectively) suppress the contribution from one component over another.

In previous work, Hieftje and co-workers (22) had employed rhodamine 6G-impregnated Nafion films attached to the distal end of a fiber-optic probe as humidity sensors. During the course of this work it became apparent that rhodamine 6G in films "resided" in at least two distinct environments within the Nafion polyanion matrix (film). Unfortunately, the time resolution with their instrument was not high enough to allow resolution of the individual components. Figure 9 shows the three-dimensional plot of phase-resolved fluorescence as a function of emission wavelength and detector phase angle for a "concentrated" rhodamine 6G-impregnated Nafion film. The modulation frequency for this data set is 50 MHz and laser excitation is at 457.9 nm. The total collection time is approximately 1 hour. Obviously, the information content for such a

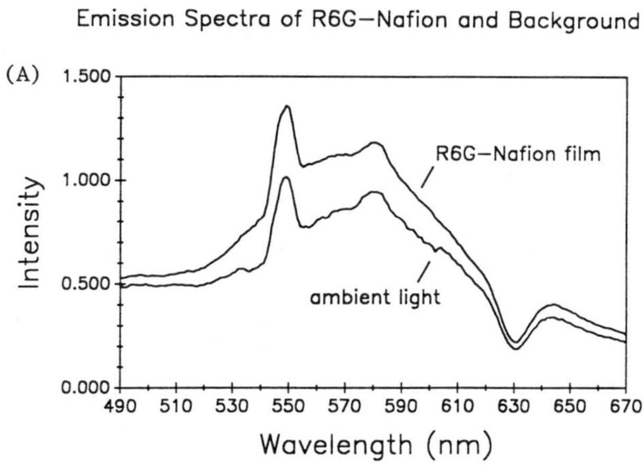

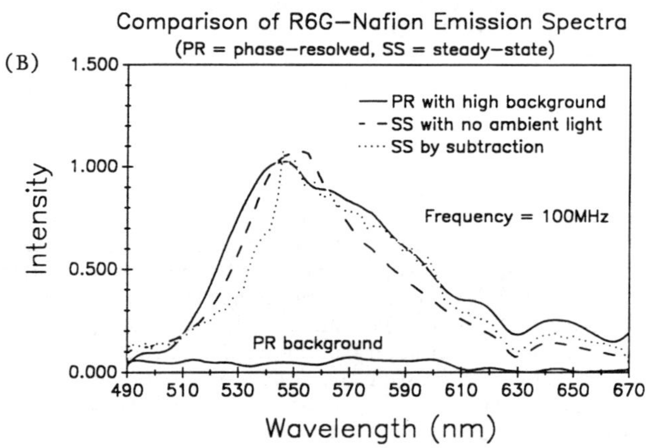

Figure 6. (A) Steady-state emission scans for a rhodamine 6G-impregnated Nafion film in the presence of room light background. (B) Phase-resolved spectra (—) of the same sample shown in A. Laser excitation at 457.9 nm; modulation frequency 100 MHz.

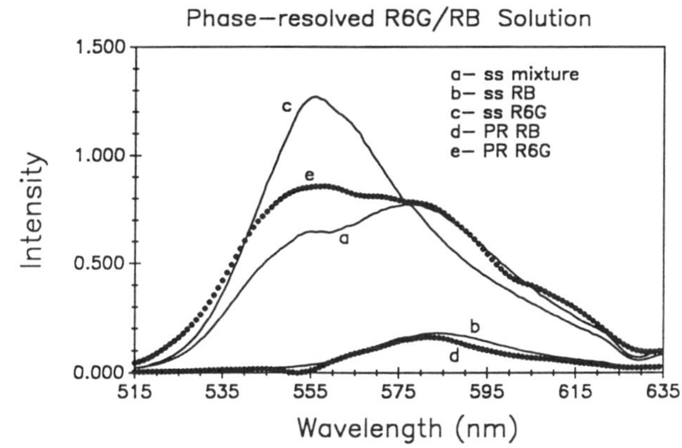

Figure 7. Steady-state and phase-resolved emission spectra for 1:1 binary mixture of rhodamine 6G and rhodamine B. Laser excitation at 457.9 nm. Curve a: steady-state emission spectrum for mixture; Curve b: steady-state emission spectrum for rhodamine B; Curve c: steady-state emission spectra for rhodamine 6G; Curve d: phase-resolved emission spectrum nulling rhodamine 6G signal; Curve e: phase-resolved emission spectrum picked to slightly enhance the rhodamine 6G component over the rhodamine B, but not null it.

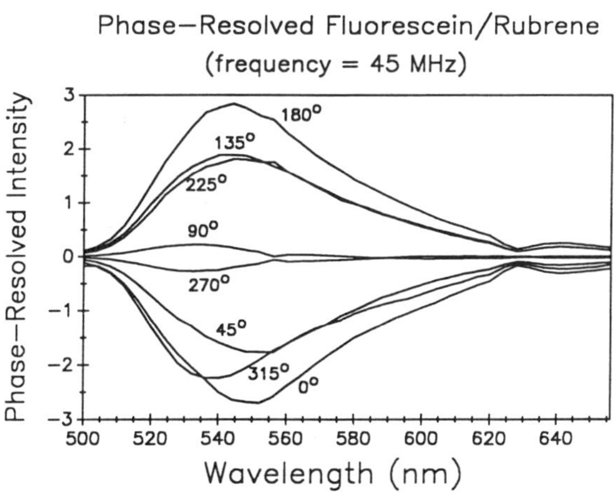

Figure 8. Phase-resolved fluorescence emission spectra for 1:1 mixture of fluorescein and rubrene showing the dependence of detector phase angle setting (curve numbers) and the selective enhancement of one component over another. The spectra are complicated by energy transfer from fluorescein to rubrene.

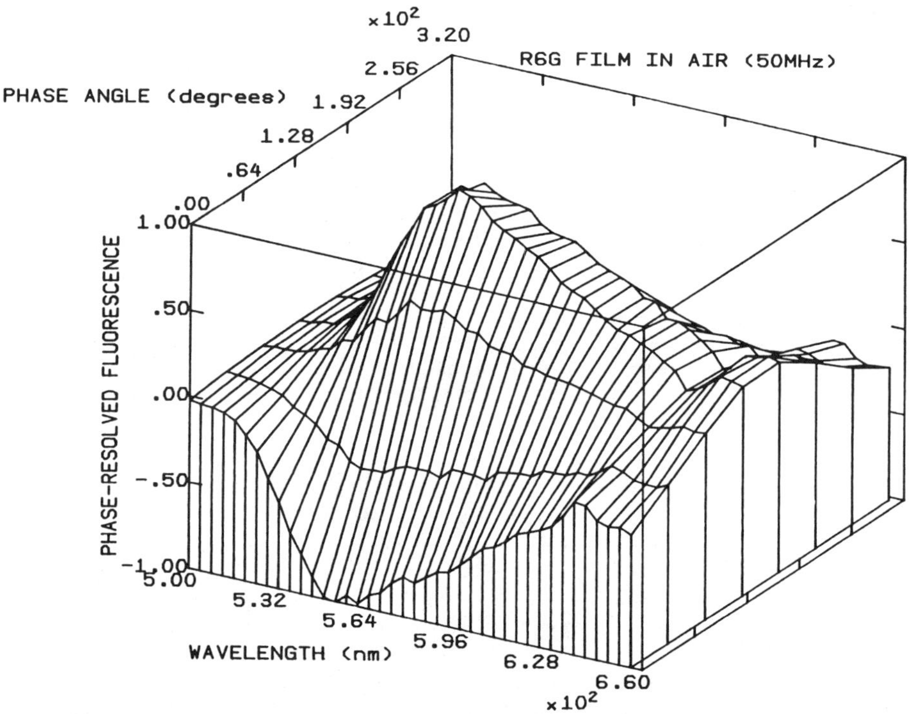

Figure 9. Three dimensional plot of phase-resolved fluorescence versus detector phase angle versus emission wavelength for a typical rhodamine 6G-impregnated Nafion film. The film was impregnated using a 1 m<u>M</u> aqueous rhodamine 6G solution. Laser excitation at 457.9 nm; modulation frequency is 50 MHz.

data set is quite high, but it is not apparent that the emission is made up of multiple components. However, a look at a narrow portion of the detector phase angle plot (Figure 10) for the same sample collected using a 100 MHz modulation frequency provides clear evidence for the presence of two distinct (9° curve). If the emission were due to a single emitting species the position of the emission maximum would remain constant as the detector phase angle is varied and any spectrum could not exhibit both positive and negative PRI regions (7). Conversely, when the emission maxima shifts as the detector phase angle varies and a sign change in the spectrum is seen one can only attribute this to emission from a second component (7). Figure 11 shows the individual phase-resolved emission spectra recovered for the rhodamine 6G-impregnated Nafion film. These spectra were recovered by suppressing the fluorescence at 500 nm and scanning the entire spectrum and by suppressing the contribution at 600 nm and scanning. Again, if the emission were from a single component both spectra would have been identical. In this case, the spectra are not identical. In fact, the dashed vertical line in Figure 11 can be used to determine the relative difference in emission maxima between the two components. In this particular case, the difference between the two component maxima is approximately 7 nm. Resolution of this complex emission process was impossible prior to the development of our fiber-optic-based multifrequency instrument.

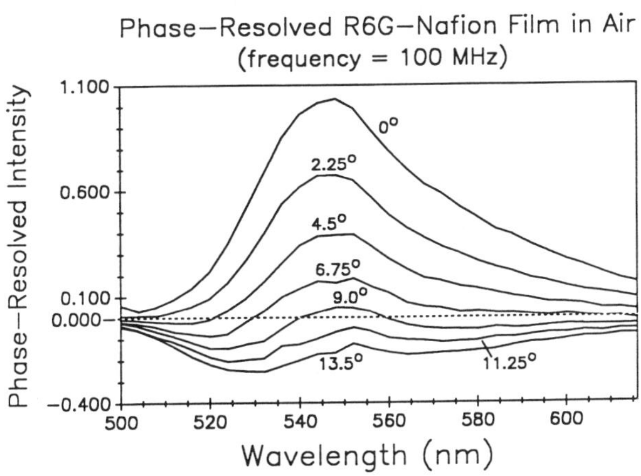

Figure 10. Phase-resolved fluorescence emission spectra for the same sample shown in Figure 9 with a narrow detector phase angle region scanned (0 to 14°). Note that the emission maxima shifts with detector phase angle and the sign of the phase-resolved intensity changes (curve 9.0°). Both these observations are indicative of the presence of two components. Laser excitation at 457.9 nm; modulation frequency is 100 MHz.

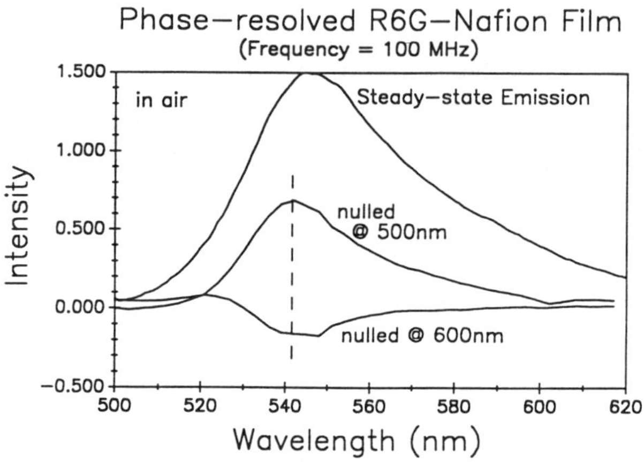

Figure 11. Phase-resolved fluorescence spectra for the individual rhodamine 6G components "contained" in the rhodamine 6G-impregnated Nafion film. The dashed vertical line is used to judge the difference in the emission spectra (7 nm).

So, to better understand the rhodamine 6G-Nafion film dynamics, lifetimes of the "dilute" and "concentrated" films were measured. Using the fiber optic, the "dilute" film was found to have a single emitting component of 3.84 ± 0.18 ns which corresponds well to free rhodamine 6G in solution (3.7 ns). However, the "concentrated" film results were always best described by more complex decay process.

Since no polarizers are used with the fiber optic, the second emitting component in the "concentrated" film could have been attributed to a different vibronic state of the rhodamine 6G. This was disproved by determining the lifetimes of the film species through the fiber and in the sample chamber where the polarization was accurately controlled (0° excitation, 54.7° emission). The results for both approaches were the same. Both configurations exhibited phase shifts greater than 90°, which is indicative of an excited-state process (23). Using global analysis, the first lifetime (fixed at 4.0 ns) was typically 30 to 33% of the fluorescence. The second lifetime depended critically on the humidity surrounding the film. Generally, it decreased from a maximum of 15 ns to 5 ns as humidity increased. We are presently performing additional experiments to help elucidate this complex decay process.

Acknowledgments

The authors would like to acknowledge Jingfan Huang for aiding in the collection of several of the spectra shown herein. This work was supported by BRSG S07 RR 07066 awarded by the Biomedical Research Support Grant Program, Division of Resources, National Institutes of Health, by the Donors of the Petroleum Research Fund, administered by the American Chemical Society (Grant # 19620-G4), the Health Care Instruments and Devices Institute at SUNY-Buffalo, by a Non-Tenured Faculty Grant from 3M, Inc., by a New Faculty Development Award from the New York State/United University Professions, and by the Center for Advanced Technology (SUNY-Buffalo).

Literature Cited

1. Warner, I. M.; Patony, G.; Thomas, M. P. Anal. Chem. 1985, 57, 463A.
2. Bright, F. V. Anal. Chem. 1988, 60, 1031A.
3. Weber, G. Nature 1961, 190, 27.
4. Ainsworth, S. J. Phys. Chem. 1961, 65, 1968.
5. Warner, I. M.; Christian, G. D.; Davidson, E. R.; Callis, J. B. Anal. Chem. 1977, 49, 564.
6. Warner, I. M.; Fogarty, M. P.; Shelly, D. C. Anal. Chim. Acta 1979, 109, 361.
7. McGown, L. B.; Bright, F. V. C.R.C. Crit. Rev. Anal. Chem. 1987, 18, 245.
8. Knutson, J. R.; Davenport, L.; Brand, L. Biochem. 1986, 25, 1805.
9. Bright, F. V. Appl. Spectrosc. 1988, 42, 1245.
10. Angel, S. M. Spectroscopy 1987, 2(4), 38.
11. Seitz, W. R. C.R.C. Crit. Rev. Anal. Chem. 1988, 19, 135.
12. Bright, F. V.; Monnig, C. A.; Hieftje, G. M. Anal. Chem. 1986, 58, 3139.
13. Bright, F. V. S.P.I.E. Conf. Proc. 1988, 909, 23.
14. Bright, F. V. Appl. Spectrosc. 1988, 42, 1531.
15. Lakowicz, J. R. Principles of Fluorescence Spectroscopy, Plenum Press, New York 1983.
16. Spencer, R. D.; Weber, G. Ann. N.Y. Acad. Sci. 1969, 158, 361.
17. Jameson, D. M.; Gratton, E.; Hall, R. D. Appl. Spectrosc. Rev. 1984, 20, 55.
18. Spencer, R. D. Ph.D. Thesis, University of Illinois, Urbana, IL 1970.
19. Beecham, J. M.; Gratton, E. S.P.I.E. Conf. Proc. 1988, 909, 70.
20. Lakowicz, J. R.; Cherek, H.; Balter, A. J. Biochem. Biophys. Meth. 1981, 5, 131.
21. Bright, F. V. Ph.D. Thesis, Oklahoma State University, Stillwater, OK 1985.
22. Zhu, C.; Bright, F. V.; Wyatt, W. A.; Hieftje, G. M. Proc. Symp. Chem. Sensors 1987, 87-89, 476, Turner, D. R., Ed., Electrochem. Soc.
23. Lakowicz, J. R.; Balter, A. Biophys. Chem. 1982, 16, 117.

RECEIVED March 9, 1989

INDEXES

Author Index

Aizawa, Masuo, 129
Alarie, Jean-Pierre, 318
Angel, S. M., 345
Arney, Larry, 143
Arnold, Mark A., 303
Ballantine, David S., Jr., 157,222
Barnard, Steven, 252
Baur, John, 114
Bowyer, James R., 318
Bright, Frank V., 380
Brinker, C. J., 208
Brown, R. S., 331
Burgess, Lloyd, 143
Buttry, Daniel A., 237
Chaniotakis, N. A., 26
Chau, Lai-Kwan, 283
Coury, Louis A., Jr., 78
Dessy, Raymond E., 140,143
Dyne, K., 331
Frye, G. C., 191,208
Galiatsatos, Christos, 78
Geise, Robert J., 65
Giuliani, John F., 364
Hafeman, Dean G., 46
Heineman, William R., 22,78
Hill, H. Allen O., 105
Hougham, B. D., 331
Hoyt, Arthur M., 318
Huber, Edward W., 78
Ikariyama, Yoshihito, 96,129
Janata, Jiri, xi
Jarvis, N. L., 157
Jones, Thomas P., 283
Klein, Napthali P., 105
Kristensen, Eric, 114
Krull, U. J., 331
Kulp, T. J., 345
Langry, K., 345
Lasky, Steven J., 237
Leszczyszyn, David, 114
Litwiler, Kevin S., 380
Luo, Shufang, 252
Martin, S. J., 191,208
May, Leslie J., 114
McConnell, Harden M., 46
Meyerhoff, M. E., 26
Munkholm, Christiane, 252
Murray, Royce W., 1
Murthy, A. Surya N., 105
Myrick, M. L., 345
Niemczyk, T. M., 191
Parce, J. Wallace, 46
Park, S. B., 26
Petersen, James, 143
Porter, Marc D., 283
Pranitis, D. M., 26
Psalti, Ioanna S. M., 105
Ricco, A. J., 191,208
Ridley, M. N., 345
Seitz, W. Rudolf, 248,273
Sepaniak, Michael J., 318
Shimada, Naoki, 96
Sigal, George B., 46
Sponaugle, Sandra, 78
Stole, Scott M., 283
Tanaka, Masaru, 129
Tromberg, Bruce J., 318
Vandenberg, E. T., 331
Vo-Dinh, Tuan, 318
Walt, David R., 252
Wangbai, Ma, 273
Wightman, R. Mark, 114
Wohltjen, Hank, 157,222
Yacynych, Alexander M., 65
Yamauchi, Shigeru, 96
Yim, H. S., 26
Yuan, Ping, 252
Zellers, Edward T., 176
Zhujun, Zhang, 273

Affiliation Index

Geo-Centers, Inc., 222
Indiana University, 114
Inorganic Chemistry Laboratory and Oxford Centre for Molecular Sciences, 105
Iowa State University, 283
Lawrence Livermore National Laboratory, 345
Microsensor Systems, Inc., 157,222
Molecular Devices Corporation, 46
National Rehabilitation Center for the Disabled, 96
Northern Illinois University, 157
Oak Ridge National Laboratory, 318
Rutgers, The State University of New Jersey, 65
Sandia National Laboratories, 191,208
State University of New York at Buffalo, 380
Tokyo Institute of Technology, 129
Tufts University, 252
U.S. Naval Research Laboratory, 364
University of Central Arkansas, 318
University of Cincinnati, 22,78
University of Iowa, 303
University of Michigan, 26,176
University of New Hampshire, 248,273
University of North Carolina, 1
University of Tennessee, 318
University of Toronto, 331
University of Utah, xi
University of Wyoming, 237
Virginia Polytechnic Institute and State University, 140,143

Subject Index

A

Acetaminophen, response of electropolymerized films, 70,71*f*,*t*,72
Acetylcholine receptor, function as neurotransmitter receptor, 333
Acetylcholine-receptor-mediated optical transduction from lipid membranes
- experimental procedures, 334–335
- fluorescence spectrum of labeled receptor, 335,336*f*,337
- future research, 341,343
- incorporation of *N*-(7-nitrobenz-2-oxa-1,3-diazole-4-yl)-*N*-phosphatidylethanolamine, 338–339,340*f*
- linear approximation of fluorescence enhancement, 341,342*f*
- material preparation, 334
- measurement of enhancement, 335
- reproducibility, 335,337*t*
- role of fluorescent lipid component in fluorescence enhancement mechanism, 338–339,340*f*
- selective receptor mediation of enhancement process, 337–338

Acoustic plate mode device
- acoustoelectric coupling, 196,203
- applications in chemical sensing, 192
- calibration of mass sensitivity, 201,202*f*
- comparison of mass sensitivity to that of other devices, 201,203

Acoustic plate mode device—*Continued*
- comparison to quartz crystal microbalance, 197
- comparison to surface acoustic wave device, 197
- detection of aqueous Cu^{2+}, 203
- experimental reagents, 200
- fabrication, 199–200
- frequency of coupling between transducer and substrate, 192
- fundamentals, 192–193
- instrumentation, 200
- instrumentation of sensors, 197,198*f*,199
- liquid entrainment, 195–196
- mass sensitivity, 193,195
- mode resolution, 200–201
- schematic representation, 193,194*f*
- sensitization to species in solution, 192
- surface derivatization, 200
- viscous coupling, 203,204*f*,205

Acoustic wave sensors, description, 191
Acoustoelectric coupling acoustic plate mode device, 196,205
Affinity reagent phases, examples, 318–319
Alamethicin, response of photocurrent amplitude to step in bias potential, 60,62*f*,63
Alkaline phosphatase, catalysis of reaction, 305
Amplification technology for sensors
- examples of sensors, 260–261,262*f*

Amplification technology for sensors—*Continued*
optimization, 259–260
procedure, 256,257*f*
Analytical bioelectrochemistry, development, 106
Anion-selective membrane components, use of metal–ligand complexes, 27
Anion-selective membrane electrodes
design of potentiometric gas-sensing systems, 38–43
deviation from Hofmeister behavior, 29
effect of pH on response, 35,37*f*
electrodes using bis(diethyldithiocarbamato)mercury(II), 35–38
Hofmeister pattern, 29
potentiometric selectivity coefficients, 30,32,34–36
response properties, 29
response vs. metalloporphyrin, 30,32,33*f*
selective membranes using metalloporphyrins as ionophores, 30–35,37
Antibodies, generation, 319
Antibody-based fiber-optic chemical sensors
critical instrument design considerations, 319
fluoroimmunoassay principles, 319–320
instrumentation, 320–321,322*f*
optical configuration, 321,322*f*
sensing based on competitive-binding fluoroimmunoassay, 321,323,324*f*
sensing based on direct immunoassay, 323,325,326*f*
sensing terminal designs, 321,324*f*
total internal reflection, 320–321,322*f*
Aqueous Cu^{2+}, detection by acoustic plate mode device, 203
Ascorbic acid, effect on response of catechols to poly(*N*-vinylpyrrolidone)-modified electrodes, 86,87*f*,88
Attachment of sensing reagents to fiber tip
amplification technology, 256,257*f*,259–260
indicator derivatization, 258–259
polymerization, 259
surface activation, 256,258

B

Band microelectrode
advantages, 107
collection current vs. electrode gap, 107,111*f*,112
enzymatic reaction vs. collection current, 107,110*f*
BET analysis, description, 210
Biocatalytic-based biosensors
description, 304
examples, 304
general schematic, 304,306*f*
Bioelectrochemistry at microelectrodes
advantages, 106–107
band microelectrodes, 107,110–111*f*,112
disc microelectrodes, 107,108–109*f*
Bioreceptor-based biosensors, examples, 304
Biosensors
applications, 65
classes, 304
definitions, 303–304
desired qualities for thermally responsive sensors, 143
fiber-optic-based enthalpimeter, 143–149
ideal configuration, 66
interferences, 65–66
oxidation of hydrogen peroxide, 65
polymer-film-based enthalpimeter, 149–155
preparation with amplification technology, 261,263,264*f*
schematic representation of surface morphology, 68,69*f*
use of electropolymerized films in construction, 66–76
Biosensors based on lipid membrane electrochemistry, limitations for analytical applications, 332

C

Carbon–fiber microelectrode
chemical heterogeneity of neurotransmitter storage and release in rat brain, 123–127
concentration profile at outlet of chromatographic loop injector, 115–119
concentration profile in chromatographic columns, 118–123
preparation, 114–115
Catechols, determination by poly(*N*-vinylpyrrolidone)-modified electrodes, 82,85*f*,86
Cellulose acetate films
advantages for use in sensors, 285,288
base hydrolysis, 289
characterization of microstructure and permselectivity at glassy carbon, 290–295,297
construction of multicomponent pH sensor, 298,299*f*,300
dye immobilization, 289
electrochemical characterization, 291,292*f*,293
fabrication as optical sensor, 285
hydrolysis scheme, 295,297*f*

Cellulose acetate films—*Continued*
instrumentation, 290
IR external reflection spectroscopic characterization, 293,294*f*,295,297*f*
model for structural evolution, 295,297*f*
permeability vs. molecular composition, 295
preparation, 288–289
reagents, 290
scope of sensor fabrication for immobilized direct dyes, 295–300
Chemical heterogeneity of neurotransmitter storage and release in rat brain
description of carbon–fiber electrode used for measurement, 123–124
discussion, 123
pattern of heterogeneity, 124,126*f*
preparation of rats for study, 124
results from single stimulation, 125,125*f*
Chemical inhomogeneities, measurements with chemical sensors, 114
Chemical sensor(s)
advantages of use of fiber optics, 129
definition, 1
measurements of chemical inhomogeneities, 114
selectivity vs. reversibility, 178
uses of piezoelectric materials, 7–12
Chemical sensor based on acoustic wave devices, applications, 191
Chemical sensor based on electrical phenomena
conductivity, 2–3
electrochemical reaction currents, 307
potentiometry, 3
Chemical sensor based on optical fibers and waveguide
coupling of light-transmitting medium to chemical medium, 12,14
design of indicator phases, 15–16
elicitation and detection of optical response, 14
nature of indicator phase, 14–15
optical fibers, 12
schemes for coupling indicator phases with optical elements, 12,13*f*
Chemical sensor design
impact of technology on research, 2
lipid bilayer measurement, 17
thermal sensors, 17
Chemical transducer, *See* Chemical sensor(s)
CO_2 sensor, preparation with amplification technology, 260–261
Coated surface acoustic wave sensor
apparatus, 181
description, 178
experimental procedure, 181–182
oscillator frequency vs. film deposition, 178–179
Coated surface acoustic wave sensor—*Continued*
response vs. vapor–coating interaction, 179
use of *trans*-directing ligands as reagent, 179–181,183*f*
vinyl acetate vapor measurement, 182–189
Coating deposition methods, surface acoustic wave device, 164
Competitive-binding assays, description, 320
Competitive-binding fluoroimmunoassay, sensing for fiber optics, 321,323,324*f*
Conductivity, electrical, measurements by chemical sensors, 2–3

D

Design of indicator phases
immobilization of polymer binding, 16
monolayers, 15
partitioning polymers, 16
trapping membranes, 15–16
Detectors with enhanced gas sensitivity based on differential gas-sensing membrane electrode
advantages of using two electrodes, 39
calibration curve, 42,43*f*
determination of differential cell potential, 41–42
schematic representation, 41,43*f*
Diffusion coefficients, determination with surface acoustic wave devices, 210–211
Diffusion in polymer films
advantages of surface acoustic wave device techniques, 220
diffusivity values, 218,219*f*,220
frequency transients, 218,219*f*
Direct assays, description, 320
Direct dyes
structures, 296
utility as optical pH sensor at hydrolyzed cellulose acetate, 296,297*f*,298
Direct fluoroimmunoassay, sensing for fiber optics, 323,325,326*f*
Disc microelectrode
limiting size, 107,108*f*
use in determination of electrochemically coupled enzymatic oxidation of glucose, 107,108–109*f*
Dispersion at outlet of loop injection
current–time traces after introduction of dopamine, 115,117*f*
time required to reach half of steady-state current vs. position of electrode, 118,119*f*
tube used with microvoltammetric electrode, 115,116*f*

Dispersion in chromatographic columns
characterization of radial dispersion, 118,120–123
chromatograms measured at radial positions, 120,121*f*
geometry of electrochemical cell, 120,121*f*
normalized radial concentration distribution, 120,122*f*
normalized retention time and reduced axial plate height, 120,122*f*
separation efficiency, 120
Dopamine, detection in brain, 123–124,127
Dual-wavelength-absorption fiber-optic sensor for trichloroethylene and chloroform
apparatus, 346,347*f*
calibration curves, 349,350*f*,351,352*f*
future research, 351
instrumental setup, 346,348*f*,349
measurement of groundwater contaminants, 346
reagents, 346
response vs. exposure time, 349,350*f*,351,352*f*

E

Elastic modulus, determination by surface acoustic wave device, 235
Elastic properties of thin polymer films by using surface acoustic wave devices
background, 223–225,228*f*
calculation of glass transition temperature, 226
description of surface acoustic wave devices, 226
determination of glass transition temperature, 231,233–234
determination of melting point, 231,233–234
effect of temperature on frequency, 227,228–230*f*
frequencies of delay lines, 226
polymer coatings and physical parameters, 225*t*
slope of frequency–temperature curves, 227,231,232*f*
Elastic waves, types generated on piezoelectric materials, 157
Electrical conductivity measurements, chemical sensors, 2–3
Electrochemical characterization of cellulose acetate films
electrochemical charge vs. hydrolysis time, 291,292*f*,293
Electrochemical characterization of cellulose acetate films—*Continued*
molecular dimensions of electrochemical probes, 291*t*
Electrochemical potential, measurements by chemical sensors, 3
Electrochemical reaction currents
measurement by chemical sensors, 3–7
schemes for coupling enzyme reactions, 4,5*f*,6
Electrochemical sensor(s)
development for use in biological systems, 78
enhancement of chemical selectivity by polymer films, 6
Electrochemical sensor design, impact of miniaturization of electrodes and electrochemical cells, 6–7
Electrochemical sensor with porous polymer film, applications, 284
Electrochemically coupled enzymatic oxidation of glucose
dependence of catalytic current on glucose concentration, 107,109*f*
effect of collection current on glucose concentration, 107,111*f*,112
effect of reaction on collection current, 107,110*f*
reaction sequence, 107,108*f*
Electrochemiluminescence of luminol
calibration curve in presence of H_2O_2, 135,137*f*
cyclic voltammogram, 130,134*f*
differential pulse polarogram, 134*f*,135
effect of dissolved oxygen, 135,137*f*
fabrication of fiber-optic electrode, 130,132*f*
luminescent intensity vs. concentration, 135,136*f*
luminescent intensity vs. negative-potential scan, 135*f*
pulse interval optimization for excitation, 135,136*f*
Electrochemistry of biological molecules
developments, 105
problems with enzyme studies, 106
Electrochemistry of protein–protein complexes, electron transfer, 105–106
Electrode(s), schemes for coupling enzyme reactions, 4,5*f*,6
Electrode fouling, resistance to poly(*N*-vinylpyrrolidone)-modified electrodes, 88,89,90*f*
Electrode surfaces, advantages of modification with polymer films, 78–79
Electropolymerization, advantages, 67
Electropolymerized films for biosensor construction
apparatus, 67

Electropolymerized films for biosensor construction—*Continued*
calibration curves, 72,73*f*
comparison of polymer films, 74–75*f*,76
electropolymerization, 70
experimental procedure, 68
flow-injection analysis system, 68,69*f*
interference, 70,71*t*
long-term response to acetaminophen, 70,72
materials, 67–68
response–recovery time, 72,74
response to acetaminophen, 70,71*f*
schematic representation of biosensor surface, 68,69*f*
Energy-transfer fibers, description, 263,265
Enhancement, measurement, 335
Enthalpimeters, biosensor applications, 143–156
Enzyme(s)
coupling to electrodes, 4,5*f*,6
problems with electrochemical studies, 106
Enzyme-based optical biosensors, preparation with amplification technology, 261,263,264*f*
Enzyme electrodes, use of discrete–macroscopic membranes, 66
Excitation wave form, definition, 381
Extrinsic sensor, description, 253

F

Ferrocene derivatives, use as mediator in bioelectrochemistry, 106
Fiber optic(s)
advantages for use in chemical sensors, 129
application as chemical sensors, 143
Fiber-optic ammonia-probe-based biosensors
calibration curve for glutamate, 315,316*f*
membrane phases involved in response, 312,314*f*,315
Fiber-optic-based biocatalytic biosensors
development, 304
example of internal enzyme biosensor, 312,313–314*f*
for NADH, 305,307–311
for *p*-nitrophenyl phosphate, 305,306*f*
immobilization of enzyme at ammonia probe, 312,314*f*,315
problems with biocatalyst, 309,312*f*
Fiber-optic-based enthalpimeter
calibration curve, 146,148*f*
comparison to polymer-film-based enthalpimeter, 154,155*f*,156
instrument design, 144,145*f*,146
interferences, 149
Fiber-optic-based enthalpimeter—*Continued*
interferometer output, 146,147*f*
modeling of heat transport, 146,149
schematic representation of instrument, 144,145*f*
sensitivity to environmental noise, 149
Fiber-optic-based fluorescence sensing, approaches, 381
Fiber-optic chemical sensor(s)
advantages of poly(vinyl alcohol)-based indicators, 273
applications, 318
function, 129
Fiber-optic chemical sensor for continuous monitoring
advantages, 252–253
classifications, 253
examples prepared with amplification technology, 260–264
excitation and emission spectra of fluorescein, 253,255*f*,256
methods for attaching sensing reagents to fiber tip, 256–260
schematic representation, 253,254*f*
sensors based on fluorescence energy transfer, 263–267
sensors based on irreversible chemistry, 267,268,269*t*,270*f*
Fiber-optic chemical sensor with poly(vinyl alcohol)-based indicators
apparatus, 274
Mg(II) sensor, 275,277–282
pH sensor, 275,276*f*
procedures, 274–275
reagents, 274
Fiber-optic electrode
characterization, 130,133*f*
fabrication, 130,132*f*
schematic representation of configuration, 130,132*f*
use in electrochemiluminescence of luminol, 130–137
Fiber-optic internal enzyme biosensor
ethanol calibration curve, 312,314*f*
reaction scheme, 312
schematic representation, 312,313*f*
Fiber-optic sensors
field applications, 345–346
laboratory applications, 345
types, 345
Flow-injection analysis system, schematic representation, 68,69*f*
Fluorescein
copolymerization of molecules into polymer attached to fiber-optic tip of sensors, 256,257*f*
excitation and emission spectra, 253,255*f*,256

Fluorescence
determination of receptor function, 333
phase resolution, 383
Fluorescence lifetimes, definition, 382
Fluorescence spectroscopy
advances due to fiber-optic development, 381
advantages, 380
limitations, 380–381
Fluorescent lipid component, role in fluorescence enhancement mechanism, 338–339,340*f*
Fluoroimmunoassay
classifications, 319–320
measurement protocols, 320
Formate ion, colorimetric detection, 368,370*f*
Frequency-dependent demodulation, definition, 382
Frequency-dependent phase, definition, 382
Frequency-domain lifetime determinations, theory, 381–383

G

Gamma (γ) radiation immobilization of polymer films
advantages, 80
description, 79–80
dosage profile vs. radius, 83,83*f*
experimental procedures, 80,81*f*,82,83*f*
irradiation apparatus, 80,81*f*
Gamma (γ) rays, description, 79
Gas flow system, use in surface acoustic wave devices, 212,213*f*,214
Gas-sensing systems, potentiometric, *See* Potentiometric gas-sensing systems
Gas sensors, preparation with amplification technology, 260–261,262*f*
Gasoline sensor, preparation with amplification technology, 261,262*f*
Generic transduction mechanism for biosensor development, basis, 331
Geothermal-well measurements
calibration curve of fiber-optic sensor, 358,359*f*
instrumentation, 358,360
temperature fiber-optic sensor spectral response, 358,359*f*
temperature profiles of fiber-optic sensor, 360–361*f*
Glucose
detection by quartz crystal microbalance, 237–244
electrochemically coupled enzymatic oxidation, 107–112
Glucose—*Continued*
real-time determination by pulse voltammetric biosensing device, 96–103
schemes for detection in liquids, 237
Glucose oxidase
immobilization in poly(*N*-vinyl alcohol)-modified electrodes, 91,92*f*,93
reaction for glucose measurement, 91
Glutamate biosensor
glutamate response curve, 315,316*f*
reaction scheme, 309
response curve, 309,313*f*
Gramicidin D, effect on lipid bilayer resistance, 60,62*f*

H

Half acceptance angle, determination, 320
Hofmeister pattern, description, 29
Homogeneous immunoassay based on electrochemical immunoassay
principle, 129–130,131*f*
use of fiber-optic electrode to improve sensitivity, 130–137

I

Immobilization of glucose oxidase in poly(vinyl alcohol)-modified electrodes
effect of γ irradiation dose, 91,92*f*,93
response to glucose, 91,92*f*
Immunoglobulin G, structure, 319
Indicator of sensors, derivatization, 258–259
Indicator phase
definition, 12
design, 15–16
Intrinsic sensor, description, 253
Ion-selective electrodes
advances in development, 26–27
evaluation of selectivity of polymer membrane electrodes, 17
See also Anion-selective membrane electrodes
Ion-selective membranes
selective detection of ammonia in solution, 27
use in static and flow-through gas-sensing configurations, 27,28*f*
IR external reflection spectroscopic characterization of cellulose acetate films, compositional evolution vs. hydrolysis, 293,294*f*

L

Lactate dehydrogenase, catalysis of NADH reaction, 307
Lifetimes, frequency domain, 381–383
Light-addressable potentiometric sensor (LAPS)
  alternating photocurrent amplitude vs. bias potential, 49,51*f*
  amplitude of insulator charging current through external circuit, 49
  analytical applications, 46
  capacitance, 47
  electrical measurements on phospholipid bilayer membranes, 59,61*f*
  equivalent electrical circuit with phospholipid bilayer, 50,51*f*
  experimental materials, 53
  formation of phospholipid bilayer membranes, 55
  generation of hole–electron pairs, 47,49
  ionophores and ion channels, 55–56
  lipid bilayer membrane apparatus, 53,54*f*
  measurement
    membrane capacitance, 50,52
    membrane resistance, 52–53,54*f*
    photocurrent with lipid bilayer membrane apparatus, 55
    transmembrane potential, 50
  monitoring of chemically sensitive surface potentials, 49–50
  partial equivalent circuit, 48*f*,49
  response of photocurrent amplitude to step in bias potential in presence of alamethicin, 60,62*f*,63
  schematic diagram, 47,48*f*
  sensor photocurrent amplitude and phase vs. capacitance, 56,57*f*
  specific conductance of lipid bilayer membrane
    vs. gramicidin D concentration in solution, 60,62*f*
    vs. valinomycin concentration in solution, 59–60,61*f*
  theory, 47–53
  time constant of sensor photocurrent response to step in bias potential vs. resistance, 56,58*f*
  use with phospholipid bilayers, 46
Lipid bilayer(s), measurement by chemical sensors, 17
Lipid bilayer membrane apparatus
  cross-sectional view, 53,54*f*
  measurement of photocurrent, 55
Lipid matrices, ability to act as generic transducers, 332–333
Lipid membrane(s), function, 331–332
Lipid membrane perturbations, fluorescence spectroscopic analysis, 332
Liquid entrainment, acoustic plate mode device, 195–196
Liquid-phase sensors, acoustic plate mode devices, 191–205
Loop injector, dispersion of output, 115
Luminol, electrochemiluminescence, 130–137

M

Mammalian brain
  heterogeneous structure, 123
  neurotransmission, 123
Mass-sensitive applications of piezoelectric materials, schemes for using quartz crystals, 8,9*f*
Mass sensitivity
  acoustic plate mode device, 193,195
  surface acoustic wave device, 159,160*t*,161*f*,162
Membrane capacitance, measurement by LAPS, 50,52
Membrane resistance, measurement by LAPS, 52–53,54*f*
Metalloporphyrins
  potentiometric selectivity coefficients, 30*t*
  structures, 30,31*f*
Mg(II) sensor
  absorption spectra of eriochrome black T, 277,278*f*
  effect of eriochrome black T on response, 281*f*,282
  fluoresceinamine, 275,277,278*f*
  fluorescence intensity vs. added Mg(II), 277,279*f*
  indicator emission spectrum, 280,281*f*,282
  intensity vs. time, 280*f*
  reactions involving eriochrome black T, 277,278*f*
Microbiosensing, performance characteristics for glucose, 103*t*
Microelectrodes
  advantages, 106–107
  band, 107,110–111*f*,112
  disc, 107,108–109*f*
Microfabricated chemical sensors, advantages for industrial-hygiene-monitoring instruments, 177
Microfabrication, advantages, 96
Miniaturization of electrodes and electrochemical cells, effect on electrochemical sensor design, 6–7
Monitoring of toxic chemical exposure
  inadequacy of conventional sampling methods for contaminant collection, 176–177
  limitations of direct-reading instruments, 177

Monitoring of toxic chemical exposure—*Continued*
problems, 176
Multidimensional fluorescence using simple fiber-optic-based probe
detection limits, 386,387*f*
elimination of stray light in measurements, 389,390*f*
experimental materials and procedures, 385–386
instrumentation, 383,384*f*,385
lifetime(s), 381–383
lifetime determination, 386,387–388*f*
phase resolution, 383
phase-resolved fluorescence vs. emission wavelength vs. detector phase angle, 389,392–393*f*,394
phase-resolved fluorometric measurements on remotely located samples, 389,391*f*
temporal resolving power, 386,388*f*,389
Multielement optical-waveguide-sensor arrays
block diagram of sensing system, 373,376*f*,378
configurations, 373,376*f*
oscilloscope traces of colorimetric response, 377*f*,378

N

Near-IR surface-enhanced Raman spectroscop
advantages, 353
future remote research, 358
geothermal-well measurements, 358,359*f*,360,361*f*
potential applications, 353
spectra of 3-chloropyridine, 353–354,355*f*
spectra of 3-chloropyridine and picoline, 354,355*f*
spectra of pyridine, 356,357*f*
spectra of tris(2,2-bipyridine)-ruthenium(II), 356,357*f*,358
use of diode laser, 354,356,357*f*,358
Nicolsky equation, determination of selectivity of polymer membrane electrodes, 27
N-(7-Nitrobenz-2-oxa-1,3-diazole-4-yl)-*N*-phosphatidylethanolamine, fluorescence enhancement, 338–339,340*f*
p-Nitrophenyl phosphate, fiber-optic-based biocatalytic biosensors, 305,306*f*
p-Nitrophenyl phosphate fiber-optic biosensor, response curve, 305,306*f*

O

Offset term, definition, 118
Optical biosensors, description, 129
Optical fiber(s), functions, 252
Optical-fiber- and optical-waveguide-based chemical sensors
coupling of light-transmitting medium to chemical medium, 12,14
design of indicator phases, 15–16
elicitation and detection of optical response, 14
nature of indicator phase, 14–15
optical fibers, 12
schemes for coupling indicator phases with optical elements, 12,13*f*
Optical sensors with porous-polymer-film
characteristics for chemical analysis of aqueous solutions, 284–285,286–287*t*
influence of immobilization on reactivity of colorimetric reagents, 284–285,286–287*t*
molecular-level descriptions of microstructure, 285
response time vs. utility, 285,286–287*t*
strategies for design, 284
use of cellulose acetate films, 285–300
Optical-waveguide chemical sensors
application to selected vapors, 365,367*t*
development, 364
dynamic response of thiolene sensor to acetone vapors, 371*t*,372*f*
fabrication, 368,371*t*,372*f*
geometries, 365,366*f*
multielement arrays, 373,376–377*f*,378
schematic representation, 365,366*f*
solution-phase small-molecule detection, 367,368*t*,369–370*f*
vapor–polymer solubility interactions, 371,373,374–375*f*
Optoelectronic enzyme sensor, construction with luminescence-generating enzyme, 129
Optrodes
definition, 129
See Fiber-optic sensors
Oscillator circuitry
frequency shift vs. mass loading per area of film, 209
use in surface acoustic wave devices, 212

P

Peclet number, definition, 115
pH sensor
construction from coimmobilization of two dyes at hydrolyzed cellulose acetate, 298,299*f*,300
excitation and emission spectra for fluoresceinamine, 275,276*f*
excitation emission intensity, 275,276*f*

pH sensor—*Continued*
factors influencing response time and sensitivity, 275
preparation with amplification technology, 260,262*f*
Phase resolution, fluorescence, 383
Phase-resolved fluorescence spectroscopy, description, 383
Phase-resolved intensity, definition, 383
Phospholipid bilayer membranes
electrical measurements by LAPS, 59,61*f*
formation, 55
Photocurrent, measurement with lipid bilayer membrane apparatus, 55
Photolithographic techniques, applications, 96
Photolithography, oligomer polymerization, 368,371
Piezoelectric devices, use as active component in detection systems, 238
Piezoelectric materials
generation of elastic waves, 157
mass-sensitive applications, 8,9*f*
precise positioning of microelectrode, 7
properties, 7
scanning tunneling microscopy, 7
use as microscopic sensors, 7–8
use for quartz crystal microbalance, 8,10
use in surface acoustical wave devices, 10–12
Plasticization effects, determination by surface acoustic wave device, 235
Platinization technique, microfabrication of biosensor, 96–97
Polymer film(s)
advantages for modification of electrochemical sensors, 78–79
γ radiation immobilization, 79–83
Polymer-film-based enthalpimeter
calibration curves, 151,155*f*
comparison to fiber-optic-based enthalpimeter, 154,155*f*,156
fabrication, 149–150
instrument design, 150–151,152*f*
modeling of heat transport, 151,154
output, 151,153*f*
Polymer-film-modified electrodes
applications, 66
formation, 66–67
Polymer films on chemically modified electrodes, enhancement of chemical selectivity of electrochemical sensors, 6
Polymer membrane electrodes, selectivity evaluation, 27
Polymer modification of electrodes, strategies, 79
Polymeric delivery systems, description, 267
Polymeric materials, need for characterization of properties, 222
Polymerization for sensor preparation, procedure, 259
Poly(vinyl alcohol)
description, 91
structure, 84*f*,91
Poly(vinyl alcohol)-based indicators
advantages, 273
Mg(II) sensor, 275,277–282
pH sensor, 275,276*f*
use in fiber-optic pH sensors, 274–282
Poly(vinyl alcohol)-modified electrodes, immobilization of glucose oxidase, 91,92*f*,93
Poly(vinylidene fluoride) films, use in enthalpimetric sensor, 150
Poly(*N*-vinylpyrrolidone)
description, 82
structure, 82,84*f*
Poly(*N*-vinylpyrrolidone)-modified electrodes
effect of ascorbic acid on response to catechol, 86,87*f*,88
enhanced response to catechols, 82,85*f*,86
resistance to electrode fouling, 88,89–90*f*
Pore size distribution, definition, 210
Porous samples, characterization with surface acoustic wave devices, 210
Potentiometric gas-sensing systems
detectors with enhanced gas sensitivity based on differential gas-sensing membrane electrodes, 39,41–42,43*f*
Severinghaus design, 38
sulfur dioxide sensing based on sulfite-selective membrane electrode, 38–39,40*f*
Potentiometry, measurement by chemical sensors, 3
Protein–protein complexes, electrochemistry, 105–106
trans-$PtCl_2$(ethylene)(pyridine), behavior in ligand substitution reactions, 179–180
Pulse voltammetric biosensing device for real-time determination of stationary-droplet glucose sample
calibration curve for real-time determination of glucose, 100,101*f*
performance characteristics, 103*t*
schematic representation, 97,98*f*
schematic representation of electrochemical fabrication technique, 97,98*f*
schematic representation of sensing principle, 97,99*f*
sensor output vs. sample volume, 100,102*f*,103
transient responses, 100,101*f*

Pulse voltammetric biosensing device for real-time determination of stationary-droplet glucose sample—*Continued*
typical responses, 97,99*f*,100

Q

Quartz crystal microbalance(s)
calculation of maximum frequency change for glucose binding, 244
comparison to acoustic plate mode device, 197
control of solution temperature in cell, 240
description, 8,238
diagrammatical representation of crystal, 240,241*f*
enzyme immobilization, 238
measurement of glucose by using hexokinase, 238
oscillator circuit, 240
response to different glucose concentrations, 242,243*f*
response to pure buffer vs. glucose in buffer, 240,241*f*,242
Sauerbrey equation for frequency changes, 242
schematic representation of instrument, 238,239*f*,240
use of piezoelectric materials, 8,10

R

Radial velocity of carrier stream, definition, 118
Rayleigh surface acoustic wave
advantages, 157
See also Surface acoustic wave
Real-time determination of glucose, pulse voltammetric biosensing device, 96–103
Reduced nicotinamide adenine dinucleotide (NADH) fiber-optic biosensors
response curve
bioluminescence biosensor, 309,311*f*
lactate, 307,308*f*
pyruvate, 307,310*f*
response times, 307
schematic diagram of probe, 305,308*f*
schematic representation based on bacterial luciferase, 309,310*f*
Regenerable antibody-based fiber-optic chemical sensors
configuration of sensor, 325,328
direct measurement of tetrahydrobenzo[*a*]pyrene, 327*f*,328
preparation, 325
Reversibility, chemical sensors, 178
Rhodamine 6G–Nafion film dynamics, phase-resolved fluorescence, 389–394

S

Salicylate-selective membranes using metalloporphyrins as ionophores
potentiometric response, 32,34*f*,35*t*,37*f*
potentiometric selectivity coefficients, 30*t*
Sandwich assays, description, 320
Sauerbrey equation
conditions for application, 242,244
description, 242
Scanning tunneling microscopy, description, 7–8
Selectivity, chemical sensors, 178
Sensors based on fluorescence energy transfer
advantages, 265,267
criteria, 263
energy transfer, 263,265
pH response, 265,266*f*
phenol red absorption, 265,266*f*
Sensors based on irreversible chemistry
approach based on dual excitation–emission reagent, 267,268*f*
approach based on mixture of reagents, 267,268*f*
configuration, 269,270*f*
polymeric delivery system, 267
response time, 269*t*
selection of sensing reagent, 267
Shear modulus, definition, 224
Sol–gel film characterization
adsorption isotherms, 214,215*f*
advantages of surface acoustic wave device techniques, 217–218
effect of aging time on film porosity, 214,217*t*
pore size distributions, 214,216*f*
Solution-phase small-molecule detection, optical-waveguide chemical sensors, 367,368*t*,369–370*t*
Spectroelectrochemistry, description, 130
Sulfite–bisulfite-selective electrode based on bis(diethyldithiocarbamato)mercury(II)
calibration curve, 35–36,37*f*
mechanism of ionophore action, 36,38
potentiometric selectivity coefficients, 36*t*
Sulfur dioxide sensing based on sulfite-selective membrane electrode
flow-through gas-sensing arrangement, 38–39,40*f*
response toward sulfite standards, 39,49*f*
Surface acoustic wave, effect on changes in substrate surface, 208

Surface acoustic wave device
advantages, 223
applications, 10–11,158
array sensor system configuration, 170,171*f*
background, 209–211,213
characterization
diffusion in polymer films, 218,219*f*,220
porous samples, 210
sol–gel films, 214–218
thin-film properties, 208–209
coating deposition methods, 164
comparison to acoustic plate mode device, 197
description of operating principles, 223
design, 162
determination of diffusion coefficients, 210–211
determination of glass transition temperature, 231,223–234
dual-delay line sensor configuration, 163,165*f*
effect of temperature on frequency, 227,228–230*f*
effect of temperature on sensitivity, 166,168,169*f*
effect of wave–polymer film interaction on glass transition temperature, 224–225
elastic modulus determinations, 234
experimental procedure for thin polymer films, 225–226
factors influencing properties, 158
factors influencing response, 223
features, 157–158
film selection, 11–12
future applications, 234–235
future developments, 174
gas flow system, 212,213*f*,214
instrument design, 211
layout of dual-sensor chip, 170,171*f*
limitations and capabilities of sensor array, 173
mass sensitivity, 10,159–162
noise level, 209
operating frequency vs. mass sensitivity, 159,160*t*,161*f*,162
oscillator circuitry, 212
parameter values, 223,224*t*
plasticization effects, 235
response of coated sensor to different vapors, 168,169*f*
response patterns of array sensor to different vapors, 170,172*f*,173
response time of sensor, 166
scaling laws, 159,160*t*
schematic representation as feedback element of oscillator circuit, 209,213*f*
sensitive and selective coatings, 163–164
sensitivity as vapor sensor, 222

Surface acoustic wave device—*Continued*
sensor response, 164,165*f*,166
sensor response vs. solubility interactions 223–224
shear modulus of polymer, 224,228*f*
signal determination, 158–159
simplified delay line configuration, 158,161*f*
sorption isotherm of sensor, 166,167*f*
theory of operation, 158–162
thermal activation energies, 234–235
thermal expansion coefficients, 234
thin-film formation, 211–212
use of piezoelectric materials, 10–12
Surface activation, procedure, 256,258

T

Thermal activation energies, determination by surface acoustic wave device, 234–235
Thermal expansion coefficients, determination by surface acoustic wave device, 234
Thermal sensors, description, 17
Thin film(s), formation, 211–212
Thin film of organic polymers, use in construction of electrochemical and spectrochemical sensors, 283–284
Thin-film properties, monitoring with surface acoustic wave devices, 208–220
Thiocyanate-selective membranes using metalloporphyrins as ionophores
potentiometric response, 30,32
potentiometric selectivity coefficients, 30*t*
Time-dependent fluorescence, definition, 381–382
Toxic chemicals, problems with monitoring, 176
trans-directing ligands
back-bonding, 180
description, 180
effect on rate and mechanism of reaction, 180
factors influencing reaction rates, 180–181
olefin-substitution reaction pathway, 180,183*f*
Transmembrane potential, measurement by LAPS, 50

V

Valinomycin, effect on lipid bilayer resistance, 59–60,61*f*

Vapor detection, surface acoustic wave microsensors, 157–174
Vapor–polymer solubility interactions, measurement by optical-waveguide chemical sensors, 371,373,374–375*f*
Vinyl acetate
  health effects, 179
  problems with characterization of occupational exposures, 179
Vinyl acetate vapor measurement by using coated surface acoustic wave sensor
  dual-oscillator configuration, 182,183*f*
  experimental procedures, 181–182
  interaction with coating film, 182,184
Vinyl acetate vapor measurement by using coated surface acoustic wave sensor—*Continued*
  response
    vs. exposure time, 186,187*f*
    vs. temperatures, 186,187*f*
    vs. vinyl acetate concentration, 184,185*f*,186
  selectivity of sensor, 188*t*,189
Viscoelastic polymers
  curves of idealized modulus vs. temperature, 224,228*f*
  glass transition temperature, 224–225
Viscous coupling, acoustic plate mode device, 203,204*f*,205

Production: Rebecca Hunsicker
Indexing: Deborah H. Steiner
Acquisition: Robin Giroux

Elements typeset by Hot Type Ltd., Washington, DC
Printed and bound by Maple Press, York, PA